Overcoming Ptolemy

AsiaWorld

Series Editor: Mark Selden

This series charts the frontiers of Asia in global perspective. Central to its concerns are Asian interactions—political, economic, social, cultural, and historical—that are transnational and global, that cross and redefine borders and networks, including those of nation, region, ethnicity, gender, technology, and demography. It looks to multiple methodologies to chart the dynamics of a region that has been the home to major civilizations and is central to global processes of war, peace, and development in the new millennium.

Recent Titles in Series

Overcoming Ptolemy: The Revelation of an Asian World Region, by Geoffrey C. Gunn
Poetry and Terror: Politics and Poetics in Coming to Jakarta, by Peter Dale Scott with Freeman Ng
Memory, Reconciliation, and Reunions in South Korea: Crossing the Divide, by Nan Kim
Peace in the East: An Chunggun's Vision for Asia in the Age of Japanese Imperialism, edited by Yi Tae-Jin, Eugene Y. Park, and Kirk W. Larsen
Neonationalist Mythology in Postwar Japan: Pal's Dissenting Judgment at the Tokyo War Crimes Tribuna, by Nariaki Nakazatol
The Bonin Islanders: Narrating Japanese Nationality, by David Chapman
Heritage Politics: Shuri Castle and Okinawa's Incorporation into Modern Japan, 1879–2000, by Tze May Loo
Ethnic Capital in a Japanese Brazilian Commune: Children of Nature, by Nobuko Adachi
Yokohama Street Life: The Precarious Career of a Japanese Day Laborer, by Tom Gillfethnic
Visualizing Modern China: Image, History, and Memory, 1750–Present, edited by James A. Cook, Joshua Goldstein, Matthew D. Johnson, and Sigrid Schmalze
Refining Nature in Modern Japanese Literature: The Life and Art of Shiga Naoya, by Nanyan Guo
The "Other" Karen in Myanmar: Ethnic Minorities and the Struggle without Arms, by Ardeth Thawnghmung
Malay Kingship in Kedah: Religion, Trade, and Society, by Maziar Mozaffari Falarti
A Localized Culture of Welfare: Entitlements, Stratification, and Identity in a Chinese Lineage Village, by Kwok-shing Chan
Radicalism, Revolution, and Reform in Modern China: Essays in Honor of Maurice Meisner, edited by Catherine Lynch, Robert B. Marks, and Paul G. Pickowicz
Water: The Looming Crisis in India, by Binayak Ray
Filling the Hole in the Nuclear Future: Art and Popular Culture Respond to the Bomb, edited by Robert Jacobs
Rising from the Flames: The Rebirth of Theater in Occupied Japan, 1945–1952, edited by Samuel L. Leiter
Japan in the World: Shidehara Kijūrō, Pacifism, and the Abolition of War, Volumes I and II, by Klaus Schlichtmann
A Matter of Honour: Being Chinese in South Africa, by Yoon Jung Park
Imperial Subjects as Global Citizens: Nationalism, Internationalism, and Education in Japan, by Mark Lincicome
The Chinese in Cuba, 1847–Now, by Mauro García Triana and Pedro Eng Herrera
Windows on the Chinese World: Reflections by Five Historians, by Clara Wing-chung Ho
Lake of Heaven: An Original Translation of the Japanese Novel by Ishimure Michiko, by Bruce Allen
Tommy's Sunset, by Hisako Tsurushima
The Culture of Fengshui in Korea: An Exploration of East Asian Geomancy, by Hong-Key Yoon
China's Unequal Treaties: Narrating National History, by Dong Wang
Precious Steppe: Mongolian Nomadic Pastoralists in Pursuit of the Market, by Ole Bruun
Yoko Tawada: Voices from Everywhere, edited by Douglas Slaymaker

Overcoming Ptolemy

The Revelation of an Asian World Region

Geoffrey C. Gunn

LEXINGTON BOOKS
Lanham • Boulder • New York • London

Published by Lexington Books
An imprint of The Rowman & Littlefield Publishing Group, Inc.
4501 Forbes Boulevard, Suite 200, Lanham, Maryland 20706
www.rowman.com

6 Tinworth Street, London SE11 5AL, United Kingdom

Parts of chapter eight were previously published as Gunn, Geoffrey C. 1991. "Anglo-French Rivalry over the Spratlys (1930–1937)." 262–283. In R. D. Hill, Norman G. Owen and E. V. Roberts, eds. *Fishing in Troubled Waters: Proceedings of an Academic Conference on Territorial Claims in the South China Sea*. Hong Kong: Centre of Asian Studies, University of Hong Kong, 1991.

British Library Cataloguing in Publication Information Available

Library of Congress Cataloging-in-Publication Data Available

ISBN 978-1-4985-9013-6 (cloth : alk. paper)
ISBN 978-1-4985-9014-3 (electronic)

∞™ The paper used in this publication meets the minimum requirements of American National Standard for Information Sciences—Permanence of Paper for Printed Library Materials, ANSI/NISO Z39.48-1992.

Printed in the United States of America

Contents

Acknowledgments

The research for this monograph was carried out with the support of the Academic Research Grant from the Cultural Affairs Bureau of the Macao Special Administrative Region Government. I am obliged to this institution for general guidance and initial text reviews. I am also obliged to two anonymous Lexington Press reviewers for sage comments upon the conceptualization and scope of this work besides sparing me errors of commission and omission.

Looking back, my triple majors as an undergraduate at Melbourne University in geography, history-political science, and East Asian studies were strategic to this study. So were several youthful trips across the Afro-Eurasian heartland including some close inspection of North African Greek and Roman ruins, ancient Cyrene in particular. I have also navigated and/or sailed in some of the waters and seas that I describe—the Malacca Straits, the South China Sea, the Indian Ocean and Red Sea included. In other words, unlike the French *géographes de cabinet* or armchair geographers, I have mostly traveled the map terrain that I describe. I personally know few individuals working in this field and none in Japan but thanks are due to those who also improved my previous books (series editor Mark Selden in particular). Over the years I have also profited from consultation with Toulouse-based geographer Frédéric Durand, author of map collections on the Malay world area.

Some paragraphs and sentences in chapter 8 are adapted from my conference paper, "Anglo-French Rivalry over the Spratlys (1930–1937)," in R. D. Hill, Norman G. Owen, and E. V. Roberts, eds., *Fishing in Troubled Waters: Proceedings of an Academic Conference on Territorial Claims in the South China Sea* (Hong Kong: Centre of Asian Studies, University of Hong Kong, 1991: 262–83).

Conventions

As a rule, I do not interject modern or national names or spellings such as Okinawa, Melaka, and Sri Lanka into historical contexts. Similarly, I avoid the names of invented nations such as Indonesia, Burma/Myanmar, Malaysia, unless I am addressing a modern context. Even Taiwan/Formosa raises naming issues from a 500-year perspective. Nevertheless, in the case of China, I seek to match modern equivalents to ancient Chinese names. Taking zero longitude as understood by Ptolemy as my point of reference, unless otherwise specified all other regions of Eurasia are eastward, as with Near or Middle East, Southeast, Far or East-Northeast Asia.

Abbreviations

BnF	Bibliothèque nationale de France
GPS	Global Positioning System
PRC	People's Republic of China
VOC	Verenigde Oost-Indische Compagnie (Dutch East India Company)
UNCLOS	United Nations Convention on the Law of the Sea

Introduction

In the present age the notion that all parts of the world are interconnected is self-evident. The recognition that global solutions alone can address mounting environmental concerns in the contemporary era is part of this paradigmatic shift in thinking. But for most of human history this was not the case. Allowing for the rise and fall of empires across the Afro-Eurasian landmass since antiquity, the "discovery" and conquest of New Worlds was a feat accomplished by Western Europe. Nevertheless, understandings of the universe in Renaissance Europe were layered upon the classical foundations of ancient Greece and Rome in turn drawing upon even earlier traditions of mathematics and astronomy, some with Asian origins. As exemplified by Claudius Ptolemy (ca. 90–168 CE), the Alexandrian school mapped the heavens and placed the Earth at the center of the universe. Importantly, in acknowledging a spherical world, Ptolemy bequeathed a masterpiece in geometrical geography offering instructions on how to delineate the Earth on a plane surface or map. Although long lost to Latin Europe, his reconstructed *Geographia* also included instructions on how to complete a map of the world according to his projection. In setting down coordinates for over 6,000 places in the ancient world, he also signaled the importance of careful astronomical observation. As a mathematical model featuring a grid of lines of latitude and longitude imposed upon a sphere, it is important to acknowledge that Ptolemy passed on the now familiar convention of dividing the globe into 360 degrees (east or west of a prime meridian) further divided into minutes and seconds in the same way that we divide hours of time into minutes and seconds. Without it, we would be lost. Ptolemy may thus be credited with setting down the first recorded treatise on geo-positioning.

Until the first translation of *Geographia* from Greek into Latin appeared in 1407, all knowledge of these coordinates had been lost in the West.

Notwithstanding new empirical truths revealed by such travel accounts of Asia as those of Franciscan missionaries, Venetian merchants, and Muslim contacts, newly redrawn Ptolemaic maps entering print versions in the late fifteenth century stubbornly exaggerated the extremities of the Eurasian landmass eastward. As Ptolemy wrote in *Geographia* in an allusion to China, "The inhabited part of our Earth is bounded on the east by the Unknown Land which lies along the region occupied by the easternmost nations of Asia Major, the Sinae and the nations of Serice." This land of silk or Serica, as referenced in a number of classical Greco-Roman texts, was said by Ptolemy to be 119.5 degrees east of Alexandria.

Unquestionably the Ptolemaic legacy embraces diverse and complex elements. These include celestial observation, the science of astronomy, geography, both cartographic and descriptive, and, as alluded, the mathematical modeling of a spherical Earth on a plane surface surrounded by a longitudinal and latitudinal grid. With its multivalent associations across the ages and reaching into the present, the Ptolemaic corpus can be confusing for the uninitiated. As a grand cosmological system, its various elements or components are simply not of a kind and have to be unpacked accordingly. We should also be alert to a long transitional period in the history of even "post-Ptolemaic" cartography. For example, as Margaret Small (2013: 154–55) points out, distinctions should be made between, on the one hand, cartographic techniques and developments that may or may not be mathematical, and textual geographies which often ignore maps, on the other. In the Western classical tradition, this distinction reaches back to Pliny and is revived in the late Renaissance period but examples can be found in all the major civilizations.

In large part this book is concerned with the enduring influence of Ptolemy's *Geographia* across the centuries notwithstanding challenges prompted by new data and understandings that would eventually lead to the overcoming of certain of the Alexandrian's key tenets especially relating to the size and configuration of the continents and encircling seas. Here we distinguish between *Geographia* with its concern for geography and map projections and Ptolemy's other major work *Almagest* coming to dominate Western astronomical thinking for over a thousand years. We understand that the tri-continental Ptolemaic template sourced from *Geographia* actually set a trap for the first wave of New World discoverers (as with Columbus). But we also allow that, long prior to the entry of the first European voyages into the Indian Ocean, the Ptolemaic concept of an enclosed Southern Ocean had been disproven by Arab seafarers. In this light, it will be useful to set down the major features of the received Ptolemaic world leading to myriad misunderstandings and corrections down through the centuries.

One of Ptolemy's mistakes was to underestimate the size of the world. Dividing the world into 360 degrees (a convention holding to the present), he calculated each degree to be 50 miles (80.5 km) rather than 70 (111.3 km) as correct. He not only exaggerated the length of the Mediterranean by about 30 percent, but also that of the Eurasian landmass, thus greatly reducing the distance between the western tip of Spain, and the east coast of Asia. Ptolemy's estimate yields a circumference of 28,800 kilometers, just over a quarter smaller than the correct circumference (40,070 km). Thus, the further east in the Ptolemaic map template the more the errors accumulated. Spanning 180 degrees of longitude from the Canary Islands, his world ended somewhere in the middle of China. Today we know Beijing at around 116 degrees east of the Greenwich Meridian (0° longitude).

As summarized by historian Donald F. Lach (1965: 69), other elements or actually flaws in the Ptolemaic template relating to Asia include the insertion of a *Sinus Magnus* or Great Bay at the eastern end of the southern land bridge otherwise obscuring the sea route from India to China. Besides elaborating a truncated India, an oversized Ceylon, and an elongated Malay Peninsula, Ptolemy produced a highly generalized rendition of the space between India and China. Dubbed *India Extra Gangem*, this vast region broadly encompasses what is known today as Southeast Asia including Yunnan and Tibet. Making only vague allusion to *Seres/Serica*, sometimes *Sinae* (China), Ptolemy sowed confusion to navigators and mapmakers alike by adding such toponyms as Zaba and Cattigara believed to be key ports in Southeast Asia but of uncertain location.

In his magisterial study of the reception of Ptolemy in Florence at the end of the fourteenth and beginning of the fifteenth century, Patrick Gautier Dalché (2007: 285) advises that "rather than being a single incident in the history of mapmaking, the translation of Ptolemy and the diffusion of his work took place in an intellectual and cultural context within which complex and varied motivations were at play." It is hard to disagree with his contention that the whole must be considered within the wider context of the intellectual trends at work through to the sixteenth century. But how long did the weight of the Ptolemaic legacy survive? Scholars are divided. As Anthony Grafton (1995: 225) explains, between 1450 and 1700 a revolution in the form of knowledge and expression took place in Europe. The textual canons of the fifteenth century "inevitably lost their air of cohesion and completeness" especially as the world came to be known directly. But that is not to say that tradition was swept away. Rather, in contradictory ways, the ancient texts served "both as tools and obstacles for the intellectual explanation of the new worlds." In his discussion of "alternative geographies," John Rennie Short (2000: 39) also acknowledges that Ptolemy's model for the celestial universe (such as elaborated in *Almagest*) scarcely survived the Copernican Revolution (and in any case disproven by medieval Islamic astronomers and mathematicians). Geography, however, was a different matter; "Ptolemy's basic model of describing the earth was improved rather than overturned; the improvements included extending the known world and refining the measurement of latitude and longitude."

As addressed in a dedicated chapter, improving on Ptolemy's measurements was also a project adopted by the French Sun King, Louis XIV (r. 1643–1715). India, Siam (Thailand), and the Portuguese-administered territory of Macau in southern China ceded by the Ming all loom large in this account as sites where arriving Jesuit missionaries fed back to Europe vital new geographical information. In the case of China, their accounts were practically the first since the semi-fictionalized Cathay of Marco Polo's description. Later-arriving Jesuits would go on to serve the Qing court in the actual trigonometric surveying of the empire, thus calculating the eastern extremity of Eurasia and the oceans that surround it. We could say that, more than any other single European enterprise (and setting aside the endeavors of Muslim geographers dating back to the ninth century), it was the Jesuit surveying and mapping of China that finally corrected the distortions formulated by Ptolemy as to the full eastern extent of the Eurasian landmass and the actual circumference of the planet.

For entirely different reasons, Philip II of Spain (r. 1556–96) also sought to geometrically map his New World conquests, subsequently extended into the Pacific Ocean (and with only about half the Earth's surface surveyed with any detail 100 years later). In writing of the Spanish king's dual concern to dovetail chorographic or city-scape maps with geographic area maps, as Barbara E. Mundy (1996: 4) contends, "Ptolemaic concepts have colored how we think of spatial representation up to the present." To be sure "coloring" is a good metaphor and we should be alert to how such a taint or legacy actually influenced the concerned elites when it came to attitudes, actions, and interactions, especially in the wake of Columbus.

WORLD REGIONS

With a focus upon Asia referenced against Europe, we wish to know how the regions—world regions—of the vast Eurasian landmass have been conceptualized and configured globally, especially via cartographic and other representations of space. Although not the central concern of this book, we should also note that the historical construction of boundaries of modern nation-states continues to be a source of sometimes conflictual nationalistic claims and counterclaims. Especially we wish to know how European conceptions of China changed over the centuries, from vague imaginings as a source of silk practically back to the age of Ptolemy, to the more exoticizing accounts of late medieval European (and Muslim) travelers, to confusions as to the location of the *Sinus Magnus*, down to firsthand accounts of Portuguese arriving on the coast of China in the late Renaissance period.

The European ancestry of the Asian world-region system as we know it today is deeply embedded in Western intellectual tradition. Given the weight of millennium-old Asian civilizational verities, this might strike a reader in, say, India or China, as presumptuous or even fallacious. The unknown Asia referenced against the known world of Europe obviously has a long pedigree reaching back to the ancient Greeks and Romans and, even before them, the Phoenicians. The European "priority" in spatializing the globe becomes even clearer in the light of the rediscovery and virtual reification of the classical template by late Renaissance cartographers and later-arriving conquistadors and colonial trailblazers. In other words, the spatial configuration of the Asian world by Europe as represented in text and maps and, more often than not reflecting the views and biases of powerful scholarly elites, calls for a very special interrogation. Neither do we apologize for sweeping 2,000 years of contacts across Eurasia, especially given the ancient origins of regional framings. At the extremities of the known world, not only were ancient Egypt and Han China (206 BCE–220 CE) broadly contemporaneous, but they were linked by the sinews of long-distance trade.

Still, we need to test or relativize this Eurocentric view of world-region formation against other civilizations, allowing as well Islamo-centric, Indo-centric, Sino-centric, and other culturally or civilizationally bounded views of the world. At the same time we cannot ignore millennium-long contestations, clashes, and other encounters across the Eurasian landmass from the time of Alexander the Great (336–323 BCE), to the rise of Islam in the seventh century, to the Arab conquests including direct Islamic contacts with Tang China (618–907 CE), and to the Mongol empire (1206–1368 CE). Nor can we ignore complex civilizational interactions in the eastern Mediterranean engaging, variously, the Byzantine Eastern Roman empire, rising Ottoman Turks, the Mamluks, along with Venetians and Genoese, including the events surrounding the Christian Crusades (1095–1291 CE) against Islam. While Latin Europe and the Islamic Other suggests a cultural binary, there was also an intense two-way transmission of ideas, inventions and scholarship mostly in Europe's favor. Without imposing a Eurocentric reading of macro history, or even suggesting the inevitability of a European priority alongside an Asian stagnation, the argument does assume 1500 as a turning point heralding the Spanish and Portuguese conquests of New Worlds and oceans. In this spirit—and at the risk of overdetermination—I seek to demonstrate through cartographic history just how Western Europe came to impose its political and ideological dominance upon most of the rest of the world, only challenged in the age of decolonization, as new nation-states (bounded territories) emerged.

Recalling the classical past to comment upon the present is not a new theme, but the novelty is in the presentation of meta-geographical representation across long time.

THE GLOBALIZATION TROPE

Scholars disagree as to definitions of globalization and, especially, the question of origins or beginnings. Nevertheless, the new world history perspective has advanced our understanding of both origins, periodization, as well as the scope and scaling of the global.[1] The new world history has also unsettled certain verities as to the inevitability of the rise of the West versus a stagnant East. Not even the recent ascent of China and India to mega-economy status has allayed this bias or centrism in thinking. Building upon my own two books as to the origins of globalization back to the European discoveries, namely, *First Globalization: The Eurasian Exchange, 1500–1800* (2003), with its emphasis upon East-West intellectual transactions, and *History without Borders: The Making of a Southeast-East Asian World Region, 1000–1800* (2011), with an acknowledgment of even deeper East-East alongside West-East reciprocal exchanges, this study seeks to stimulate some rethinking on the global with attention to the historical construction of region. Taking the *longue durée* as a standard, we come to realize how our sense of the global or planet earth has been a long, fluid, and even tortuous adventure of exploration, contestation, and revelation on the part of mankind, best expressed as civilizations or identities forged around beliefs, shared languages, cultures, and especially religious identities.

As suggested, reaching back to ancient Greece and Rome along with Han China, civilizational centers across the Eurasian heartland were in communication, carried through by the advent of Islam, and the Mongol conquests, alongside the rise of ancient maritime and terrestrial "silk roads." While not always well documented, ancient civilizations knew each other's strengths, resources, and approximate locations. Alongside texts, route guides, sailing directions, and physical maps, undoubtedly mental maps also existed of these distant polities and trading centers, however dimly apprehended. Yet, it is a fact of history that nation, region, and civilization do not always neatly coincide. To be sure, nation and region tend to be more bounded than civilizations, but they are also crosscut by cultural-religious-linguistic zones. On the other hand, regions sometimes survive the rise and fall of civilizations and physical geography hardly changes at all (though desertification in parts of Afro-Eurasia, siltation, and sedimentation in tropical zones, accompanied by climate change, have indeed had national and regional impacts). More broadly, the ebb and flow of empires across Eurasia from Alexander the

Great, to the Mongols, to the Ottomans, to the Soviet Union and its break-up, renders fixed conceptions of nation-region-space as entirely misleading.

Europe, as the ancients were well aware, was not Asia but merely the western edge of the vast Eurasian continent and, not being Asia, Europe mixed fear with curiosity of the Asian "other." The installation of an Islamic caliphate in Spain (ca. 1000 CE), the Reconquista (ca. 718 to 1492), and the Crusades did not dim Europe's outreach to the furthest ends of the Asian continent. But it would be the mental revolution associated with the Renaissance that would actually spell the death knell to the Ptolemaic model of a geocentric world such as stubbornly adhered to by the Catholic Church. The Copernican Revolution, the new age of print dissemination combined with new knowledge derived from the discoveries, and the conquest of new lands heralded a truly "first globalization" of cultures, leading to a cornucopia of East-West and West-East exchanges, both economic and ideological. Globalization so understood also called up a new discourse on space especially at the hands of schools of cartography, not excepting a later generation of Enlightenment philosophers when it came to pondering humankind's place in the universe.

THE ORIENTAL CONTRIBUTION TO THE RISE OF THE WEST

The historical Ptolemy with his base in Alexandria was undoubtedly heir to broader civilizational transfers reaching back to Babylonia, yet with great acclaim his recovered texts were appropriated by early Renaissance Europe. The non-Western reader of this book might well ask, but doesn't the Ptolemaic inheritance legitimize a Eurocentric framing of the world? Even if not, wasn't it the Ptolemaic imaginary that led the conquistadors into world conquest? By positioning the Mediterranean at the center of his known world, wasn't Ptolemy the one who set up an enduring East-West polarity? Certainly we gain this impression in consideration of the long shadow cast by Ptolemy upon European cartographers even through the age of "discoveries," although not necessarily if we study practical navigation and scientific advance. Neither can we ignore the ways that the Ptolemaic corpus was received by different civilizations and traditions, as with the Islamic which both borrowed and modified the Alexandrian heritage, and the Chinese who were impressed but skeptical at the same time.

As John Hobson (2004: 3) signals, to impose strict binaries upon a "stagnant" East and a "pristine" and dynamic West in the sweep of history might be false when we consider such eastern contributions to Europe as papermaking, printing, the compass, gunpowder, and the lateen sail. Setting aside the transfer of Islamic science to China leading to the construction of celestial observatories, Muslim geographers and astronomers also mathematically disproved Ptolemy's Earth-centric view of the universe. Whatever one's opinions as to the salience of Ptolemy in shaping our worldview, it is hoped that this work will stimulate thinking around the origins of science as a grand cross-civilizational affair where many traditions and ideas mingled.

THE INTERROGATION OF SPACE AND PLACE

Through the ages travelers have resorted to route guides, sailing directions, star charts, or maps for journeys and pilgrimages, both real and imaginary. This is fortunate as their legacy (where it survives on parchment, paper, or even stone) has supplied a veritable cascade of interpretations down until the present. In its time, the eighteenth-century wave of utopian literature, certain with a travel vogue, called down "no place." In the present age we travel vicariously via digital media going simultaneously every place we wish while actually going nowhere. Globalization has also spawned the phenomenon of "no place" especially relating to sites transformed by mass migration, transitory experiences, and dislocations. Present-day Macau has been so defined (Wong 2014: xxxvi), but it is easy to think of other candidate cities in Asia so transformed.

A cultural studies approach to maps and spatial representation offers yet another optic. As defined by French anthropologist Marc Augé (1995: 77–78), "If a place can be defined as relational, historical and concerned with identity, then a space which cannot be defined as relational, or historical, or concerned with identity will be a non-place." Post-structuralists as with Giles Deleuze and Felix Guattari (1972) have also entered the discussion on the territorialization/deterritorialization of space. In their analysis, territorialization fits the pattern of mapping space, as much the colonizers need to impose fixity over conquered space. Deterritorialization may mean to take the control and order away from a land or place (territory) that is already established. It is to undo what has been done.

To be sure, as Japanese media analyst Ema Ota (2014: 1) summarizes, "place is not an immobile truth . . . our ideas of place are always in flux and mediations generate even greater flux, we cannot cling to an ideal of place, the sanctity of the local or the demon of the global (or vice versa)." For her, "given the simultaneity of multiple places," space ultimately defies location. Given that the experience of the world is based upon perception and cognition, the craft of mapmaking, it follows, is a "mediated experience" and, after Jacques Derrida, with language central in this mediation. In this interrogation, as Ota (2014: 1) continues, both embeddedness and deterritorializion/reterritorialization pose questions as to empowerment or loss. In turn, she points to

Henri Lefebvre's landmark *La Production de l'espace* (1974), especially his sense of the changing discourses of space from the mathematical, to the mental, to the artistic, in fact allowing multiple spaces; economic, geographic, demographic, ecological, national, regional, and global (as indeed borne out in the text). If then, space is mentally conjured (imaginary), constructed, or socially produced, then it lies outside of fixed coordinates. Anthropologists, notably, understand deterritorialization as referring to a weakening of ties between culture and place.

Together, territorialization/deterritorialization can be seen as sequential outcomes of the historical processes of European expansion, colonization/decolonization, and postmodern hybridity under globalization. A range of Asian regions as with India, Malay(sia), Indochina, the Indonesian archipelago, and China under the "unequal treaties" system underwent this process of deterritorialization/deterritorialization. For some thinkers, globalization has exacerbated the process of territorialization of peoples from their local cultures, especially under the impacts of migration and commodification. In this argument, something new evolves through a process of reterritorialization as local culture becomes enveloped in global culture.

This brings to the discussion the question of maps which, over the centuries, have sought to territorialize or totalize representation of space. As Peta Mitchell (2012: 3) explains, by the age of Enlightenment the concept of cartography as science fed into the tradition of map as a "purely scientific document." Maps stood as representations of geographical territory or even as a "mirror of nature" or, in other words, "unproblematic."[2] Noting the influence of Albert Korzybski in the field of general semantics, she is not alone in explaining how, in the early decades of the twentieth century, a shift appears away from the Aristotelian tradition to an understanding of representation and use of the map as metaphor (albeit with the critique only becoming mainstream in the 1980s).

According to geographer David Harvey (1991: 240–249), the European conception of space has gone through many transformations from feudal days (imbued in legal political and social relations), and Renaissance (rooted in a new objectivity and stripped of fantastic elements). In fact, our text will reveal an even more complex array of map transformations. Semiologists, as with Roland Barthes and Martin de Certeau, have seen in modern maps a scientific element that actually negates representation and imposes a power relationship, and with the new reality actually subverting the true picture. J. B. Harley (2001) has gone further in defining a map as a "social construction" and therefore subverting the traditional, "positivist" model of cartography.

Whether or not we agree with the above interrogation of spatial representation, and my own bias is firmly planted in a positivistic unfolding of map history, we nevertheless cannot ignore the subjective element in the production of three-dimensional geographies. The production of faux maps and other dystopias in eighteenth-century Europe illustrates this truism, a tradition harking back to the ancient Greeks as with the invention of Atlantis. To take an example from the European Middle Ages, Christocentric map renditions served feudal princes and overlords to better understand their enemies. On its part, late Renaissance cartography enriched by the revelations of the European discoveries literally staked out vast territories of control, hyperbolic as with the Papal division of the world into Iberian spheres. Early colonial cartography fenced in and roped off new domains, as armies advanced on the terrain, map in hand. The post-Westphalian nation-state system embraced such fixity of boundaries, domains, and identities with alacrity. More than ever, for powerful centralized states vying for "ownership" of resources and waters including even oceanic reefs and rocks, space is a construction. But today, under the agency of globalization, the hybridization of cultures, along with the role of new media and human mobility, the fixity of boundaries are increasingly blurred and reterritorialized, just as we see new identities and associations embedded in even newer conceptions of spatiality. Stated another way, we witness the remediation of new transnational space in ways that could not have been imagined by mapmakers of old. Even in the age of scientific geography as with the setting down of coordinates or the current use of Global Positioning System (GPS), we still have to navigate issues of freedom versus control.

This book offers a range of map and other images explaining how cartographers first reproduced, then corrected and/or imaginatively embellished the rediscovered Ptolemaic global template including its sectional maps of Asia. As explained, this was a centuries-long process engaging not only Latin Europe but the Islamic world, itself in contact with Tang dynasty, China, via both terrestrial and maritime silk roads. Still, with Islamic mapmaking entering a long stasis following the Mongol advance westward across Eurasia and the withdrawal of the great Ming-era voyages into the Indian Ocean realm, we are also alert as to the global shifts occurring in the sixteenth century heralded by the European "discoveries."

Another feature of this book is a concern to explain the technical innovations wielded by navigators in enabling their advance. While many civilizations dating back to ancient Babylonia and Egypt experimented with such instruments as the astrolabe to calculate the position of the Sun and prominent stars with respect to both the horizon and/or the meridian or latitude of a ship at sea, the prospect of open sea navigation also called down a need for constant improvement in navigational devices. As demonstrated, it would take long centuries to perfect instruments like the quadrant and the sextant to allow celestial objects to be accurately measured relative to the horizon, to develop accurate shipborne timepieces essential to calibrate longitude, to master magnetic declination to avoid shipwreck, and many other technical details, often becoming national priorities. More than anything as the history of cartography reveals, it was the practical construction of sailing guides and marine charts by mariners of many traditions that first revealed the true dimensions of the continents and oceans.

Some of the maps embellishing our narrative are practically iconic, as with the worldmap of the Muslim geographer Muhammad al-Idrisi (1154 CE); the Catalan Atlas (1375); the Fra Mauro map (ca. 1450); the Catino planishere of 1502, summarizing the Portuguese voyages; the Waldseemüller map (1507), adding the American continent; and the atlas by Abraham Ortelius (1570). Still, these maps have to be explained in the context of their times. They also have to be relativized with mapping and astronomical traditions of other civilizations. To say the least, the mathematically modelled Ptolemaic tradition of map coordinates that came to be esteemed by Renaissance and post-Renaissance makers of globes and inventors of map projections was far from universally accepted across various cartographic traditions and civilizations, at least until the universalization of science itself. Paradoxically, as it were, corrections to the Ptolemaic template of the Eurasian landmass as with the Jesuit survey of China, Korea, and Tibet (d'Anville, 1737) could not have been made without engagement with Arab, Persian, Chinese, Korean, Japanese, and other Asian traditions of mapmaking.

WORLD REGIONS AS PERIPHERIES

But just as historical kingdoms inspired and celebrated monumental centers, so they supported or exploited peripheries, satellites, and regions. The Roman empire was the "classical" example, building roads and fortifications, measuring distance to outlying provinces as with the Roman mile, entering into long-distance trade with other civilizations, India included, just as monuments and artifacts of its territorial reach abound. In the Sinic world, the center acknowledged tributary states linked by an elaborate ritual of "tribute trade" missions, a system enduring over millennium, only eclipsed within recent recorded time. Christians, Muslims, Hindus, and Buddhists also developed special spatial topographies. But Chinese provincial mapping, measured in *li* or Chinese miles, was always ahead of anything comparable in Europe. In my argument, the making of world regions, not excluding the Islamic world or even the Mongol empire across the vast Eurasian space, were dynamic constructions outside of classical enumerations or predictions.

We see in the sweep of history the triumph of those civilizations which, alongside armies and fleets of galleys, were supported by a literate caste of philosopher-scribes including astronomers and cosmographers, matched by a tradition of preservation of knowledge. Greco-Romans, Byzantines, medieval and Renaissance Europeans, Arabs, Persians, Indians, Turks, and Chinese civilizations all shared these attributes. Typically, all maintained "colonies" or outliers with varying degrees of autonomy versus integration. Typically, the centers displayed their wealth ostentatiously, such as vested in monuments, grand works, and bureaucracies. The more mercantile were serviced by merchant communities entering into external commerce—maritime or terrestrial—to further leverage or arbitrage that trade. Some of them, as with Venetians and Ottomans, even hosted or dispatched envoys or diplomats. This was a commerce not only in commodities or precocities traded over long distance but also in the transmission of new ideas, technologies, foods, and further knowledge of distant regions (alongside disease and other contaminations). Necessarily, even the earliest civilizations and empires required a meta-geographical sense of the known world at the least to better understand their enemies.

Another version of regional globalism is supplied by Martin Lewis and Kären E. Wigen (1997) who argue that our present-day assumptions as to continental division and meta-geographical regions are actually cultural concepts developed from classical times. As they argue, such "unconscious spatial frameworks" have to be unpacked, offering the example of the Urals becoming the division between the so-called continents of Europe and Asia. While not denying the roots of Eurasian regional divisions back to Alexander the Great and the importance of physical geography as with mountain ranges, deserts, and oceans, our focus is upon Ptolemaic distortions attending south, southeastern (mainland and maritime), eastern, and northeastern Asia, especially with respect to location and identity. Allowing that the classical template as applied to South Asia (India) more closely matches the modern reality, confusions over *Sinae/Serica* and China, the *Sinus Magnus* and the South China Sea, *India Extra Gangem* and Southeast Asia, and Tartary versus Manchuria, Mongolia, Siberia, and the Russian Far East, took long centuries to untangle. The disambiguation of Australia as a continent from a putative Great South Land is another confusion and the last great Ptolemaic puzzle to be solved. In a similar vein, broader cultural constructs of Orient or Occident practically reaching back to Ptolemy also require unpacking. Viewed in this light, we cannot take the accepted meta-geographical divisions of the world for granted, but should be cognizant that region-making has always been an ideological exercise and with civilizational and cultural hegemons largely in control.

Such a view is also consistent with our reading of the Ptolemaic legacy. However, the notion that world regions can be simply read out of procrustean classical renditions of geography is false, especially given the fluidity of historical change in the millennium postdating ancient Greece and Rome.

Notably, the shift from pastoral or nomadic to settled societies over much of recorded history has been tumultuous, especially across the Eurasian continent. The explosive "Barbarian" Arab, Mongol, Turkish, and Russian-Cossack expansions across Afro-Eurasia added a new dynamic to the classical rendition of the "settled world." The existence of non-state peoples still found in many isolated places around the earth, such as described by ethnologists and others, often unsettles mainstream cultures and their construction of state.

To emphasize the fluidity of cross-civilizational contacts across the Eurasian heartland, the Muslim Arab expansion from the ninth century, overridden by the Mongol conquests of the following centuries and, in turn, the Ottoman Turkish eclipse of Christian Byzantine drew and redrew the classical map of the known and unknown worlds. The expansion of Han China (206 BCE–220 CE) out of the Yellow River heartland was hardly known in ancient Europe and with Song China (960–1279) unknown until the advent of long-distance Arab seagoing voyages. Scarcely comprehended in Europe, the Ghana empire linking with the cross-Sahara trade from an even earlier origin commenced to knit new regional configurations in north Africa. Also unknown to Europe, the Incas and Aztecs would create their own empires in the thirteenth and fourteenth centuries. Contemporaneous seaborne migrations across the Pacific and Indian Oceans islands, as with the "Austronesian expansion" bequeathed a cultural, in part, linguistic "unity" across a swathe of oceanic space from Madagascar in the west to Easter Island in the east. Completely outside of world trading systems and major civilizational influences, a range of non-state peoples as with Australian aborigines devised their origin myths or "dream worlds."

Although Muslim geographers and astronomers had been testing and revising Ptolemaic theory for centuries, it is surprising that northern European cartographers remained stubbornly conservative. This is all the more so in the light of rich empirical evidence of the sea silk roads reaching to China collected at such places as Venice (enabling the production of revisionist world images). Although it is now understood that Copernicus drew upon a body of Islamic knowledge in formulating his heliocentric or Sun-centered model, the Catholic Church in particular was laggard in jettisoning the Earth-centered model such as handed down from the ancient Greeks. As Eugenio Lo Sardo (2013: 59) explains, Aristotelian cosmology was actually upheld by the Counter-Reformation as a bulwark against heresy. Eventually, he declaims, it was the Jesuits of the China mission who led the way in overseeing "the collapse of the last remnants of mythological and religious beliefs in such things as the existence of a Terrestrial Paradise in the southern hemisphere and in the presence of a vast subterranean Hell."

Resisting the post-modernist temptation of simply reading back history, we seek instead to instance the pioneering labors of individuals and institutions (schools of mapmakers and geographers) in actively shaping the image of regions, continents, and future countries. As such, this study places a positivistic spin upon what was then an exciting new age of geography and frontier science, to which many brilliant individuals and civilizations contributed. In other words, my images are time bound allowing of course new understandings of cartographic representation. As such, the present study is not strictly concerned with "spatial structures" outside of a long historical time frame. However, it is alert to culturally-specific or relativized understandings of the universe and the Earth across the vast terrain that this work traverses.

THE "PTOLEMAIC TRAP"

Acknowledging that traditions of astronomy and compilations of geographical knowledge predate ancient Greece and Rome, unquestionably the greatest impact upon Renaissance and post-Renaissance cartographic knowledge was the Ptolemaic view of the world. For the ancients, cosmology was science. Notably, the Alexandrian school of geography, as epitomized by Ptolemy, raised the study of the known world to new heights as with his rendition of the Earth as a sphere, upon which were added representations of the Afro-Eurasian landmass, alongside ocean, dressed in a matrix of coordinates that we now know as parallels of latitude and lines of longitude. Armed with the astrolabe, the astronomical instrument devised by the Greeks, Ptolemy's method of ascertaining and testing measurable data stood the test of time, notwithstanding the shortcomings. Informed by his Hellenic predecessors, Ptolemy's achievement was to place the study of the world on an empirical basis. For Ptolemy, to establish principles of cartography, knowledge had to be tested through scientific observation (as with astronomic calculations) in a systematic manner.

Unique among Phoenician and Hellenic forebears—and their legacy cannot be dismissed—Ptolemy's principles of cartography survived across millennium, just as longitude and latitude determined scientifically are essential navigational principles in use unto today. Carried through in the Islamic world, even allowing for Byzantine embellishments to surviving Greek manuscripts translated into Arabic, the Ptolemaic tradition which also included a geocentric view of the universe was recovered by late medieval and Renaissance Europe, entering the print revolution and the new "discovery" literature of the late sixteenth century. To be sure, centralized bureaucratic polities as with ancient China also observed the heavens, produced star charts, recorded tributary trade missions with near and distant kingdoms, authored geographies, and constructed spatial representations. However, we owe to the

Alexandrian "school" the first mathematically informed and systemically reproducible spatial geographies that had no peer.

While eastward-sailing navigators rounding the Cape of Good Hope in the late fifteenth century and entering the Indian Ocean would soon put to rest the Ptolemaic notion of an enclosed ocean, the identification of a discrete *Sinae* or China as opposed to the Khanate visited by Marco Polo remained in the future. Westward-sailing navigators as with the Magellan-Delcano circumnavigation of 1511–12 and those in his wake would also produce their own scaling of land and sea, especially by taking the measure of the vast Pacific Ocean. As explained in some detail, *Os Lusíadas* (The Lusiads), the masterpiece composed by Portuguese poet laureate Luís Vaz de Camões in the 1560s and 1570s, was entirely redolent of this age of transition, mixing classical erudition with "discovery" tropes which he mostly experienced firsthand.

In an age when the astronomical measurement of longitude was still experimental, it would take at least two centuries before the easternmost extent of the Eurasian landmass was scientifically calibrated, an endeavor commencing with astronomical observations conducted by the first-arriving Jesuit missionaries in Asia in the 1580s and with the French Sun King also investing in the venture one hundred years later. But even for the navigators, the accurate measurement of longitude at sea would not be achieved at least until the invention of the chronometer at the end of the eighteenth century.

At least one of the best-known examples of the weight of classical thinking bearing upon westward-sailing navigators, namely Columbus, was his refusal to believe that he had actually discovered a new continent and that he had navigated to Asia. Discussions on the location of the *Sinus Magnus* relative to Cattigara and *regio Sinarum* or China also entered the conference of the Junta of Badajoz-Elvas in 1524 on the true location of the antemeridian. While, to be sure, the navigators and those who produced marine charts would quickly relegate such classical referents to irrelevance, Ptolemy's *Geographia* nevertheless still retained its intellectual as well as practical spell. This was especially apparent in its method of geo-referencing names on a globe or two-dimensional map surface. In the hands of such practitioners as the Paris-based Cassini dynasty of astronomers and Jesuit emissaries to India and China, the enterprise of measuring the Earth and logging coordinates would carry on through the centuries. This book examines each of these themes.

As taken up in a number of chapters of this work, received cosmographic ideas influencing Columbus and his contemporaries concerned both the circumference of the globe, and the number of miles to a degree, as well as the size of the continental landmass of Europe, Africa, and Asia, in relation to the whole circumference. Another problem with Ptolemy especially relating to his book *Geographia* is the imprecision of his recorded coordinates the further the distance from the Mediterranean. As his modern translators J. Lennart Berggren and Alexander Jones (2002: 22) point out, "At the eastern edge, where the lands represent central China and Southeast Asia, it is virtually impossible to identify any of the features on Ptolemy's maps with real counterparts."

Also, as mentioned, geographers of the ancient world greatly overestimated the eastward extent of Eurasia across 180 degrees of longitude from the western extremity of Europe to the Far East. As map historian Helen Wallis (1992) explains, this was a distortion that increased with distance eastward, seriously misplacing China. Through his underestimation by one quarter the true circumference of the Earth, Ptolemy actually set the "trap" for westward-sailing navigators in search of the source of Oriental "spices," code for trade with India and the East. Because of these distortions, Ptolemy maps reappearing in Europe in the fourteenth century set back by centuries the true revelation of Eurasian space and region.

We find that the major European schools of map production in the seventeenth and eighteenth centuries as with France and the Low countries often lagged the discoveries. In many cases they refused to entirely abandon Ptolemaic stereotypes especially when it came to renditions of the Malay Peninsula, the *Sinus Magnus*, *Terra Australia Incognita*, and China as opposed to "Tartary" or what would become known as the Russian Far East. Even so, setting aside stagnations and reverses, some map productions stand out as bold—even breathtaking—as with revealing the divide of oceans and continents. Accordingly, it is important to acknowledge the key cartographic innovators, their schools, their influences, and their impacts.

An important part of the overall narrative concerns the actions of arriving Jesuit missionaries in China in serving the Kangxi Emperor (r. 1661–1722) in mapping the dimensions of the Chinese empire. With China mapped through a trigonometric method, and with the results feeding back to Europe under the enterprise of official French mapmaker Jean Baptiste Bourguignon d'Anville (1697–1782), it could be said that the ancient Ptolemaic configuration of *India extra Ganges* along with *Sinae* such that had deluded Columbus and others could be laid to rest. Nevertheless, not even d'Anville and emulators could resist drafting post-Ptolemaic maps (as of India), a practice practically continuing into the early twentieth century.

THE ISLAMIC BRIDGE

Allowing that "regional globalism" can be tracked back to the Hellenic world, we should also acknowledge that in an age described as the European

Dark Ages, a truly multiethnic and cosmopolitan civilization came into being, namely, the world of Islam or Dar al-Islam. Appearing like a starburst on the world stage in the eighth and ninth centuries, Islamic *ummat* or community not only spanned the northern Afro-Eurasian heartland but also embraced swathes of southern Europe, surviving in parts of Andalusia until 1492, with outliers across the terrestrial silk roads to China and, via the Arab maritime trade to India connecting up with parts of present-day Indonesia. As Janet Abu-Lughod (1989) established, the *pax Mongolica* actually served to weld an Islamic bridge bringing, importantly, new knowledge to China where Persian astronomers set up an observatory. As a co-fraternity of believers across the bounds of race, Islam did not extinguish nations and regions. To the contrary, it bound believers with a shared identity and with the Arabs bequeathing a new and complimentary global topography/chorography, as with the Maghreb (western North Africa) and its companion, Mashriq, signaling Arab countries to the east of Egypt.

With Islam as a bridge, and with the Caliphate of Cordoba in Andalusia (929–1031) as a prime center of learning, as with translating ancient Greek texts into Arabic, Latin, and Hebrew, Europe regained the intellectual losses of the Dark Ages following the barbarian conquests of the old Roman system. With the *Reconquista*, Catholic Iberia set the scene for the age of discovery, not just as a physical projection of Europe upon a new world, but for the cartographic revelation of the great unknown entering the Gutenberg revolution in Western Europe. We recall that Columbus set sail just after Islamic Grenada fell to Castile in 1492.

Correcting Ptolemy became a small industry on the part of new schools of mapmakers, just as late Renaissance science came to challenge his Earth-centered view of the universe. In a word, European discovery globalism from Polo, Vespucci, Columbus, da Gama, to Magellan (and their respective chroniclers) was epochal to the fate of millions. This is not only an allusion to the victims of Europe's New World conquests, but the material and nonmaterial dividends supplied to Europe. Even setting aside the wholesale plunder of New World gold and silver, Europe received vital new botanica, pharmacopoeia and food sources, alongside new knowledge on non-European cultures, including earth science, navigational lore, and commercial intelligence, seemingly ushering in an age of European exceptionalism (although more enlightened historians today would not so readily dismiss the scientific, material, and artistic achievements of non-European civilizations).

The Gutenberg revolution in printing, both north and south of the Alps, also boosted the dissemination of new information gleaned from the discoveries, some factual, some fantastic. As frequently referenced in world history, it was the European Enlightenment that positioned relativism as a credible optic upon the world outside of religious-cosmographic orthodoxy which, in Europe, was the church. In my design, the appearance of universal histories and encyclopedias in various European centers marked the dawn of a new age of rational learning and critical reflection based upon observation. As alluded, the announcement and gradual acceptance of a heliocentric world was also prophetic of this sea change in thinking as the scientific revolution took off.

INTERPRETATIONS, INTERROGATIONS, AND METHOD

Without question, the royal patrons, elites, and commercial classes across Europe were the first consumers—and critics—of the new geographical information entering print in vernacular language editions, as with Ramusio, Linschoten, Hakluyt, Harris, Churchill, and the even more accessible Enlightenment-era collections, and there was much to wonder, review, and interrogate. As with any historical interpretation, my rendition stands in a long line of interpreters: some ancient, some modern, and others postmodern. An early standard still worth consulting is *Essai sur l'histoire de la cosmographie et de la cartographie pendant le moyen-âge*, a multivolume work, accompanied by an atlas of 117 world maps published in Paris between 1848 and 1852 by Portuguese diplomat-author Manuel Francisco de Barros e Sousa, the second Viscount of Santarém. In pioneering this field, Visconde de Santarém (to adopt his Portuguese title) literally gave new meaning to cartography (Webster's dictionary dates the first use of this term to ca. 1847). Especially worthy of mention is the elucidation of the influential Dieppe school of mapmakers by the French cleric Albert Anthiaume in his *Cartes marines*. Published in Paris in 1916, his method is surprisingly modern. A major influence upon Anthiaume was the Finnish Arctic explorer Adolf Erik Nordenskiöld (1832–1901), known for his facsimile-atlas of the early history of cartography (1889). The various publications of Armando Cortesão on Portuguese cartography and navigational techniques certain dating to the prewar period are outstanding (1935). Leo Bagrow's *History of Cartography* (1951) is another benchmark in this field. Founder of the journal *Imago Mundi*, Bagrow's iconoclastic text was also revisionist with respect to his interpretation of Ptolemaic map renditions.

Allowing a revival of interest in global meta-geography, my method is both constructive and deconstructive. Two seemingly contradictory trends are at work here. One is the endurance of the Ptolemaic world template such as recovered in early Renaissance Europe notwithstanding revisionist astronomy, voyages of discovery, new science, and new data including that supplied by local Asian map traditions. The other concerns the erosion of the Ptolemaic legacy, first his astronomy as the scientific revolution advanced and, in long stages, his geography (excepting of course his mapping system and concept of latitude and longitude). As such, the book unfolds a fourfold agenda. First, it looks back to classical points of reference, essentially to highlight the weight of Ptolemaic representations of the Eurasian world as it was received in late medieval Europe from Arabic sources. Second, it tracks the gradual erosion of the Ptolemaic template in the face of empirical truth revealed by medieval travel accounts and, more dramatically, via the revelations of the Columbian, da Gama, and Magellan revolutions. Third, through selected examples, as with India, Southeast Asia, and China, it seeks to expose textual and cartographic adjustments to the classical models flowing from the scientific revolution. Fourth, through an examination of Jesuit astronomical observations conducted in Malacca, Thailand, Macau, and other points in China, as backed by the French crown, it demonstrates how Eurasia was actually measured and sized with respect to its true longitudinal coordinates such had deluded Columbus and even succeeding generations. As argued, by overcoming the "Ptolemaic trap" the constructions of place produced by schools of European mapmakers and establishments also shadowed the age of European colonialism, and with their legacy even surviving the age of decolonization and the shaping of new nation-states. In short, my project is to problematize the creation of geographical knowledge, to raise awareness as to the historical construction of region in Asia, and, as a more latent agenda, to sound an alert as to the perils of overdetermination in the construction of boundaries such as exemplified by international legal questions besetting "ownership" of the South China Sea.

THE BOOK

Referencing the Hellenic tradition and setting the scene, chapter 1 elaborates upon the historical Ptolemaic system including both its mathematical and descriptive components, the tri-continental geographical template to which all neoclassical studies would refer back to. It also examines the written record, namely, the ancient tradition of periplus or navigational guides alongside the two seminal world histories of ancient Rome, respectively those of Pliny the Elder and Strabo as they relate to Eurasia, also becoming reference points for European geographers and cartographers practically into the modern age.

Chapter 2 highlights the custodial role of Islamic scholars in retrieving the "lost" Alexandrian tradition, making it accessible to late medieval Europe alongside broader Islamic knowledge of the Eurasian world. A first section of this chapter seeks to establish the importance of the Islamic crossover with Latin Europe, especially with reference to the transmission of knowledge, both philosophic and scientific, not excepting transmissions of Ptolemy to the Arab world. Second, the chapter elaborates upon the seminal contributions of the twelfth-century Sicily-based Arab geographer Muhammad al-Idrisi, especially his framing of Asia, including China. A third section examines Arab primacy in both theoretical astronomical and practical navigational knowledge of the Indian Ocean far ahead of Europe. A fourth section discusses Arab/Muslim geographies of Asia, while a final section draws attention to Islamic encounters and exchanges across the terrestrial and maritime silk roads to China reaching their apogee during the Mongol-Yuan dynasty with its capital in Khanbalik.

The process of Ptolemaic revival and reception in Renaissance Europe is carried through in chapter 3. To this end, it investigates the transitional status of astronomical and cartographic knowledge during this period with respect to its classical lineages. It also references Catholic Europe's outreach to the Orient noting the primacy of Venice in connecting up Latin Europe with the Eurasian silk and spice roads. It then evaluates the impact of the Ptolemy-revival upon navigation and the Columbian discoveries, and reciprocally examines revisions made by the leading schools of European mapmakers in the light of practical information supplied by the discoveries.

The impact of the Iberian discoveries upon the Ptolemaic stereotype leading to its eventual disaggregation and abandonment is assessed in chapter 4. More broadly, this chapter sets down the main lines of the Iberian encounter with Asia, especially with respect to the construction of regions and how knowledge of this meeting across civilizations was received back in Europe. A first section is concerned with Portuguese and Spanish navigational advances in the age of discovery. A second section develops the themes of Iberian "expansion" with specific reference to the Columbian and Magellan revolutions, while a third section focuses upon Iberian mapmaking in the wake of Columbus/Magellan.

Chapter 5 seeks to make sense of how the European discoveries were incorporated into textual accounts entering print. As per the Ramusio collection of the late sixteenth century we are concerned to highlight the precocious role of Venice in disseminating travel narratives as well as in creating its own spatial framing of the New World. With respect to Portugal, we highlight Luís Vaz de Camões's eternal classic *Os Lusíadas* as a transitional text, informed by the discoveries but warped into a classic mold. By contrast, in *Peregrinação*, Camões' Portuguese countryman Fernão Mendes Pinto melds fact with fiction in his swashbuckling account but also redolent of a sense of place. Finally, Englishman Peter Heylyn sets out the English priority, mixing fact and fantasy over hundreds of pages, yet never entirely divorcing himself from the classics.

Chapter 6 describes a Ptolemaic world-in-motion, and with scholarly elites increasingly fed by Enlightenment science and the application of reason. Neither does it ignore the longevity of Chinese astronomical traditions with respect to the arrival of Jesuit astronomers on the coast of China from the last decades of the sixteenth century offering to Europe the first convincing post-Ptolemaic imaging of China's actual space since the age of Marco Polo. This chapter leads with a brief discussion of the (European) scientific revolution as it came to revise classical and medieval conceptions. It then offers the example of the Jesuit challenge to Ming Chinese astronomical traditions, alert as well as to Jesuit revelations of the size and shape of China and the Eurasian landmass finally putting paid to the Ptolemaic stereotypes. A third section examines the ancient Chinese astronomical tradition as confronted by the first-arriving Jesuits in China. A final section highlights official French attempts to harness Jesuit astronomical prowess to service their academic ambition to measure the globe.

Chapter 7 reveals how, working with the Kangxi Emperor, the Jesuit mapping of China brought to Europe, at the hands of French cartographer d'Anville, new conceptions of the old Cathay. First, this chapter positions d'Anville's scientific mapmaking with reference to classical renditions of Asian space. As a foil to d'Anville's neoclassical pretensions, a second section discusses the new relativist thinking on global geography such as presented by French Enlightenment philosophers.

A final chapter examines the clash besetting Western and Oriental concepts of power, state, and territory, as exemplified by post-Columbian explorations of the Pacific and the historical geography of the South China Sea. A first section introduces early Spanish constructions of China and the China Seas. A second section turns to attempts by a range of nations and discoverers entering the southern Pacific Ocean to reveal Ptolemy's Great South Land. A third section turns to an examination of the Western legal concepts of *Mare clausum* (closed sea) versus *Mare liberum* (open sea). Then follows a fourth section bringing to the fore French/English mapping, surveying, and measurement of the South China Sea. The chapter concludes with a discussion on what I view as a clash of concepts besetting colonial interlopers and age-old Asian dynastic priorities, with echoes down unto the present.

A conclusion examines post-Ptolemaic constructions of empires and nation-states. While the central focus of this book has been upon the way Europe imposed its conceptual renderings of space upon "Asia"; here we seek answers to the question as to Asian cognizance of space and region, or how

"Asia" was valorized locally in such disparate culture areas and regions as China, the Russia Far East, Japan, India, Southeast Asia, and parts thereof?

NOTES

1. See the author's "East-Southeast Asia in World History," *TRaNS: Trans – Regional and – National Studies of Southeast Asia* vol. 2, no. 1 (2014), 5–30.
2. To this end, Mitchell (2012: 3) examples the (re)-discovery of linear perspective (in art) by Filippo Brunelleschi (1377–1446), just fifteen years after d'Angelo's translation of Ptolemy's *Geographia.* Citing Edgerton (1975: 120), author of a dedicated study on this subject, this was an application of Ptolemy's Euclidean geometrical model pure and simple.

Chapter 1

The Classical View of the Afro-Eurasian World

While a truly scientific understanding of the universe would await Galileo, Copernicus, Kepler, Newton, and others, the sense of the sphericity of the world based upon celestial observation and measurement in general dates to Greek cosmology in Asia Minor, carried through especially with the Alexandrian school of geographers associated with Claudius Ptolemy. Granted that other civilizations also developed their own understandings of the cosmos, the Hellenic tradition is exceptional. Not only did it present a mathematical representation of the earth, but also as it also came to influence Islamic conceptions of geography, fitted in with medieval Christocentric views of a geocentric universe and, with the neoclassical revival of the European Renaissance, would gain new credibility as manuscript versions including maps entered print fixity. More generally, as this book develops, it would take much experimentation and critical testing of knowledge before post-Ptolemaic representations of the Asian world gained credence in European elite circles.

In the way of setting the scene, this chapter opens with a discussion on the Alexandrian tradition of celestial observation back to its Greco-Roman and Babylonian antecedents. Second, it examines the written record, namely, the ancient Roman tradition of *periplus* or navigational guides, alongside the two seminal world histories of ancient Rome, respectively those of Pliny the Elder and Strabo. Especially as we need to know how the very first renditions of Ptolemaic space applied to India and Eurasia in general, a third section focuses upon the Ptolemaic achievement and method of geography. Not simply an academic exercise, a fourth section explains the rediscovery of Ptolemy in early Renaissance Europe and, with the invention of print, the dissemination of Ptolemaic knowledge including map images on the cusp of the age of discoveries.

THE GRECO-ROMAN ASTRONOMICAL TRADITION

Just as the ancient Greeks drew upon anterior astronomical traditions, so many civilizations gazed the heavens, produced star charts, or made other observations and calculations certain relating to calendars and navigational almanacs. Depictions of the skies range from aboriginal Australian cave paintings, to the civilization that decorated the Lascaux cave paintings in France, to evidence from prehistoric humans in China and to ancient Egypt. In the earliest Greek cosmological systems, the Earth was envisaged as a shield-like disc floating on water and surrounded by the mythical streams of Ocean. This is the view presented around 800 BCE in Homer's *Iliad* and *Odyssey*, and with the *Iliad* illustrating a "shield of Achilles" depicting a "river ocean," enshrining the idea of an ocean sea encircling the continental land and with seas, Sun, Moon, and stars all part of the center of the disk. As map historian Thomas Suárez (1999: 60) summarizes, the Greek world of sea and islands undoubtedly helped to frame this conception of space. Over time, highly schematic and circular conceptions of the world emerged in the Greek tradition, more generally distinguishing between the landmass of the *ecumene* or known or inhabited world, and the unknown, and were surrounded by a ring-shaped ocean.

As Walter Burkert (1992: 128) explains, by around the eighth century BCE, the Greeks had entered into intensive cultural exchanges with the high cultures of the Semitic (or Mesopotamian world) and adopted and transformed what they had received and learned. Aside from a massive import of goods into ancient Greece during this "orientalizing" period, there was also a direct import of written culture, as with the alphabet, the writing tablet, the scroll, and the book format. In a word, he concedes, the Greeks were the most easterly of the Westerners. As better understood today through archaeological research, the key to this exchange were the contributions of Babylonian astronomers. As Francesca Rochberg-Halton (1998: 752) explains, Babylonian lunar and planetary theory became the foundation for further refinement by Greek astronomers Apollonius, Hipparchus, and Ptolemy.

Although the paradigm of a spherical earth is sometimes linked to Pythagoras (sixth century BCE), in 330 BCE Aristotle was one of the first Greek thinkers to propose a spherical Earth on the basis of physical theory and observational evidence and, as mentioned below, his legacy is duly acknowledged by Ptolemy. The ancient Greeks stand apart from other civilizations, China included, for their development of mathematical astronomy. Greek geometry especially came to express precise relationships within real and conceptual geometrical bodies.

While the division of a circle into 360 degrees is believed to have originated in Babylonia, the concept was taken up by the Greeks as with Euclid (mid-third century BCE), and Apollonius of Perga (second century BCE). In the second century BCE, Crates of Mallus devised a terrestrial sphere that divided the Earth into four approximately symmetrical continents, two in the north and two in the south, separated by great rivers or oceans, with people presumed living in each of the four regions, including a land of the Antipodes (Thrower 2008: 22). Carried on by Ptolemy or more generally the Ptolemaic tradition, a belief existed among the ancients that the Indian Ocean was enclosed on the south by land, and that the lands of the northern hemisphere should be balanced by land in the south.

Living three centuries before Ptolemy (and a contemporary of Crates), Hipparchus of Nicaea (present-day Iznik in Turkey) also absorbed Babylonian astronomical understandings especially as to the motions of the planets. He also divided the circle into 360 degrees of 60 arc minutes. Hipparchus and his predecessors used various instruments for astronomical calculations and observations, including the astrolabe used for the measurement of geographic length (longitude) and width (latitude) of the observed stars from any part of the Earth. Together with the gnomon or sundial used to calculate time, and the armillary sphere to demonstrate the motion of the stars (and with the compass added), they became basic navigational devices during the European age of discovery. Besides constructing star charts and celestial globes, Hipparchus devised an imaginary grid of equally spaced parallels and meridians crossing each other at right angles. He also determined latitude and longitude by astronomical observation.

Marinus of Tyre is often upheld as the founder of mathematical geography active in 100–150 CE. He was also an influential figure especially in the way that he assigned to each place its proper latitude and longitude, and introduced improvements in the construction of his maps (*Encyclopedia Britannica,* Vol. 17, 1919). According to Norman Thrower (2008: 22–23), Marinus attempted to enrich the world map by incorporating information from new sources. These contributions directly influenced Ptolemy, a later critic of Marinus (see Figure 1.1).

The Alexandrians

Consensus holds that Claudius Ptolemy was established in the port city of Alexandria ca. 146 CE–170 CE (taking his name from the ruling Ptolemy dynasty). Then a Roman outpost, Alexandria had been founded in 332 BCE by Alexander the Great whose army put an end to a long period of Persian domination in Egypt, a reference to his siege in 332 BCE of the ancient Mediterranean city-state of Tyre (in present-day Lebanon). With a new Macedonian order firmly established, Alexandria emerged as a major capital of the ancient Hellenistic world and remaining so for almost a thousand

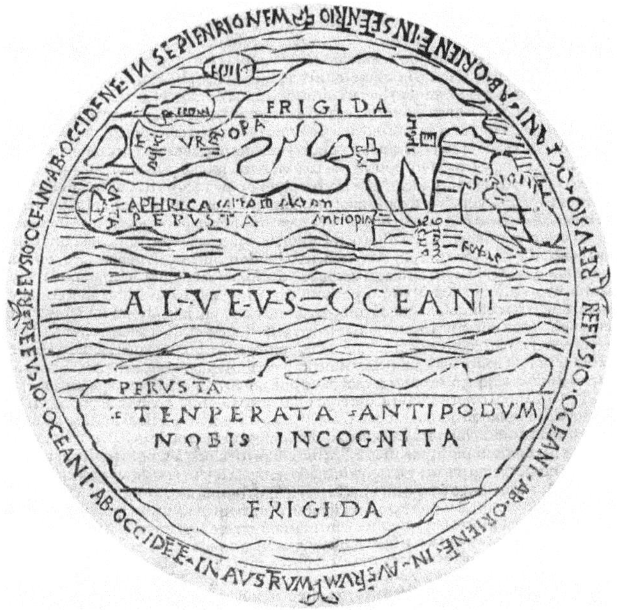

Figure 1.1 Roman-era World Map after Macrobius. Ambrosius Aurelius Theodosius Macrobius (AD 395–423). *Commentarii in Somnium Scipionis,* 1492. Europe, Africa, and Asia are shown in the upper hemisphere. A vast southern continent lies in the lower hemisphere. An ocean called Alveus separates the continents. *Source*: State Library of New South Wales.

years until falling to Muslim conquerors in 641 CE (Goodchild 1981: 17–18; 37–38). Under the Ptolemies, the Egyptian dynasty of that name, Alexandria became the new cosmopolis in the eastern Mediterranean, reputedly hosting one of the largest and most significant libraries and centers of learning in the ancient world. There, in Ptolemaic Egypt, the Egyptian tradition merged with Greek astronomy and Babylonian astronomy. Ptolemy and others of the Hellenic world were also heir to the Phoenician seafaring tradition that spread across the Mediterranean from 1550 BCE to 300 BCE, including celestial navigation making use of the stars and constellations. The ancient Greeks also employed the astrolabe.

Preceding Ptolemy at the head of the library at Alexandria, Eratosthenes of Cyrene (present-day Libya) (ca. 276–196 BCE) postulated a spherical Earth divided into zones. The Earth's circumference was first determined around 240 BCE by Eratosthenes (and with some accuracy as borne out by modern measurements). He also preceded Ptolemy in creating the first map of the known world incorporating parallels and meridians and based on the available geographical knowledge of the era. In assuming a spherical Earth, he also understood the 360 degree division. As discussed below, this would be revealed in *Geographia* after its recovery and translation into Latin in 1409. Dubbed the "father" of geography, he believed that the known world was surrounded by a continuous sea (a conception carried forward by other Greek scholars). He also allowed that Africa was separate from a southern continent. According to Thrower (2008: 20), Eratosthenes's measurement of the Earth and his method should be regarded as one of the great achievements of Greek science. Drawing upon the legacy of Alexander the Great, he also placed India on his world map, actually represented as a quadrangular shape and thus capturing the peninsular character of south India. The large island of Taprobane (Ceylon) is recorded (Lach 1965: 9–11; Faussett 1993: 10).

Excursus on the Spherical World

As Bradley E. Schaeffer (2005) has queried, is it plausible to date celestial globes back to the time of Hipparchus? He answers in the positive. Ptolemy's *Almagest* (Book VIII, chap. 3) gives a detailed discussion on the construction of solid globes for showing stars. The concept of star globes was common in Greek times as evidenced by the existence of many Greek and Roman coins and engraved gems that show such globes. In particular, a small bronze coin from Roman Bithynia depicts Hipparchus seated in front of a globe resting on a table. But the primary evidence that star globes date back at least to Hipparchus is that Ptolemy specifically states the matter in *Almagest* (Book VII, chap. 1). Notably, as taken up in chapter 4, even Copernicus was obliged to acknowledge such astronomer predecessors as Hipparchus, Ptolemy, and al-Battani (Albategnius), the medieval Raqqa (present-day Syria)-based Arab mathematical astronomer known for correcting the solar year and some of Ptolemy's calculations. Celestial mapping in the Islamic tradition, it should be mentioned, primarily adopted the Ptolemaic conceptualization of the skies, adding elements from pre-Islamic Bedouin influences in the Syrian-Iraqi heartland from the seventh century (Savage-Smith 1992: 70).

Little physical evidence of Greco-Roman renditions of the universe as a sphere remains outside images on certain coins and other fragments. In this sense, the so-named Farnese Atlas at the National Archaeological Museum in Naples is exceptional. A second-century Roman marble copy of a sculpture of the Titan Atlas kneeling with a celestial globe weighing on his shoulders, the Farnese Atlas is the oldest known representation of the celestial sphere, also displaying an accurate representation of the ancient Greek constellations. Standing 2.1 meters tall, and with the globe 65 centimeters in diameter, the Titan Atlas globe records forty-one constellations (as distinguished by Ptolemy) accurately placed against a grid of reference circles, including the Equator, tropics, and colures (either of two great circles on the celestial sphere) the Arctic and Antarctic circles. It is simply the oldest surviving depiction of the original Western constellations (actually matching descriptions from Hipparchus's only surviving work, *Commentaries*, according to some).

THE CLASSICAL WORLD REGIONAL VIEW AS RECEIVED IN STRABO AND PLINY

As attested by modern archaeology, various Roman objects such as coins have been recovered on the two coasts of India, at points on the Malay Peninsula, and at Funan (Oc Eo), site of a major Indianized civilization in the western Mekong River delta (present-day Vietnam). Nevertheless it was India that was best known to the ancients including Ptolemy through trade contacts reaching the Mediterranean. As befits an empire, the accounts bequeathed by a handful of Roman authors are seminal to any understanding of this world and its external connections. Among them are Strabo, Pliny, and Ptolemy, and the anonymous author of the Periplus of the Erythraean, as discussed below. Neither should we ignore the Alexandrian legacy, a reference to the wealth of knowledge brought to ancient Greece by the eastward thrusting armies of Alexander the Great. Writing at different periods straddling the advent of the common era, they were all heir to a shared Greco-Roman tradition, itself building upon precursors who had earlier stamped their authority upon the eastern Mediterranean region. To be sure, as Lach (1965: 12) remarks, these were elite conceptions of distant realms, of which locals were practically oblivious. In any case, it would take time before the notion of map projection would evolve in the sense of acknowledging regions of the world along with relative position and size.

Separate from the tradition of celestial observation, the Romans developed their own practical tradition of rendering space, especially for administrative and defense purposes. According to Claude Nicolet (1991: 95–98), during the Augustan period (27 BCE–AD 14), there were a series of undertakings related to geography, cartography, and administration, along with the publication of cadastral books, a census, and maps. Notable from this period is the map of the known world by Roman statesman Vipsanius Agrippa, including an oversized India. Aggripa supplied measurements in length and width. However, unlike the Ptolemaic tradition, he eschewed coordinates. Neither was his map aligned north and south. Augustus was also a road builder and the Roman metric of miles was a standard across the empire, feeding into *peroploi.* Aggripa's was a simple cylindrical projection, as opposed to conical projection in which the meridians are drawn in terms of converging lines (such as discussed by Ptolemy). Especially, Aggripa's conception of political-administrative regions or *provincia* fed into Pliny (and perhaps Strabo) among other authors down until medieval times.

Aggripa's legacy may be seen in the *Tabula Peutingeriana* or Peutinger Table believed to be a thirteenth-century copy of a Roman map dated to around 300 CE including segments representing India, Taprobane, and even China. An elongated parchment scroll 0.34 meters high and 6.75 meters long and lacking scale, the map was rediscovered in Germany and once owned by Konrad Peutinger, a sixteenth-century scholar. Notably, it traces the *ecumene* (as known to the Alexandrians) in great detail. For instance, the port city of Muziris on the Malabar coast of India (present-day Kerala) is marked Templ(um) Augusti or Temple of Augustus, highly suggestive of Roman trade links (McLaughlin 2010: 10–11).[1] Mapmaker Abraham Ortelius commissioned manuscript copies of the original, which were made in 1598.

Alexander the Great

Epochal in the sweep of ancient world history, between 326 BCE and 324 BCE, Alexander the Great (356–323 BCE) drove his armies eastward through Persia until the Indus Valley leaving a mixed legacy of conquest although returning to Greece with a vastly superior knowledge of the inhabited world. As heir to Greek civilization, the Macedonian world conqueror was undoubtedly informed as to the possibilities of reaching an ocean at the end of the eastern world. As Lach (1965: 5–6) explains in a learned discussion on "India and the Greek tradition, 600–100 BCE," the "father of history" Herodotus (ca. 484–425 BCE), had formed a conception of areas east of Mesopotamia with the notion of "India" occupying the easternmost space of the known world. Around 300 BCE, the Greek historian Ctesias of Cnidas wrote treatises called *Indica* and *Persica* seeking to correct Herodotus, though injecting even more imaginative elements into his description, many lingering on through Pliny and others to the European Middle Ages. Aristotle (384–322 BCE) had also postulated a vast Asian continent.

As Suárez (1999: 61) comments, Alexander's decision to split his armies into three divisions for the return journey, respectively, inland, coastal, and ocean, greatly expanded Greek knowledge of central Asia and Indian subcontinent, along with the western Indian Ocean. His conquest also initiated more regular trade with India in spices or, more generally, exotic products in demand. Still, having reached the Indus, and apprising the Indian subcontinent, he failed to make even indirect contact with lands beyond India (or the Ganges). The historical Alexander, as known today, owes much to Arrian of Nicomedia's *Indica* (ca. 150 CE), a work which mostly describes what Alexander's lieutenant Nearchus witnessed on the return voyage from India. According to Lach (1965: 8), Arrian also brought new knowledge of the

Ganges and the eastern and western parts of the Indian Ocean. Nevertheless, the legacy of Alexander the Great's expedition would continue to be reinterpreted down until modern times, especially as it both distorted and fed into Ptolemaic and post-Ptolemaic renderings of Indian space.

With Rome replacing Greece as the new hegemon over Egypt and the Red Sea littoral, the ancient trade routes gained a new impetus. To note, an envoy from Ceylon arrived in Rome at the time of Augustus in 26 BCE. Some Roman subjects visited India during this period and with Roman missions reaching China in the second and third centuries. During this period both land and maritime routes were wide open, a window not again ajar until the Portuguese entered these seas in the sixteenth century (Lach 1965: 13).

Emerging after a period of internal dissension following the death of Alexander, Emperor Seleucus Nikator (around 300 BCE) failed in another military adventure to the Indus. He was, however, more successful in diplomacy, dispatching an emissary, Megasthenes, to cross India to the Maurya court in Pãtiliputra (Patna) along the valley of the Ganges. Megasthenes's work *Indica* expanded Greek knowledge to embrace not only the Ganges River system and its origins in the Himalayas, but Ceylon, Tibet, the shape of the Indian continent, the tropical zone with its unique flora and fauna, and, importantly, the monsoon system facilitating long-distance travel across the Indian Ocean. In Lach's (1965: 8) words, although lost, Megasthenes's work would "set the image of India for the Greek world."

The Periplus of the Erythraean Sea

Route descriptions were undoubtedly a feature of Mediterranean sea voyaging of great ancestry, although few remain. One that has survived the ages is an anonymous account of the maritime trade in the Red Sea area connecting with the monsoon trade. Dubbed the *Periplus of the Erythraean Sea*, it is dated from 40 to 70 CE.[2] In the words of Raoul McLaughlin (2010: 9), author of possibly the first detailed study of Rome's eastern trade, no other surviving Roman text offers such detail on commerce and eastern markets, and without the *Periplus* our understandings would be considerably impoverished.

As with other Roman texts, certain descriptions of places can be matched with present locations; but others have spawned considerable debate and even a small literature. The *Periplus* is also valuable in mentioning many of the key trade goods driving the east-west maritime trade, but some trade goods are obscure as well.

> This whole voyage as above described, from Cana and Eudaemon Arabia, they used to make in small vessels, sailing close around the shores of the gulfs; and Hippalus was the pilot who by observing the location of the ports and the

> conditions of the sea, first discovered how to lay his course straight across the ocean. For at the same time when with us the Etesian winds are blowing, on the shores of India the wind sets in from the ocean, and this southwest wind is called Hippalus, from the name of him who first discovered the passage across. (para. 57) (Schoff 1912)

Hippalus, a Greek navigator and merchant living in the first century BCE, is credited with the revelation that the prevailing winds in the Indian Ocean reverse direction seasonally. He also hypothesized that, out of sight of land, these seas could be traversed and, with the reverse of the winds, a return voyage could be undertaken. As also recorded by Pliny, Hippalus successfully sailed from the mouth of the Red Sea to the delta of the Indus River and thence to the Malabar or western coast of India (para 58) (Suárez 1999: 61).

India is well attested in the *Periplus*, along with Ethiopia, Eritrea, Somalia, the Hadraumaut coast, and the East African or "Swahili coast," plausibly matching present-day Tanzania. But trade with the Indian harbor of Barygaza (identified with Baruch in Gujarat) is especially referenced, along with a long list of trade items. The Chola kingdoms, the Tamil country, and the enigmatic Chryse are also mentioned. By the second century CE, as described in the *Periplus*, Roman sailors were frequenting the eastern coast of India, just as Asian commodities were arriving in larger quantities in the eastern Mediterranean, including (Puteoli) Puzzuoil, near Naples. Again, the Roman connection with the two coasts of India, whether direct or indirect, is matched by archaeological evidence.

According to Enlightenment philosophe Guillaume-Thomas Raynal (Abbé Raynal) (1784: 96–99), under the ruling Egyptian Ptolemies, "Alexandria was the mart of the merchandise that came from India, and the port of Berenice on the Red Sea, was put into a state ready to receive them." As validated during the Napoleonic interlude in Egypt, a Nile canal linked to the sea had been constructed to better facilitate this trade. According to Raynal—although offering no source—boats entering this trade not only plied the Red Sea but sailed south down the east African coast and, with others sailing toward the Persian Gulf and further east to the mouths of the Indus, the Malabar coast, and Taprobane/Ceylon. In turn, "All the seafaring and trading nations in the Mediterranean resorted to the ports of Egypt to purchase the productions of India."

Evidence of Trade with China

In this age, trade was staged moving from one merchant group to another and it was by this means that trade goods from as far away as China connected with the ancient Greco-Roman world. A range of Chinese items unearthed in archaeological sites in the Mediterranean world suggest that the trade was considerable, as with Chinese earthenware and bronzes. But, outside of such trophies, a small catalog of Asian perishables also reached the Roman world, from Indian slaves to exotic animals, to pearls, spices, and, importantly, silk from distant China. With Rome via Syria and Egypt connected up by the terrestrial "silk roads" spanning the Eurasian landmass, alongside coveted silk, so "spices"—a blanket term embracing a range of eastern exotics—arriving from India and beyond were the main staple of the maritime trade (at least alongside silver going the other direction). So considerable was the trade, that the drain upon Roman silver aroused the scorn of Roman moralists (Lach 1965: 15). Suárez (1999: 61) is undoubtedly correct to suggest that along with the trade link came expanded geographical knowledge.

Strabo (64/63 BCE–ca. 25 CE)

We detour to profile Strabo, first to acknowledge him as a pioneer geographer and, second, to prefigure a future Strabo revival in the Renaissance, meaning that his legacy reacquired a new importance among geographers along with a Ptolemy revival, as discussed below. Born in Amasya, present-day Turkey, Strabo also traveled widely in the Hellenic world, Egypt included. Moving to Rome in 44 BCE, he lived in the time of Augustus and the early part of the reign of Tiberius (Hamilton and Falconer 1854–57, vol. III: v). In his seventeen-volume work *Geographica*, Strabo presented a descriptive history of people and places from different regions of the world known to his era, embracing a survey of the topographical, historical, and political characteristics of the principal regions of the Roman world. Besides history, *Geographica* also engaged philosophy, political theory, mathematics, earth science, and history. Among a long list of predecessors in the study of geography, Strabo cited the antique Greek astronomers Eratosthenes and Hipparchus ("Learn celestial phenomena from Hipparchus"), but it is "the illusions of Homer," he eulogizes, "while scoring the factual data of Herodotus" (Lach 1965: 13). While acknowledging the astronomical and mathematical efforts of such giants toward geography, he also claimed that a descriptive approach was more practical. According to his modern translators, H. C. Hamilton and W. Falconer (1854–57, vol. III: vi), written in Greek from Greek sources, *Geographica* may not have had great impact in his age. Neither, according to McLaughin (2010: 9), did he update events in India occurring in his own lifetime despite the records of thousands of Roman merchant travelers sailing eastward every year.

Strabo also assumed that the Earth is spheroidal, offering the proof that those who sail cannot perceive lights at a distance when placed at the same level as their eyes (Hamilton and Falconer 1854–57, vol. III: x). In paying tribute to Homer, "the founder of geographical science," Strabo acknowledged his truism that the Earth was completely surrounded by oceans. He also described countries by name, as with Libya, Ethiopia, Arabia, countries further east (India, and Siam), and west (Iberia), along with Isles of the Blest (Canary Islands) (although Homer was in fact not aware of these distant places). Similar understandings of land meeting ocean in the north or the Arctic and to the south (Arabian Gulf; Cape Comorin) was also implied. He also acknowledged a communicable Atlantic Ocean (not interrupted by isthmus). His understanding was of a circumnavigable sea. He acknowledged tidal movements, as with the "ebb and flow of the sea, both in the increase and the diminution, is everywhere identical, or at all events has but little difference, as if produced by the agitation of one sea, and resulting from one cause." He even contradicted Hipparchus on this issue, "denying that the ocean is every where similarly affected" (Hamilton and Falconer 1854–57; Strabo Vol. I Book I: 8). In hailing Hipparchus, he stated, "the only means we have of becoming acquainted with longitudes of different places is afforded by eclipses of the Sun and Moon. . . . Everyone who undertakes to give an accurate description of a place should be particular to add its astronomical and geometrical relations, explaining carefully its extent, distance, degrees of latitude, and "climate," comparing this task with an architect building a house. "Thus it appears that the knowledge of geography is connected with meteorology and geometry, that it unites the things of the earth to things of the heaven, as though they were nearly allied, and not separated" (Hamilton and Falconer 1854–57: 13–14).

What is important to our thesis is that, as a textual geography, Strabo framed regions or country-regions, many of them enduring. Besides evocations of European countries and regions, in Book 15 Strabo describes the southern side of Asia below the Taurus with India and Persia (divided into Asiana or east Persia and Persis or west Persia). Book 16 describes the western half of south Asia (Assyria with Babylonia, Mesopotamia, Syria, Phoenecia, Palestine, the Persian and Arabian Gulfs and the coast of Ethiopia, and, lastly, Arabia), and with Book 17 dedicated to Africa. Strabo's *Geographica* may have circulated throughout the Byzantine Empire in manuscript form but all copies are ancestral to a now-lost copy discovered in the European Middle Ages. According to a study by Falconer (preface; Hamilton and Falconer 1854–57 Vol. 3: xviii), in the mid-nineteenth century, some twenty manuscripts then survived in European libraries.

Pliny the Elder

The Roman administrator and writer Pliny the Elder is best known today for his encyclopedic *Naturalis Historia* or *Natural History* published ca. 77–79

CE (and with the historical Pliny ending his life in in the eruption of Vesuvius in 79 CE). It is also one of the largest single works to have survived from Roman times. The number of extant manuscripts is about 200. *Natural History* was one of the first ancient texts to be printed, appearing in Venice in 1469, and with the first translation appearing in English in 1601 (see Holland 1601). Nevertheless, Pliny's *Natural History* replete with such fantastical elements as sciapods or umbrella feet men, cynocephalu or dog heads, Amazons, and so on, continued to cast its spell into the late Middle Ages or even beyond. More generally, his evocation of anthropophagi such as entering the island realms of the nether reaches of the Earth even came to influence late Renaissance thinking, perhaps as reflected in early Spanish attitudes toward the natives of the Canary Islands and the Caribbean.

After an initial survey of cosmology and geography also announcing a range of European locations in Book 5, Pliny turns to North Africa; the Eastern Mediterranean and Asia Minor. Book 6 covers countries from the Black Sea to India; Persia; Arabia; Ethiopia; and the Nile valley. Having described in some detail the Black Sea, and the Armenia region, Pliny's description of Asia is suggestive, but highly deceptive. Notably, a description of silk making adds an ethnographic touch and verisimilitude to the existence of China. Following Horace Rackham's (1949–54) translation:

> XX. The first human occupants are the people called the Chinese [Seres], who are famous for the woolen substance obtained from their forests; after a soaking in water they comb off the white down of the leaves, and so supply our women with the double task of unraveling the threads and weaving them together again; so manifold is the labor employed, and so distant is the region of the globe drawn upon, to enable the Roman matron to flaunt transparent raiment in public. The Chinese, though mild in character, yet resemble wild animals, in that they also shun the company of the remainder of mankind, and wait for trade to come to them.

Obviously the geography of China was not known as a bounded region, at least not in the way of Persia and India, and it would take practically a millennium before medieval travel accounts by Christians and Muslims would improve upon Pliny.

Pliny also explains his sources and the state of knowledge on India going back to Alexander. He is no less expansive on the island later becoming known as Ceylon, explaining (XXIV), "We however have obtained more accurate information during the principate of Claudius, when an embassy actually came to Rome from the island of Ceylon" (Rackman trans. 1949: 54). This translation also mentions the Malay Peninsula but that is a modern gloss on what in classic versions is referred to as *Chryse*, sometimes as island as well as peninsula, a land east of India rich in gold, along with Argyre, an island east of the Ganges (possibly Burma or Arakan) rich in silver. The association of a Southeast Asian region with gold became even stronger, as with the Ophir identification leading Ptolemy to dub the Malay Peninsula, the *Aurea Chersonesus* (Savage 1984: 30; Suárez 1999: 62–63). As discussed in subsequent chapters, the Pliny/Ptolemaic distortion of the Golden Peninsula would confuse cartographers even if the first eastward-sailing European navigators were untroubled.

As Suárez (1999: 65) explains, though Strabo never reached Southeast Asia his text continued to influence European designs 1,500 years later. Not only did Strabo describe the shipping routes to India but, importantly, postulated that the ultimate source of spices—the Indies—could be attained by sailing west from Spain across the Atlantic. As such, he encouraged the use of globes, especially large size. Looking ahead to chapter 4, Strabo's ideas circulated widely at the time of the ecclesiastical Council of Florence in 1439, the failed attempt to unite the Latin and the eastern Orthodox churches, in any case dashed with the fall of Constantinople in 1453. Council participants included Paolo dal Pozzo Toscanelli, the Florentine physician and cosmographer to whom Columbus wrote in 1481. As recounted by Columbus's son Fernando (Don Hernando Colón) in a book published in Venice in 1571, among a shortlist of his father's sources favoring a short western route to India were Strabo's *Geography*, Pliny's *Natural History*, and Seneca's *Natural Questions* (Wallis 1992: 18).

The Ptolemaic Canon

Having discussed histories and source materials, many feeding into each other, and with many mythological elements stubbornly persisting, in the following section we turn to the science within the geographies, sometimes also coming under a broader rubric of cosmographies or the study of the heavens and the Earth. The seminal works in this respect are Ptolemy's *Geographia* and *Almagest* or *Syntax Mathematica,* both of which have survived through Arabic translations into Latin and modern European languages. The *Almagest* was also complimented by a companion volume *Tetrabiblos* (known in Latin as *Quadripartitum*), essentially a work of astrology or the study of the effects of astronomical cycles on earthly matters. Here, Ptolemy sets down his geocentric view of the universe (Ashmand 1822). As this book argues, while Ptolemy's astronomical theories would succumb to the heliocentric or sun-centered theories such as propounded by Galileo, Copernicus and others, his treatise on geography had considerable staying power.

As O. A. W. Dilke (1987: 177–178) points out, by the age of Ptolemy with the expansion of the Roman empire and with the silk trade with China linking up to Rome already operative, the incentive was there to modify existing maps in line with new knowledge. Allowing that Ptolemy is a "complicated figure to assess," there is also the question of the influence upon the Alexandrian of Marinus of Tyre. Alternatively, did Ptolemy just produce a synthesis of Hellenic knowledge including that of Marinus? Although Marinus's work is dated 114 CE, he may have been a near contemporary of Ptolemy. As Dilke (1987: 190) asserts, all that can be said with certainty about Ptolemy is that he provided raw material for future mapmakers to work. Another question mark surrounds the provenance of Ptolemy's maps accompanying Byzantine manuscripts, important when we consider Ptolemy-revival atlases of the early Renaissance period.

Ptolemy's *Almagest* and *Geographia*

As with *Geographia,* Greek and Latin editions of *Almagest* appeared in the fifteenth century, yet we should be careful to distinguish the works as to content and impacts.[3] While, on the one hand, *Almagest* offers Ptolemy's theoretical explanation of the cosmos, on the other hand, *Geographia* might be considered a more practical work. More particularly, as discussed below, Ptolemy's *Geographia* is a treatise on cartography or the method of mapmaking through the translation of data onto a flat or spherical surface. Together with his *Almagest,* it is also a compilation of what was known about the Roman empire of the second century CE. In *Geographia* (Books 2–7), Ptolemy catalogs some 8,000 localities and with several hundred cities marked as of particular importance. As included in the *Almagest* under the rubric Handy Tables, these were indicated by their longitudes referenced against the Fortunate or Canary Islands as zero degree, and latitudes with the Equator referenced as a zero degree parallel. Ptolemy's *Ecumene* is divided into three great continents; Europe at the center and the smallest, Libya (Africa), next in size, and Asia, the largest. Asia, in turn, is divided into twelve parts of which Southeast Asia and *Sinae* (China) comprise the eleventh and Taprobane the twelfth.

To be sure as, Suárez points out (1999: 82–83), astronomical observation may have played a part in determining these coordinates in the immediate Greco-Roman world but, further east, Ptolemy must have relied upon pilots and traveler accounts. As Suárez further remarks, the true location and identities of the Indian and Southeast Asian localities can only serve as "fodder for intelligent speculation," facts that will become clear in subsequent chapters. In *Geographia,* Ptolemy states his preference for the actual measurement of coordinates through instrumentation, but warning of the errors produced by all other methods including reliance upon travel histories and reports. This leads Suárez (1999: 83) to declare that, while impressive enough that relatively sophisticated geographical information could be compiled on the basis of such ancient long-distance sailings, "It was solely on these kinds of fallible data that Ptolemy was obliged to rely upon when constructing his map of Southeast Asia" (or *India Intra Ganges*).

Comprising thirteen sections or books, the *Almagest* is largely concerned with the motions of the stars and planetary paths. Book 1 effectively sets down Ptolemy's view of the cosmos, noting the spherical form of the heavens, with the spherical Earth at the center, with fixed stars, and with the various planets revolving around the Earth. This is the geocentric view of the cosmos, accepted in the Hellenistic world, the Islamic world and, as indicated, not disproven until by Copernicus during the Renaissance.

By contrast, *Geographia* comprises two parts: Book 1 (Theory) is a discussion of the data and of the methods used; while Books 2–5 comprise an atlas (in words). As with the model of the solar system in the *Almagest,* Ptolemy unfolds a grand scheme. The original work included maps, only reemerging in late medieval Europe but as redrawn versions, according to the template he bequeathed (and which I style "Ptolemy-revival" atlases). Although entering into Latin and Greek print versions during Renaissance times, only very recent scholarship and translations into modern European languages have made Ptolemy accessible.

Following the enumeration by classics scholar Louis Francis (1994), in Book 1 of *Geographia* (Theory) titled "The Guide to Cartography of Claudius Ptolemy,"[4] Ptolemy seeks to answer twenty-four questions. For instance, in answering the question as to the difference between the science of cartography and the more subjective chorography (or geography of any region), as Ptolemy sets down, "Cartography is the diagrammatic imitation of the known parts of the World with its unique features." By contrast, Chorography has a concern for the finest detail at the local or regional level, such as with harbors, villages, towns, and rivers courses. In sum, for Ptolemy, the role of Cartography "is to determine the nature of the Earth by showing it as one whole" [subsection 1 (§1–§3)]. As Francis (1994) interprets, Ptolemy may not be promoting Cartography as a superior science in itself to Chorography, "He is merely distinguishing between two methods of map-making, the 'global' containing outlines and large distinctive features to scale, as opposed to the 'regional' containing as much detail as possible and where relationships are more important that accurate scaling."

The Ptolemaic Presentation of Longitude and Latitude

In Book 1 (Theory), Section Two, under the rubric, "What prior considerations are necessary for the study of Cartography," Ptolemy asserts [subsection 2 (§2)], "However, before setting out to draw the inhabited World according to its true nature, we must comprehend with understanding and diligence previous scientific studies and the accounts of returning travelers."

He then introduces the reader to the concepts of longitude and latitude. Following Marinus of Tyre, and an innovation that lasts until today, Ptolemy assigned coordinates to all the places and geographic features he knew, in a grid that spanned the globe. As Francis (1994) interprets, in sub-section (§5) Ptolemy introduces the concept of fixed lines of longitude, or meridians, running north/south from pole to pole, and fixed lines of latitude, or parallels running east/west but maintaining a parallel relationship and how the coordination of the two can precisely indicate a point on Earth. Latitude was measured from the Equator, a standard observed until today. However, rather than expressed in degrees of an arc (as today), Ptolemy used the convention of "climates" or the length of the longest day (decreasing as one approaches the polar regions). He also introduces the two key astronomical instruments known to the Alexandrians, namely the astrolabe used to measure the elevation of the Pole Star in order to determine longitude, and the gnomon used to measure the angle of the Sun casting a shadow on the Earth and so allowing the determination of latitude. Above all, as in the Alexandrian tradition, Ptolemy stresses the importance of celestial observation not only to determine locations but to represent the determined coordinates spatially. As he wrote in sub-section 5 (and following Francis' translation):

> However, through observation of the heavenly bodies, accurate determination of locations can be achieved and with them how the circumference can be divided by intersecting parallel circles, (or longitudes) and then similar circles, parallel to the meridian, (or latitudes), allowing each to be inscribed diagrammatically. The two locations for any journey will therefore be represented by the intersection of longitudinal and latitudinal readings. From these it will be easily seen how much distance is required from one location to another as it might be a part of a larger circle encompassing the Earth. This does not need the mere calculation of stadia based on travelers' tales but requires an access to the map of the whole Earth.

Here, Ptolemy is setting known the basic and enduring cartographic principle of map graticule, a network of lines representing meridians and parallels that make up a coordinate system upon which a map or plan can be represented. Because of its importance, its apparent simplicity and elegance, and yet its complexities, I offer an eighteenth-century explanation.

As explained by John Harris (1764: iv), English geographer and compiler of travel accounts in the style of Hakluyt and others, by the eighteenth century "modern geographers" had made an important modification as to the way of registering longitude, either east of west of the prime meridian though conventions pertaining to parallels of latitude were well established. Harris (1764: xiv) also reserved some criticism as to Ptolemy's inaccurate measurements of latitude. The ancient Romans, he recalls, actually measured distances between places, allowing descriptions or maps of countries. He commends Agrippa's Tables in the reign of Emperor Augustus for its great accuracy. But Ptolemy adopted a different method of measurement, explained as follows:

> It is, however, true that there are many, and those many gross mistakes in the geography of Ptolemy, as for instance he places the Fortunate Islands under the same meridian, and yet gives them several degrees of longitude, and in point of latitude, he is at least ten degrees out in his computation. He speaks of a city in China, being in the latitude three degrees south, whereas we now certainly know that the most southern part of China lies in twenty degrees north latitude. He has so described the great island of Taprabane that it is in a manner unprofitable to determine, at least with certainty, whether it be Ceilon, Sumatra or Borneo. Hence some very judicious persons have doubted whether the introduction of astronomy has been any use to geography, or whether on the contrary, it has not contributed to embarrass and perplex that science, by which it is rendered much more uncertain than if the old plain geometrical method had been still followed. But in answer to this it may be very truly affirmed that the errors of Ptolemy do not arise from his use of astronomical principles, but from his following bad memoirs, and the great inaccuracy of the observations made in those times for want of good instruments. (Harris 1764: xiv)

Harris was of course correct, Ptolemy's measurement and measuring devices were not accurate, physical measurement (the geometrical method of the Romans was more functional over shorter distances and on land, such as measured in Roman miles) but the astronomical method would eventually be proven correct.

The Habitable World According to Ptolemy

Assuredly, as Francis (1994) summarizes, Ptolemy approaches his subject "from a practical rather than a scientific viewpoint." He himself did not seek to map the whole world, but to lay down practical rules for others to achieve this. As Francis comments upon Section 11, here "Ptolemy debates the size of the known, habitable world from west to east and from north to south and states the coefficients he uses." He also challenges Marinus claiming that he had "seriously overestimated these distances." He then sketches a theoretical journey from the Fortunate or Canary Islands to the Chinese capital, "which he takes to be the opposite boundaries of the known world." He also passes comment upon the merchants who traveled these routes.

Paragraphs in sub-section (§2–4) are concerned with locations in Arabia and the west coast of India, §5 is about the sea journey from India to China and the overland return journey, while §6–12 detail the coastal voyage down the eastern side of Africa. Again, in Francis's translation:

> §5. For concerning India, both about its partitioning into provinces and the inner regions themselves as far as the Golden Cheronese and from there until Cattigara, and further, that while the voyage sailing into there is towards the east, returning is towards the west, while, alongside this, acknowledging it to be irregular and uncertain of time and distance.

Besides referring to "Sera" or China, he also makes reference to an overland route to India via Bactriana and a place called Stone Tower in the Pamir desert region. Bactria was Kushan domain in inner Asia and with Stone Tower linked with Tashkurgan in Kashgar (present-day Xinjiang province of China), then a meeting point on the overland "silk roads" that distributed Indian and Chinese goods in Central Asia (McLaughlin 2010: 108). Cattigara, as discussed in a following chapter, was a received Ptolemaic name for an important port astride the sea route to China, albeit of such indeterminate location that it continued to confound mapmakers and even the likes of Columbus believing that he was en route to Zipangu (Japan) or China.

In this word depiction (because no original Ptolemy map copy survives and all Ptolemaic maps are reconstructions based upon descriptions), Africa and especially Asia exhibit greater distortions the greater the distance south (with Africa) and east (with Asia). Particularly notable is the north-south compression of India so that its western coast parallels the Equator. The southernmost locality to which Ptolemy refers is an out-sized Taprobane (Ceylon) situated on the parallel four and a quarter degrees north of the Equator. At the eastern edge comprising central Asia and Southeast Asia it is difficult to recognize any feature. Moreover, the eastern edge of Asia is folded back to join the eastern coast of Africa thereby making for an enclosed Indian Ocean. It is clear that the world known to Ptolemy did not extend far beyond the Ganges, even though Serike representing the Chinese terminus of the silk road is referenced in the *Tetrabiblos* (Berggren and Jones 2002: 19–22).

According to Francis (1994), the east/north slippage is consistent with the other objections of Ptolemy and would seem, once again, to be the result of the wrong time base. The main difference is that these locations are outside the Roman empire and, as mentioned above, Ptolemy relies on the reports of travelers for his data. In this section, as Francis (1994) points out, there is a clash between Marinus, the practical navigator, and Ptolemy, the theoretical scientist. Marinus is concerned with true distance and time taken; Ptolemy with exact, measurable distance.

In essence, as Francis (1994) recapitulates, Book I of *Geographia* is not a text book on theoretical geography "but a treatise on the art of making maps." "What evolves is a detailed analysis of map-making techniques up to his own time, followed by a synthesis of what he considers it should become and then, as a succinct summarization, a set of instructions for making it so." The remaining Books, II to VIII offer data that will—theoretically—"enable these instructions to be used to provide sectional maps of the whole of the known, habitable world." This is correct and, as discussed below, with the recovery of Ptolemy texts in late medieval-early Renaissance Europe, new maps were fancifully recreated in line with Ptolemy's text directions. Emphatically, no original maps from the classical era survived.

Ptolemy, Marinus, and his Use of Sources

Scholars of Ptolemy have also been attentive to his use of source materials and, especially, the classical heritage he transmitted. As well referenced in Ptolemy's *Geographia*, he owes much to Marinus of Tyre. We only know Marinus's geographical treatise from reading Ptolemy's critical remarks, but it is clear that he introduced improvements to the construction of maps and nautical charts. As well acknowledged by Ptolemy, Marinus's main legacy is that he first assigned to each place a proper latitude and longitude. As with Ptolemy after him, he used a "Meridian of the Blessed or Canary Islands," the westernmost land he knew, as 0° longitude and the parallel of Rhodes for measurements of latitude. His ecumene spanned 180° longitude from the Canary Islands to the eastern extremity of the Eurasian landmass (China), and about 80° latitude from the Arctic to deep into Africa. Ptolemy took over these assumptions. Still, as Ptolemy was well aware, this was only a quarter of the globe (Jones 2008).

A major issue with Ptolemy, as discussed, was the length of the Equator. Notably, Marinus estimated a length of 180,000 stadia for the Equator, corresponding to a circumference of the Earth of 33,300 km, about 17 percent less than the actual value. To the contrary, Ptolemy believed that Marinus exaggerated the eastward lying *ecumene* with China (Shera) (so identified). According to Alexander Jones (2008), the remotest places to the east of which his sources knew appear to have corresponded to China and Southeast Asia, and with parts of the east coast of Africa a few degrees south of the Equator. Greco-Roman geographical knowledge never surpassed these limits.

As Francis (1994) summarizes, in Section 17 "Ptolemy is concerned generally with the changes that have happened since Marinus wrote his commentaries and for those corrections that Marinus was unable to incorporate, presumably before his death. . . . In essence what Ptolemy argues is that Marinus consistently places locations to the east and to the north of what Ptolemy considers to be the correct position." In Section 20 concerning the nature of geographical maps by Marinus, as Francis (1994) interprets, Ptolemy passes comment upon two methods of depicting the world in mapmaking, namely on a sphere and on a flat surface. He also quotes Marinus as preferring the latter method "but of not giving due consideration to the effect of depicting a curved, spherical surface in terms of a two-dimensional, flat surfaced reproduction." Ptolemy also references Hipparchus, active three centuries earlier, to proclaim the superiority of astronomical measurement over traveler's tales. As he related, "Hipparchus alone has revealed the elevation of the Pole Star and, even then, classifying it from only so many of the places well known to geographers" (Francis 1994: Book 1, Section 4 §2). However, according to Berggren and Jones (2002: 23), notwithstanding disagreements with predecessors, Ptolemy acknowledges that the collection of data as presented in *Geographia* "substantially" owes to Marinus of Tyre, "reflecting the state of the Roman empire after the first decade of the second century CE." Additionally, we may add, Ptolemy drew upon Roman gazetteers or governance records alongside Persian and other sources. He also drew upon traveler's accounts and, undoubtedly, was also heir to a body of Roman and, before it, Phoenician knowledge. Nothing survives of the writings or maps constructed by Marinus of Tyre.

Berggren and Jones (2002: 23) also believe that Ptolemy constructed his own map based upon an earlier version produced by Marinus. Even the "provinces and satrapies" into which Ptolemy divided the known world closely follow those devised by Marinus. The chapters on longitude and latitude reveal the diversity of sources from which Marinus acquired information on lesser known parts of the world, namely, merchants mariners and soldiers. Of the known world, as these authors interpret, Ptolemy did not deviate from a tradition of accepting an inheritance of knowledge, namely the extant codification of the Roman provinces as set down by Marinus and seeing himself as a "corrector rather than creator." In fact, these two authors proceed with a series of guesses as to both Marinus's and Ptolemy's sources (only Erastosphenes is a known precursor) although doubtless they are on firmer ground in confirming the use of *itineraria* or roadmaps and *periploi*. Notably, Roman *itineraria* would list distances between localities on road networks, whether expressed in miles, leagues, or stades, but also a number relating to areas outside the Roman empire as with the Parthian empire leading to central Asia (Berggren and Jones 2002: 26–27).

Especially owing to the survival of the Periplus of the Erythraean Sea, we are more familiar with *periploi* indicating sailing routes, days of sailing, and other nautical information. As Berggren and Jones (2002: 27) confirm, Ptolemy used a source resembling the Periplus when remarking upon trade items entering the Indian Ocean trade and beyond. But writings of other historians and travelers might not have contributed much. As these authors confirm, both Marinus and Ptolemy were most comfortable with "astronomically determined latitude or longitude," albeit restricted to measurements conducted in such places as Athens, Carthage, and Alexandria.

The principle of measuring latitude by observing the Sun's noon altitude on the equinox or solstice by Hipparchus was duly acknowledged by Ptolemy (and part of a tradition shared by Erastosphenes). Rather than advancing upon this knowledge, it appears that Ptolemy was tapping into a 300-year-old database.

THE PTOLEMY REVIVAL IN EARLY RENAISSANCE EUROPE

More generally, I refer to the Ptolemy "revival" as the appearance in early Renaissance Europe of printed sets of maps purporting to be reconstructions of the ancient Ptolemaic template. The emphasis here is on both the rediscovery *and* the reconstruction of the classical template. As stated, Ptolemy bequeathed a set of instructions or guidelines on how to reproduce his representation of the known world in line with his coordinate system for locating points on a round surface. In a more misleading way, he also bequeathed name lists of terrestrial locations of the known world positioned according to celestial observations, thus setting a hard task for Renaissance geographers and successors to actually identify and correlate. Nevertheless, as discussed below, the genius of late medieval translators and Renaissance printers was to replicate this representation of the spherical globe on a flat sheet of paper. Once a printed template appeared then all others fell into a close pattern allowing for some personalization according to the artist and printing house.

Ptolemy before the Age of Print

Even so, it should be borne in mind that long before the age of print, *Geographia* was rediscovered by the Byzantine monk Maximus Planudes in 1295 and, based upon the coordinates found in the text, Ptolemy's maps were recreated. A Florentine monk named Jacobus Angelus translated Ptolemy's *Geographia* into Latin ca. 1406–07 with copies widely circulated in Europe, and with a revised version produced by Benedictine monks in 1468. As discussed below, the Angelus version would soon enter print[5] (see Figure 1.2).

Greek and Latin manuscript versions of Ptolemy's *Geographia* began to circulate in western Europe in the fifteenth century and with more than forty still extant. But these are versions and with the fate of the Ptolemy original and the actual transmission of the text from Byzantine to fifteenth-century Rome subject to conjecture. As Lach (1965: 67–68) explains, the notion that the Ptolemy original survived the ages with only minor amendments and

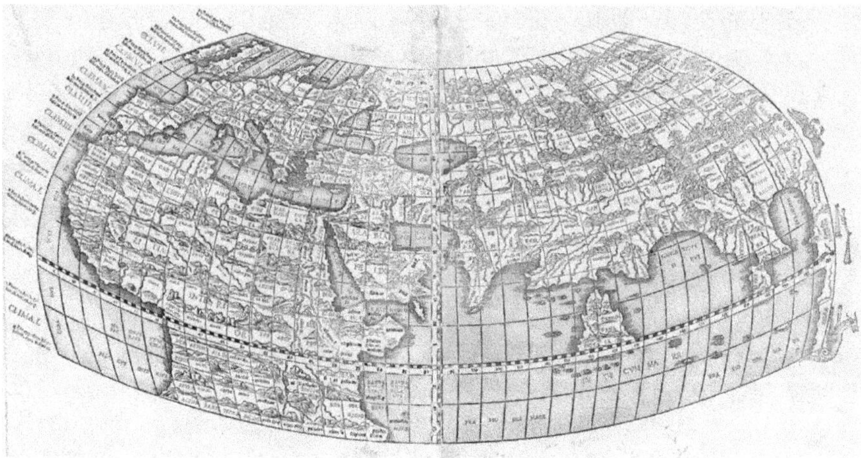

Figure 1.2 Ptolemy's Description of the Ecumene. Ptolemaic world map by Bernardo Sylvanus, Venice, 1511. *Source*: State Library of New South Wales.

revisions was severely critiqued—if not put to rest—by the work of Swedish geographer Leo Bagrow, entering publication in 1945. As Bagrow (2010: 36) argued, neither the texts nor maps passed off as Ptolemy were original, particularly on the basis that some localities referred to could not predate the eleventh century. Rather, the received versions were the works of unidentified Byzantine scholars of the tenth and eleventh centuries, and include data from Arab and Byzantine sources "gradually put together during the course of twelve centuries."[6]

During this period, late medieval world maps began to assimilate a number of Ptolemaic features. As David Woodward (1997: 316) announces, an early world map showing Ptolemaic influence, as with the closed Indian Ocean, was the Pirrus da Nonha map accompanying a manuscript of Pomponius Mela in about 1414. Starting with a print version published in Milan in 1471, Mela's *De situ orbis libri III* became a default text on global geography at the outset of the European discoveries (Romer 1998). Nevertheless, as discussed below, the recreated Ptolemies of the late Middle Ages also called down a discussion as to what exactly was the Ptolemaic standard.[7]

Still, much remains to explain regarding the reemergence/revival of Ptolemy in Rome after an absence of 1,300 years. According to Lach (1965: 16–17), it is likely that the only original part of Ptolemy's *Geographia* is the section on cartography and the list of cities and, confirming Bagrow, with the balance compiled by unknown Byzantine scholars of the tenth and eleventh centuries and with the maps of even later origin. Still, it is reasonable to adopt the term "Ptolemaic" to refer to the cartographic system and body of knowledge passed down to the Renaissance world.

Map Projections

It is important to note that the reconstructed Ptolemy world maps such as entering fifteenth-century printed texts incorporated his three primary projections, defined by Oxford as "a method for converting the three-dimensional nature of the world into a two-dimensional representation as a map." These were the conic or first-projection; the curvilinear or the second projection; and the perspective or third projection. The first is simple, in which the lines of longitude are straight and meet at a point lying north of the pole, the lines of latitude are curved. In the second projection, both lines of latitude and longitude curve. The third projection produces an oval map of the world. Because of its theoretical clarity, the conic was the most popular of Ptolemy's three projections together with the trapezoidal projection used for representing the content of his Handy Tables in map-form.

Needless to say, the subject of map projections is highly technical if not abstruse, and with European mapmakers engaged in a great deal of experimentation and adaptation over the centuries. According to Evangelos Livieratos (1996), although better simulating the spherical shape of the Earth, the curvilinear projection is difficult to execute, requiring advanced skills and experience. By comparison, the trapezoidal projection better fits the construction of maps of the known world or areas represented in Ptolemy's Tables. In *Geographia*, Ptolemy presents twenty-six tables in total, covering the known world at that time in the three continents: Europe, Asia and Africa. Smaller scales are used in the case of Tables with large ground coverage (scarce place density) as with Asia represented by four tables. Larger scales correspond to the case of tables with small ground coverage (high place density), as with Europe represented by five tables.

Maps appearing in the first of the recovered versions of *Geographia* included a single large world map along with a less detailed map adding such places as Serica and Sinae. Cattigara is located at the extreme right beyond the island of Taprobane and the *Aurea Chersonesus*. Separate and more detailed sectional or regional maps included ten maps covering parts of Europe, four of Africa, and twelve of Asia. Of particular importance to our study is the Eleventh Map of Asia (*Undecima Asiae Tabula*), because of its canonical status and frequent consultation and citation by mapmakers and navigators across the centuries, and with its depictions of *India Extra Gangem,* the Golden Chersonese, the *Magnus Sinus*, and *Sinae* (see Figure 1.3).

Ptolemy in the Age of Print

With authors and publishers of *Geographia* obliged to draw their own versions of the original maps using the information contained within the text and tables as their only guide, many of the first wave of the print Ptolemies

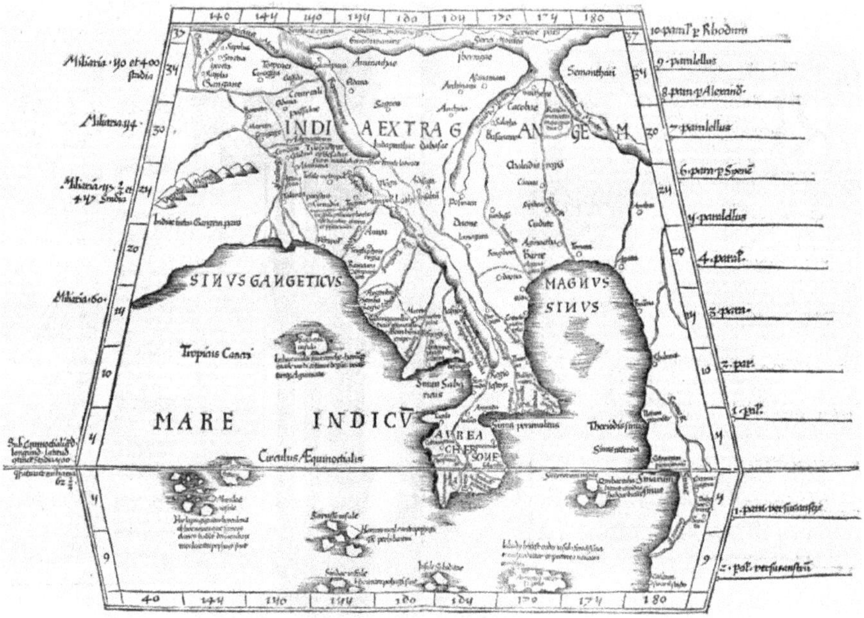

Figure 1.3 Ptolemy's 11th Map of Asia, *Tabula XI Asiae.* *India Extra Gangem.* In *Geographia*, translated by Willibald Pirckheimer with annotations by Joannes Regiomontanus. Argentoragi (i.e., Strasbourg): Iohannes Grieningerus, 1525. Scale (ca. 1:12,500,000). *Source*: Map of land east of the Ganges showing the Ganges River in the west, "Sinae" (China) in the east, along with a southeast Asian region, the Indian Ocean, and a China Sea. Map reproduction courtesy of the Norman B. Leventhal Map & Education Center at the Boston Public Library.

combine text with maps (with up to sixty-four copperplate maps).[8] Typically, as discussed in chapter 4, these maps include a full globe, the classic Ptolemaic representation, including the stereotypic *India Extra Ganges*, and Taprobane. There may be variations between printers and editions, but they conform closely to a single template. Just who struck the original map template remains unclear.

Notable was the mid-fifteenth-century Florentine map of the world based on the Jacobus Angelus Latin translation printed at Vicenza (1475), and with Venice emerging as an early print capital in Europe. A version with maps appeared three years later engraved by the celebrated artist, Arnold Buckinck (Anthiaume 1916: 22–23). The first printed version of Ptolemy with engraved maps appeared in Rome in 1475 (or Bologna, 1477), quickly followed by other print editions in Toscana Italian (1482); and with the first printed version with woodcut maps appearing north of the Alps, at Ulm in 1482. Many other editions appeared in the following centuries in either Italian or Latin such as at Cologne (1597; 1608); Arnheim (1617), and Padua (1621). On its part, the *Tetrabiblos* was printed in Venice in 1484 based upon a twelfth-century Latin version produced in Barcelona, in turn taken from the Arabic. With the first Greek text printed in Basel in 1533 by Erasmus, Greek editions of Ptolemy's *Geographia* continued to be published down until the nineteenth century.

CONCLUSION

Undoubtedly the Alexandrian heritage stood upon the shoulders of such giants of the Hellenic worlds as Pythagoras, Aristotle, Eratosthenes, and Crates of Mallus, not to mention an even earlier Phoenician (and Babylonian) tradition. Citing the ancient tradition of *periplus* along with the two seminal world histories of ancient Rome, respectively Pliny the Elder and Strabo, we sought in these texts to highlight the foundational framing by the ancient world of Asia. In other words, we sought to know how the ancients and, more specifically, the Alexandrians represented space, especially Eurasian space. In particular we found that the text versions of the histories actually survived the ages and so, as precocious meta-geographical summaries, became cardinal reference works over the ages. This is important to acknowledge because, as demonstrated in the following chapter, Pliny and Strabo were not even supplanted by empirical knowledge entering obscure and unpublished Arab manuscripts dating from the tenth to the thirteenth centuries (and with Strabo's *Geographica* in print in Latin by 1469). Moreover, as Lach (1965: 69) explains, and a theme we have more broadly embraced, Ptolemy's work was widely admired in Renaissance Europe as a treasure of ancient Greek learning and its reception was not immediately challenged by the revelation of new knowledge as with the voyages of Marco Polo, navigational maps based on compass directions, and other evidence. In the words of map historian Wallis (1992: 17), the recovery of Ptolemy's *Geographia* was simply "the outstanding cartographic event in the fifteenth-century." We could add that the real importance of this event was it continued to impact geographers, humanists, and even navigators through to the succeeding century if not longer. Becoming influential texts with a rising generation of Renaissance geographers, as with the Venice-based humanist Giovanni Baptista Ramusio, schools of mapmakers also sought to both adapt and honor Ptolemy.[9]

As Mitchell (2012: 42) sums up, Ptolemy's *Geographia* furnished the Renaissance world with a method based upon Euclidean geometry for mapping the curved surface of the Earth on a flat plane of a map. Citing Samuel Y. Edgerton (1975: 113–114), as she explains, unlike earlier European map traditions—essentially navigational maps based on compass directions—Renaissance Ptolemaic maps were outstanding in the sense of offering a geometric framework for comprehending the whole world. For the first time in the Western world all places on the globe could be fixed in relationship with each other by unchanging coordinates with their proportionate distances and directional relationships apparent. Through the method of georeferencing, the new discoveries of the late fifteenth and the early sixteenth centuries could now be traced onto a map graticule within the frame of meridians and parallels. Famed European mapmakers, such as Münster, Mercator, Ortelius, (Lorenz) Fries, and Waldseemüller, would all draw upon Ptolemy in the construction of their worldmaps as with the latter's map of 1507 situating and naming America for the first time.[10]

Even so, as Joaquim Alves Gaspar (2007: 66) points out, the translation and diffusion of Ptolemy's *Geographia* in Europe during the fifteenth century "had a relatively small impact upon nautical cartography." Rather, as embellished in separate chapters, alongside astronomical observations conducted at select sites, it would be practical navigation feeding into cartography that would revise and correct Ptolemy's latitude and longitude grid. Ptolemy's inability to even approximately measure or even represent longitude outside of the Mediterranean area also had long-reaching consequences, especially if we consider Columbus's error of believing he had arrived on the shores of India or China, the old *ecumene*, as opposed to the New World of the Americas. Astonishingly, the concept of the "other half" of the spherical world as developed by the ancient Greeks, fed into myths, speculation, and even literary utopias, down until the early modern era (Fausett 1993: 10). As discussed in a final chapter, the myth of a disproportionately sized Antipodes was only put to rest in the late eighteenth to early nineteenth centuries with advances in South Pacific navigation.

NOTES

1. See The Complete Tabula Peutingeriana—a Roman Road Map compared with a modern map.

2. Manuscript copies of the Periplus derived from a Byzantine tenth-century manuscript are found in the University Library Heidelberg (CPG 398: 40v–54v), and with a copy dating from the fourteenth or fifteenth century in the British Museum (B.M. Add 19391 9r–12r). It first entered print with the Hieronymus Froben edition of 1533. It has also attracted reprints and commentaries down until the present, as with Reinaud (1864); Schoff (1912); Casson (1989); De Romanis and Tchernia (2005); and Seland (2010).

3. The first translation of *Almagest* into a modern language was accomplished by Nicholas Halma who, in two volumes (1813; 1816), published a direct French translation from the Greek text, including detailed historical comments in a sixty-nine-page

preface. In the French translation, the *Almagest* opens with the following words, "It seems with reason, my dear Syrus, in the philosophical mind, theory is distinguished from practice" (Halma 1813).

4. In the following I have drawn from the translation and analysis of Ptolemy's *Geographia,* Books I and II, by Louis Francis (1994), sourced from a University of Oxford electronic text archive. In turn, Francis made his translation from the Greek text of *Geographia* as rendered by Karl Friedrich Nobbe (Lipsiae: 1843–1845).

5. Prior to the age of print, Ptolemy's *Quadripartitum* was used in part at least translated to French by the Norman, Nicolas Oresme (1320–82), a figure whose writings offered significant challenges to the Ptolemaic system long before Copernicus (Whitney 2004: 170).

6. Another manuscript version of *Geographia* appears to be an Arabic translation achieved by Georgios Amirutzes and son shortly after the conquest of Constantinople in 1453 on the orders of Sultan Mehmet II. This was realized from a copy of Ptolemy discovered by the conquerors in a Byzantine library. Eager to acquire scientific knowledge of the day, Mehmet surrounded himself with a group of Byzantine, Greek, and Arabic scholars, Amirutzes among them. A Greek scholar from Trabzom, Amirutzes was commissioned to draft a world map, missing from the manuscript version. This was accomplished and credibly woven into carpet. But Mehmet also learned of Latin translations available in Italy and acquired early printed versions (Bagrow 2010: 209–10). The manuscript copies of Ptolemy are kept in the Seraglio library in Istanbul. The oldest copy of the work, the thirteenth-century Codex Seragliensis GI, 57 is held in the Topkapi museum in Istanbul. The BNF has a Greek mss "Grec 1407" dated to 1438, titled Claudius Ptolemaeus *Geographia; Arrianus. De expeditione Alexandri libri VII; Historia Indica.* The Amirutzes manuscripts were reprinted in 1929.

7. Wilcomb E. Washburn (1985: 438) is one who has strenuously denied that Ptolemy himself applied the notion of an enclosed Indian Ocean, rather claiming it was a later innovation especially adopted in print editions. Thus, "this error postdated Ptolemy is, in my opinion, vastly more likely than that it derives from Ptolemy himself." More recently, scholars have sought to apply statistical computational analysis of the coordinates assigned to localities by Ptolemy and so an element of debate continues as to the transmission across the ages.

8. These can be perused via Gallica, a web portal of the BNF.

9. A good summary of the transition from manuscript to print of Ptolemy atlases can be found at MapForum.com

10. See the website of Evangelos Livieratos professor emeritus of the Faculty of Engineering, Aristotle University of Thessaloniki http://cartography.web.auth.gr/Livieratos/fil/Ptolemy.html

Chapter 2

The Islamic Crossover

Having underscored the primacy of the Alexandrian school as to its revelation of a spherical Earth, along with a method of cartographic production and replication replete with mathematically calculated coordinates, we also acknowledged that these seminal "cosmological" understandings were lost to Europe for several centuries following the collapse of the Roman Empire in 476 CE. The theme of Barbarian conquest calling down the European Dark Ages is resilient in later Renaissance writing. With the fall of Alexandria to the Arabs in 642 CE, Islamic scholars inherited classical philosophy and with many manuscripts translated from Greek into Arabic. The contrast with early Christendom, where Old Testament orthodoxy treated revisionist astronomy as heretical, could not have been greater, especially as Arab scholars in Spain under the Islamic caliphate looked up to the early Greeks as authorities in matters of natural science and astronomy. Islam also expanded east from its Arabian heartland and, especially during the period of Mongol domination of Eurasia, entered a long period of commercial, cultural, and intellectual exchange with China.

As this chapter highlights, the reception of the Ptolemaic system in the Arab world was especially important to Europe during the Dark Ages as it would not be until the Renaissance that Latin and Greek manuscript versions of the Alexandrian would find a wider audience, indeed celebrated in new print editions. Given the complexity of Muslim-Christian relations even restricted to the Arab conquest of the Iberian Peninsula and the broader Mediterranean, not excluding the impact of the Crusades, the chapter is necessarily selective. It should then be read in the light of our overarching theme as to adaptations or revisions of the Ptolemaic worldview, in this case filtered through the medieval Arab-Muslim experience, not excluding scientific contributions and new empirical knowledge of the world derived from Muslim merchant and scholarly contacts.

A first section of this chapter seeks to establish the importance of the Islamic crossover with Latin Europe, especially with reference to the transmission of knowledge both philosophic and scientific, not excepting transmissions of Ptolemy to the Arab world. Second, the chapter elaborates upon the seminal contributions of the twelfth-century Sicily-based Arab geographer Muhammad al-Idrisi, especially his framing of Asia, including China. A third section examines Arab primacy in navigational knowledge—as much knowledge of Asia—ahead of Europe. A fourth section discusses Arab/Muslim geographies of Asia. A final section on the Chinese-Muslim world exchange draws attention to Islamic encounters and exchanges across the terrestrial and maritime silk roads to China reaching their apogee during the Mongol-Yuan dynasty with its capital in Khanbalik (archaeologically attested as a site in present-day Beijing) and with its outlier in Persia.

TRANSMISSIONS FROM THE MUSLIM WORLD TO LATIN EUROPE

For 500 years Western and Eastern Christendom faced down challenges from Islam, variously the Arab-Berber conquest of Spain leading to the installation of the Umayyad Caliphate (661–750) centered on Córdoba, itself the westernmost dependency of the Caliphate of Damascus, and then under the Baghdad-centered Abbasid Caliphate at least until its destruction by the Mongols in 1258. A Muslim thrust into France was only stalled by the Battle of Tours (Poitiers) defended in 732. With the Islamic conquest of Sicily beginning in 827 and with Islamic rule commencing on the island in 902, the former Christian Byzantine outlier would only be reconquered by the Normans in the eleventh century. Under the Reconquista (718–1492), Toledo would be recaptured in 1085 and, in turn, Lisbon in 1147 but with the Catholic Christian conquest of Islamic Grenada only achieved in 1492. Within this time frame the Abbasid caliphate and Muslim culture in general recentered to the Mamluk capital in Cairo (1261).

Yet the greatest threat to the eastern Mediterranean and, indeed, central Europe, stemmed from the Seljuq Turks following the conquest by Sultan Mehmed II (r. 1444–46; 1451–81) in 1453 of Constantinople, capital of the Eastern Roman Byzantine Empire. Commencing with the proclamation in 1095 by Pope Urban II (1088–99) of the First Crusade with the stated goal of restoring Christian access to the Holy Land, Western Christendom went on the offensive with the Crusades continuing down until the fifteenth century with renewed campaigns to stem the Ottoman advance across the

Danube. Notwithstanding the religious and political zeal of the protagonists, the encounter between Latin Europe and Oriental Islam was not zero sum but involved a range of transmissions, eventually feeding into Europe's own Renaissance and even scientific revolution. As explained in chapter 6, the Crusades also reopened the eastern Mediterranean to trade and travel, especially enabling Genoa and Venice to tap the virtually uninterrupted flow of Oriental commerce.

New Knowledge from the Islamic World

As Visconde de Santarém (1852: 323–325) long ago pointed out, Arab geographies commenced to be written from the ninth century. Notably, the Abbasids gave the highest priority to the sciences, as with patronizing Nestorian scholars in Persia and Mesopotamia. Both caliphs Harun al-Rashid (r. 786–809) and al-Ma'mun (r. 813–833) actively sought out Greek works from Roman Emperor Constantine (r. 306–337 CE) and saw to their subsequent translations into Arabic thus further enhancing Arab science. As discussed below, from Córdoba the Umayyad Caliphate had already adopted this strategy. In 324 CE, Constantine founded "New Rome" at Constantinople at a time when old Rome and the provinces were overthrown by barbarian tribes.

Santarém (1852: 325–336) also drew attention to the turn in medieval Europe toward an understanding of the Muslim other, as well as in looking to Arab scholars for a recovery of science and classical learning. Whatever else, he demonstrates, the Crusades contributed greatly to the introduction into Latin Europe of knowledge of Oriental languages, knowledge also feeding back directly into medieval mapping. From the eleventh century, Christian scholars commenced to translate the works of Greek authors from Arabic to Latin. Notably, Campanus of Novara (ca. 1220–96), an Italian mathematician and astronomer, translated Euclid from Arabic. In the twelfth century, Frederick II, Holy Roman Emperor, King of Sicily (1198–1250) and crusader, patronized scholars, founded universities, and imported manuscripts from the Orient. In the thirteenth century, Raymond Llul (ca. 1232–1315), the Majorca-based scholar, actively pushed the study of Arabic, visited Tunis, and translated Al-Ghazali (1058–1111), among other works. Paris-based theologian Guillaume d'Auvergne (1190–1249), known for his Aristotelian approach to philosophy, also studied Islamic thinkers. As Santarém (1852: 326) summarizes, "the scientific base of Arab cosmography and geography was drawn from Greek systems." And, for Europe, a part of Greek science was restituted to Europe by the Arabs.

Be that as it may, scholars such as John Larner (1999: 12–13) are skeptical, asserting that Europe learned nothing from al-Idrisi and other medieval Arab scholars about the real world. "During the twelfth century, western scholars acquired a large measure of Arabic scientific learning but, for reasons which have not been fully clarified, they largely ignored what related to geography." Nothing much was derived from the Karimi merchants of Cairo who upheld extensive trade with India and the spice islands. Neither, he asserts, did European Jews filter through hard geographical knowledge from their trade contacts in India and beyond. Specifically, he mentions the manuscript written by Spanish Jew, Benjamin of Tudela (Navarre) who, between 1166 and 1171 or 100 years prior to Marco Polo, traveled as far as Al Qatif in Bahrain. Although making the first ever manuscript reference to China in Europe, however garbled, it was ignored, at least by Latin Christians.[1]

As Woodward writes (1987: 304), the influx of new knowledge from Arab sources into Europe in the tenth and eleventh centuries occurred mostly via Italy (Sicily) and, in particular, Spain, where Muslim authors achieved dozens of translations of Arabic and Greek classics on such subjects as philosophy, mathematics, astronomy, physical and natural sciences. Hitherto, very few European scholars could read ancient Greek thus closing off the early and High Middle Ages to the best classical texts. Various institutions, as with universities (Salerno, Bologna, Paris, and Oxford) developed in this period. In particular, Paris and Oxford hosted strong centers of cosmological and geographical culture, just as the study of arithmetic, geometry, and astronomy became part of the curriculum.

The Toledo School of Translators

The Arab crossover of knowledge to Europe did not stop with the Reconquista in Andalusia. Toledo under King Alfonso X of León and Castile (1252–84) continued as a major center of learning for Arab, Jewish, and Christian scholars alike. With time, a Toledo School of Translators began to make its mark, as with Gerard of Cremona (Gerardus Cremonensis) (d. 1197). Notably, the Toledan Tables edited by Gerard, containing a long list of coordinates based upon the prime meridian in the Canaries (in turn, based upon corrections of Ptolemy's computations), were then the most accurate compilation of astronomical data in Europe. According to Woodward (1987: 323), although partly based upon the work of Jabir ibn Aflah (1100–50) the Seville-based astronomer who corrected Ptolemy's *Almagest*, the principal composer was the Toledo-based astronomer al-Zarkali (ca. 1229–ca. 1287).

Besides executing a long list of translations into Latin and/or Castilian from classical Greek, Arabic, and Hebrew, including works by Archimedes, Aristotle, along with Arab works on mathematics and astronomy, the Toledo School also translated Ptolemy's *Almagest* (1175). In doing so, it also brought back to medieval Europe the hitherto lost Ptolemaic understandings of longitude and latitude. A Greek version appeared in Western Europe in the fifteenth century, followed by George Trebizond's much-critiqued Latin translation (ca. 1451).

According to Henry S. Lucas (1960: 174), while Platonic thought was well known in medieval Christian Europe, not so Aristotle. Yet Aristotelian philosophy did survive in the Arab world as with the *tafsir* (commentary interpretation) of Averroës (Ibn Rushd) (1126–98) of Córdoba. Known in Latin Europe for his enduring contributions to secular learning, his *falsafa* (philosophy) and commentaries upon Aristotle's *Physics* spawned numerous pamphlets across Europe in the 1300s and 1400s before the age of print (see Glasner 2009).

The Alfonsine Tables

Named after King Alfonso X, the *Alfonsine Tables* produced around 1252 offers yet another instance of the Arab transfer contributing to Europe's knowledge of astronomy and, in the future, navigation. As such, these were the first astronomical tables enabling calculation of eclipses and the positions of the planets for any given time to be produced in Latin Europe. Faithful to Ptolemaic theory, the *Tables* also assumed that the Earth was at the center of the universe. Based on earlier astronomical works and observations by Islamic astronomers, the new tables added observations by astronomers that Alfonso had gathered in Toledo, among them several Jewish scholars. The thirteenth-century planetary tables were relied upon by all astronomers and navigators well into the sixteenth century (*Encyclopedia Britannica* 2014b).

Transmissions of Ptolemy to the Arab World

As Marina Tolmacheva (2015: 8) summarizes, by the tenth century coincident with the Abbasid dynasty, Arabic scholarship had produced a synthesis of "received, preserved older and newly-developed knowledge." As she interprets, three recognizable, if not independent, streams then held sway in classical Islamic geography, namely (1) Greek ecumenic geography (largely, but not exclusively, Ptolemy), (2) Arabic-Islamic travel narratives, and (3) Islamic cartography. Given such understandings, as further explored below, we still must distinguish between cartography and practical navigation, on the one hand, *and* astronomy, on the other, where the Arabs were far ahead of the curve.

According to Berggren and Jones (2002: 51), the contents of *Geographia* transmitted through translation and by "less direct means" came to be received in the Arab world—Mesopotamia, Syria, Egypt, and Spain—as early as the ninth century (and apparently with translations into Arabic and Syriac). Nevertheless, no actual Arabic translation survives from the medieval period and the Ptolemaic legacy is mixed with other sources. Still, coordinates of longitude and latitude derived from *Geographia* show up in Arabic astronomical and geographical tables.

One such early Arabic adaptation was the *Kitab Surat al-'ard* (Book of the Picture of the World) ascribed to Abu Jafar al-Khwarizmi (but also linked to Abu Abdullah Muhammad Ibn Musa al-Khwarizmi, a ninth-century mathematician and astronomer). He also offers a rendition of Ptolemy's text and map but possibly through Syriac intermediaries. The al-Khwarizmi map has also been identified as the oldest representation of the Indian Ocean. According to Jean-Charles Ducène (2017: 57–64), in correcting Ptolemy he reveals an open Indian Ocean flowing into an encircling sea, a geographically correct depiction replicated over the following centuries in Islamic mapping, as with the map of Ibn Hawqal at the end of the tenth century. Baghdad-born and Cairo-based al-Masudi (896–956) and Al-Biruni writing in 1230 also accepted a continuous sea communication and so, as discussed below, did al-Idrisi in the middle of the twelfth century.

Having inherited Ptolemy's *Almagest* and *Geographia* in the ninth century, Islamic scholars should have been empowered to develop better cartography. While Islamic geography and the study of earth science did develop at widely scattered locations, it did not however translate into better mapmaking. Although duly acknowledging the Arabic contribution to Latin Europe's science, Santarém (1852: 325) is even less praiseworthy as to the Arabic contribution to the development of cartography. As he stated, "The majority of their systems are borrowed from the idea of Homer, Herodote, Socrates, Thales, Aristotle and other Greek authors." Moreover, he contended, the Arabs actually went as far as adopting the "dictates of Ptolemy." For example, al-Masudi who combined geography and history in writing his far-ranging travels actually cited Ptolemy and Marinus of Tyre. In sum, as Anthiaume (1916: 419) summarized, "the Arabs did not perfect the Ptolemaic view of the India region." Even so, this view must be revised in the light of modern scholarship, not only on Arab mapping as adverted above, but especially relating to astronomical science including measurement such as would be received in Europe in a much later age.

THE IMAGE OF ASIA IN THE CARTOGRAPHY OF MUHAMMAD AL-IDRISI

As Jan Nederveen Pieterse (2007, chap. 7; 2012: 13) adverts, the figure of Muhammad al-Idrisi (1099–ca. 1161) stands as one of the foremost names associated with premodern cosmopolitanism. This characterization is undoubtedly true and fits the multiracial embrace of Islam across the

Afro-Eurasian landmass with its important centers of religious devotion and scholarly learning. Muhammad al-Idrisi's general biography and achievements are well recorded but, as outlined below, the scope of his geographical knowledge, especially the Indian Ocean, Southeast Asian trading regime, China, and central Asia, was awesome even alongside the Marco Polo story. Born and raised in Ceuta in 1099, then under the Almoravids, the Berber dynasty with its capital at Marakeech, according to nineteenth-century researchers R. Dozy and M. J. de Goeje (1866: 2), his family lineage can be traced to Idrus II, Caliph of Malaga from a Hammoudite ruling family of this dynasty. From an early age, al-Idrisi traveled to Islamic Spain, North Africa, and Asia Minor. His sojourn in Spain also coincided with conflict and instability in Andalusia leading to an exodus of Arab scholars with a number settling in Sicily. Malaga was occupied by Granada in 1057 and with the Hammoudites going into exile. With Islamic rule commencing in Sicily in 902 following an era of contestation, Arabs of the Emirate (965–1091), formerly loyal to the Fatimid Caliphate (909–965), were overthrown by the Normans building upon their own Crusader backgrounds. By 1091, the Muslim capital of Palermo (al-Madinah) which eclipsed Córdoba as a center of learning in many respects had fallen to the Latin Christians.

More generally, the Normans tolerated and patronized certain Arab families in Sicily in exchange for knowledge. At the same time, the semi-Arabized Norman princes in Sicily ruled over an Islamic majority. Reciprocally, al-Idrisi looked up to his patron the Christian King Roger II displaying, in the words of Dozy and de Goeje (1866: 1–2), a "rare impartiality" in his text. These authors concede that little is known of al-Idrisi's stay in Sicily, although he also obviously incorporated knowledge acquired by Normans to create his book and maps. Working some fifteen years at the court of King Roger II, al-Idrisi joined mostly unknown Arab, Greek, and Byzantine scholars. Reputedly among them was Andalusian astronomer and polymath Abu al-Salt.

On the orders of Roger II, in 1154 CE al-Idrisi completed a work known in short hand as the *Kitab Rudjdjar* or *Tabula Rogeriana* or *Book of Roger*. This was a seventy-plate atlas combining Arabic text descriptions of countries and regions, or geography, along with asides on flora and fauna and local customs. Added maps show the Eurasian continent, from the Canary Islands in the west to India, central Asia, along with the Middle East and China in the east, to Britain and Scandinavia in the north, to a large swathe of Africa in the south. The Mediterranean is well developed along with the Adriatic and Black Sea, and with the Italian and Iberian peninsulas recognizable. Africa appears as the crescent-shaped landmass at the top, and with the Arabian Peninsula in the center, offering an Islamo-centric spin. Unlike Ptolemy, al-Idrisi depicted a circumnavigable Africa, just as blue sea surrounds the globe.

Although lacking much in detail, the *Book of Roger* was in fact a description of the whole known world and the most accurate, dubiously remaining so for the next three centuries. Importantly, from accumulated travel experience as well as study, al-Idrisi incorporated knowledge not only of North Africa and the Islamic world of the Maghreb, but also the Indian Ocean zone and further east, literally the world known to Muslim merchants as with the Arabo-Persian Sinbad journeys to the land of Sindh. As Hyunhee Park (2012: 90) summarizes, in creating the most sophisticated world map of his time, al-Idrisi clearly represents China at the eastern end of the world, delineating its coastline in a way to connect it up with the Islamic world.

The *Book of Roger* actually contains two world maps, one small and circular (and often reproduced), along with an atlas of seventy regional maps which, combined, complemented the miniature mappemonde. As written into a preface, al-Idrisi acknowledged a small corpus of Arab sources that he consulted relating to Africa and Spain. As Dozy and de Goeje (1866: 5) acknowledge in their annotated part translation of al-Idrisi's *Kitab Nuzhat*, although lost today these sources impart an "exceptional character" to the work. A long list of nineteenth-century Orientalist experts agree that this was the highest order compendium of geographical knowledge then existing in the European Middle Ages. In turn, the map was inscribed by Roger on a massive disc of solid silver, two meters in diameter, showing, in al-Idrisi's words, "the seven climatic regions, with their respective countries and districts, coasts and lands, gulfs and seas, watercourses and river mouths." The basis was Ptolemaic written in Arabic and divided into seven climate zones (in keeping with the Ptolemaic system) each of which is subdivided into ten sections, and with maps. Nevertheless, it is no small achievement that al-Idrisi can be credited with preserving Greco-Roman traditions during the Europe Dark Ages, such as would be revived during the Renaissance (see Figure 2.1).

The Intellectual Legacy

Alongside the role of Venice as an information and printing center, the Medici Press in Rome also played its part, especially as it published a number of influential crossover studies and translations from Arabic. Notable in this respect was an abridged version of the al-Idrisi's Arabic text published by the Medici Press in 1592 with the title *De geographia universali* or *Kitab Nuzhat al-musht* (Dozy and de Goeje 1866: 7). In George Saliba's (2007: 229) opinion, the real market of the Medici Press was European centers of learning. The first translation of al-Idrisi from the original Arabic into Latin was by the Maronites Gabriel Sionita and Joannes Hesronita published in Paris in 1619 with explanatory preface.

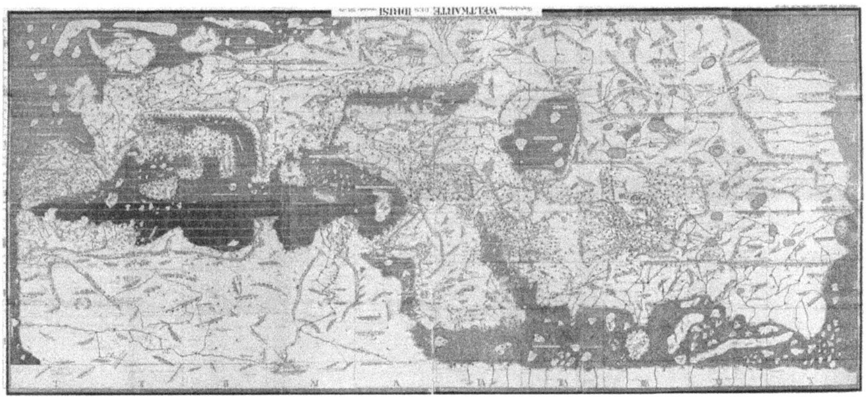

Figure 2.1 The Al-Idrisi World Map. Weltkarte des Idrisi vom Jahr 1154 n. Ch., Charta Rogeriana. World map drawn by Idrisi in 1154 and restored by K. Miller in 1927. Place names in romanized Arabic added. Originally oriented with north to the bottom. *Source*: Retrieved from Library of Congress, Geography and Map Division. Digital Id http://hdl.loc.gov/loc.gmd/g3200.ct001903.

While these printings might have awoken eyes in late medieval Europe, both came too late as direct inspirations for the great Portuguese and Spanish voyages of "discovery." That said, the fact that manuscript versions survived the ages through the activities of copyists in Egypt and Syria and elsewhere, then Arabic-reading literati also inherited the hybrid Arabo-Ptolemaic tradition. This is well understood. If the knowledge crystallized by al-Idrisi along with other Arab scholars also fed back into centers of learning as in Cordóoba under the caliphs then we are tempted to say that longitude, latitude, and understandings of spatial representation might also have been first grasped in the Iberian countries at an early date (matters taken up in separate chapter).

More generally, Santarém presents a fairly dismissive view of al-Idrisi and Arab geographers, especially for their misconceptions of the earth's sphericity. As he pointed out, several authors after the Andalusian geographer Al-Bakri (1014–1094) regarded the Earth as a fixed surface or table; others likened it to bowl cut in half; others thought that it was hollow internally. Other Arab authors contended that there were several Suns and Moons in different parts of the Earth. In al-Idrisi's system, the Earth is represented as a globe of which the regularity is interrupted by mountains and valleys on the surface. He adopts the system of the ancients, as with an uninhabited torrid zone. According to al-Idrisi, the known world only forms a simple hemisphere, half comprised of water, of which the largest part of water belongs to the encompassing Ocean environment, "in the middle of which the Earth floats like an egg in a basin" (Santarém 1852: 325).

In its time, al-Idrisi's mapmaking may also have fed into early portolan maps or may even have guided later Christian Latin crusades and conquests such as that of Ceuta and the West African littoral, but this is speculative. To suggest that al-Idrisi's legacy endured for 300 years, as sometimes claimed, requires some context. Allowing that translations and commentaries upon al-Idrisi continue in the Western and Arabic world down until the present, we still wish to know how the original was received and transmitted, and what was the intellectual reception, if it can be gauged? These are weighty questions. But we detour with reference to the bibliographical legacy.

The World According to al-Idrisi

While every text is unique, the world according to al-Idrisi is also a derivative text building upon accumulated knowledge, not only Ptolemaic, as he acknowledges, but Arabic as well. As with Latin Christian narratives of the age, he also injected some phantasmagorical elements into the text. Among the conventions that al-Idrisi bequeathed, as opposed to invented or discovered, is the view of the world as a globe, such as translated from the Arabic by early-nineteenth-century French Orientalist P. Amedée Jaubert:

> The earth is essentially round, but not perfectly owing to elevations and low-points, and water flows from one to another. Oceans surround half the globe without interruption like a circular zone, not only appearing as a half, it is for example like an egg plunged into water contained in a cup, so it is that half of the earth is plunged in the sea. The sea itself is surrounded by air and the atmosphere itself attracts or repels in ways of which we come to speak. (author's translation, Jaubert 1836–40: Vol. I: 3)

As such, al-Idrisi built upon the known. The spherical shape of the world was known since Greek and Roman times. This was also a geocentric view of the world that would only be challenged scientifically by the late Renaissance. As mentioned, he also adopted the ancient convention of "climatic zones," not be confused with climate zones in a modern sense but, rather, terrestrial latitude (as with Ptolemy), conventions adopted by such Arab and Persian geographers, as al-Biruni and al-Razi (the author of the sixteenth-century *haft iqlim* or seven climes) (Jaubert 1836–40: Vol. I: 3).

Such is obviously an important statement. Muhammad al-Idrisi acknowledges a West and an East, ipso facto placing himself in the West (whether or not as part of the Islamic Maghreb or whether as part of the western Mediterranean, or whether following Ptolemaic tradition, on which more below). The notion that maps embody conventions, certain of them idealized, or as imagined creations of their makers, or products of their time, is certainly a critical reflection coming in an age when religious allegory, Christian even more than Islamic, suffused all kinds of representations, from text to sculpture, to architecture and, not excluding cartography as with the Psalter world map conserved at the British Library ca. 1260, replete with Christian iconography. The Hereford Mappamundi dating to ca. 1285 goes further, displaying biblical events, and with Jerusalem drawn at the center and with the Garden of Eden displayed on the edge of the world.

Another convention borrowed by al-Idrisi is the notion of seven seas. Some contend that the medieval concept of the Seven Seas has its origins in Greece and Rome (Jaubert 1836–40: Vol. I: 4–8). One Arab progenitor of the "seven seas" convention was Arab geographer Ahmad al-Ya'qubi, or Ya'qubi (d. 897/8), reputed author of the *Kitab al-Buldan* or *Book of Lands* or *Book of Countries*. Whereas Ya'qubi offers a construction fitting the Sinbad or maritime trading route connecting the Gulf area with India, Southeast Asia, and Canton in China, also fitting the pattern of dissemination of Islam, such as marked by the oldest mosque in China, at Canton, al-Idrisi had more global concerns, namely, fitting the understanding of Arab and Phoenician seafarers, the Black, Caspian, Arabian, Indian, Mediterranean, and Adriatic Seas.

> The seven climates are traversed by seven seas, as it pleased God and, as mentioned below. The seven seas also include gulfs. Six of the seas are contiguous. Only one is separated. And without communication with the others. (author's translation, Jaubert 1836–40: Vol. I: 3–4)

In point of fact, al-Idrisi's addition of gulfs makes the seas difficult to count and with the sea of India and China sometimes conflated and with further reference to western and oriental seas.

> The first of the seas situated in the inhabited part of the world is the sea of China and India, of Sind and Yemen. It extends from the east of 13th latitude, above and along the equinox line, it laps China, India, the Sind, mid-Yemen, and ends at the Bab el Mandeb. (author's translation, Jaubert 1836–40: Vol. I: 4)

It is not without interest that the "China Sea" is elevated to the first sea in this discussion, suggesting the major axis of east-west maritime trade, recalling ancient maritime silk roads, invoking the "Sinbad route," such as plied by Arab navigators coasting India, the Bay of Bengal, the Malacca Straits, the Indochina coast, and the coast of China, all of which are alluded to in some (confusing) detail (and with the Bab el Mandeb the thirty kilometers strait at the southern end of the Red Sea separating the Arabian Peninsula from Africa, otherwise linking the Indian Ocean and the Mediterranean Sea).

Among the seven seas, and alongside some references to an Oriental Sea, the West Sea is described as the sea of *ténébres* (darkness) (elsewhere referenced by the translator as the Atlantic). As twice alluded, "No-one knows what exists over there." "There are two islands named islands of Fortune from where Ptolemy based his primary longitude calculation" (Jaubert 1836–40: Vol. I: 10), though it is also true that most Arab geographers placed their prime meridian on the eastern fringes of the Mediterranean. Latitude and longitude are both considered allowing a spatial vision such as revealed in the leaf maps and, more graphically, in the al-Idrisi worldmap and, although we have to believe, as with the now disappeared Norman king's silver globe.

Seasonality of Oceans, Tides, and Trade

Al-Idrisi's sense of changing wind patterns in the Indian-China seas was based upon objective knowledge or common lore among the Arab sailors who ranged these seas taking advantage of the sea currents and, critically, monsoon winds (from the Malayo-Arabic *muasim/musim* meaning season), also understood by the Greeks from the first century BCE, as with Hippalus. As mentioned in chapter 1, *The Periplus of the Erythraean Sea* credits Hippalus with discovering the direct route from the Red Sea to India along with the location of trade ports on the Indian coast (Schoff 1912). Al-Idrisi was obviously no discoverer of sea routes or the monsoon wind and sailing regime.

> The sea of China. The part of the sea of Senf to which it is contiguous. The Djaladeri Sea, as well as that of Herkend (Bay of Bengal?) and Oman, are subject to flux and reflux. Reportedly in the Sea of Oman and Fars (the Persian Gulf), the phenomenon takes place twice a year. As proven, the flux occurs during the six summer months in the Oriental Sea, while the contrary takes place in the Occidental Sea, there the reflux responds to the west during the six other months. (author's translation, Jaubert 1836–40: Vol. I: 94)

Tidal action also exercised the minds of the ancients. Muhammad al-Idrisi was not dogmatic and allows a number of opinions. Neither did he know the truth. As he mentioned, "Aristotle and Archimedes pretend that it owes to the action of the Sun combined with the action of the wind and waves (Like those occurring in the Atlantic which is an ocean). Producing the flux whereupon a reflux occurs with the wind falling away." "But Satoiois (an unknown Greek) considered the cause of the flux was linked with the Moon's successive growth and reflux accordingly" (Jaubert 1836–40: Vol. I: 94–95). In 2 BCE, the Babylonian astronomer Seleucus correctly described the phenomenon of tides and that the height of the tides depends on the Moon's position relative to the Sun. Al-Idrisi does not answer the question but tidal actions and phases of the Moon were obviously critical knowledge to mariners of any age (long prior to scientific understandings of gravity as with Galileo, Kepler, and, more especially, Newton, such as published in his *Principia* in 1687).

Immediately following his discourse on tides, al-Idrisi introduces a small passage on "China ships" entering the long-distance trade. Obviously, given the dangers of the passage as with monsoon winds and changing tides, along with sea monsters including giant roc or garudas (mythological birds) who would sweep away a ship with its talons, the quality of ships was imperative.

> All the China ships large or small which navigate in the China Sea are solidly constructed of wood. The planks placed one over the other are geometrically positioned, secure from (water) infiltration by palm fibers, and caulked with flour and fish oil. (author's translation, Jaubert 1836–40: Vol. I: 95–96)

No doubt this is an accurate description of a ship of that period, but fits that of an Arab dhow, not that of a Chinese junk which anyway would not enter interocean trade until a later era. In al-Idrisi's age the China ships were actually Arab or Persian vessels which conducted the China trade.

On China, and turning to the subject of ports of China situated on rivers and connecting with the maritime world.

> The first of the ports, as already stated, that of Khankou, is considerable. It is located on a river along which one which one passes through a large part of the country of Bahgbough, king of China and its dependencies. It is not just that he is superior but, to the contrary, all other kings in the country respect him. (author's translation, Jaubert 1836–40: Vol. I: 99)

Bahgbough is obviously an Arabism that defied translation, but Khankou is, variously, the named capital of China, such as written by the Franciscan traveler to China, Oderic de Porderone, or more likely Canton. Khankou (Canton) in turn was linked to Djankou. "Here they work Chinese glass (blown glass), as well as all kinds of silk fabrics, such as found in Djankou which is situated amidst a great river surrounding it and which reaches a great number of Chinese towns, as mentioned above" (Jaubert 1836–40: Vol. I: 100). As Friedrich Hirth and W. W. Rockwell (1911: 228n) indicate, this was the only Arab source to mention glassblowing in China, while identifying the city as Kanfu or Hangzhou (and with Kanfu its port), although Khanfou is mentioned separately by al-Idrisi. As such, these are highly laudatory statements but could also reflect a consensual view of the many stories cycled by Arab mariners returning from journeys to China.

According to Tolmacheva (2015: 13), although many later Islamic authors drew on al-Idrisi, there appears to be "a lack of systemic originality and loss of precision in world geography and cosmographical genre." This, she argues, is all the more surprising in consideration of the expansion of travel and production of new astronomical data. For example, the Idrisi sectional map system was not used at all, nor further developed by later authors, although it was not lost. Some authors contend a decline in Ptolemaic influences in later Arab maps and with a tendency to syncretism with European knowledge as in Ottoman cartography from the seventeenth century onward (Karamustafa 1992: 206).

ARAB NAVIGATIONAL PRIMACY

Granted Arab primacy in the Indian Ocean, the Gulf, and the Red Sea, it would not be surprising if Arab navigational lore did not make its impact felt on the Mediterranean as with use of the magnetic compass and the triangular-shaped lateen sail. Alongside the practical, the Arab legacy to Iberian nautical knowledge also extended to the scientific. Setting aside spherical trigonometry, such as pioneered by Muslim scholars as a navigational aid for locating Mecca, other navigational techniques crucial to the Iberian expansion include techniques for determining coordinates, as with sun-height tables, nautical charts, and basic geographical information. Once entering the Indian Ocean, the Portuguese would directly employ Arab pilots and knowledge.

Such an evolution offered Portuguese cartographers an advance over more traditional portolan charts, typically representing coastal features in line with dead reckoning such as employed in coastal navigation. Even so, Arab practical navigation based upon celestial observation did not necessarily translate into maritime charts. With its origins in Mediterranean navigation, portolan charts also incorporated magnetic compass bearings. The new Portuguese sailing charts that evolved in tandem with Atlantic explorations combined the traditional grid of magnetic courses with the introduction of scales of latitude, a development termed by map specialist Luís Filipe Barreto (n.d.: 5) as "the great technical innovation made by the Portuguese in sixteenth century cartography." Other innovations followed, such as indicating horizontally how coastlines appeared as well as records of depths plumbed.

The Haj and Arab Navigation

In Islam, the pilgrimage to Mecca is a fulfillment of religious life for those who can afford and manage the arduous and often long-distance and even cross-continental route. Millions make the journey today, the fifth pillar of Islamic obligation but, by the early decades of the last century, the numbers were closer to tens of thousands making the symbolic journey. Conducted during the last month of the Islamic year, known as *Dhu'l Hijja,* the pilgrimage required considerable planning, just as pilgrim/haj brokers would eventually emerge and one cannot entirely disengage the haj from merchant activities. As explained, four main routes were established to Mecca after the death of the Prophet Muhammad. The first, established at the turn of the ninth century by Zubayd, wife of Harun al-Rashid of the "Arabian Nights" story, was through Iraq with staging posts and reservoirs. Others routes would follow. While pilgrims were originally drawn from the Middle Eastern countries, by the eleventh century they were joined by pilgrims from Africa and even as far away as Southeast Asia (British Museum 2014).

For Muslims residing west of Egypt praying toward Mecca, in the region known as the Maghreb, and those of the Mashriq (Mashreq, Mashrek), generally denoting the region of Arab countries to the immediate east of Egypt, namely, the Levant and the broader Arabian Peninsula, and for those of the faithful living even further east as in China or the Indonesian archipelago, direction then as now is crucial for prayer as much as performing the haj. But the ancient Arabs also knew the compass as an essential maritime navigational aid, just as celestial navigation was also part of local lore in traversing vast deserts and arid regions from the Sahel to the Eurasian silk roads. Islamic mapping and navigational guides, accordingly, developed across the Islamic world to fulfill religious needs of the *ummat* or community of believers. Notably, the *ummat* knows no nationality or boundaries whatever the ethnicity or nationality. The Holy Koran is read or memorized in one language, Arabic, across the Dar al-Islam or world of Islam.

As the Portuguese scholar of scientific culture Nuno Crato (2014) observes, Muslim scholars had studied the angles of triangles on the surface of terrestrial spheres, and this study was spurred by the qibla *(kiblah)* problem, namely ascertaining the direction of Mecca. While the problem was relatively easy to solve in the vicinity of Mecca, as Islamic civilization spread to the Iberian peninsula, the African continent and the Far East, then solutions were required. As discussed below, the problem was eventually solved by the Tajikistan-born, al-Biruni, who created a precise method to determine the direction of Mecca, given the coordinates of the place in which the believer finds himself. Qibla chart making, lending to geometric construction or application of trigonometric formulas, was another aspect of this "sacred geography" (King and Lorch 1992: 189). One example offered by a British Museum (2014) exhibition was an ivory sundial and *qibla* pointer made by Bayram bin Ilyas in Turkey in 1582–83. A *qibla* compass (sometimes also called indicator or pointer) is a modified compass used by Muslims to indicate the direction to face to perform ritual prayers. This direction is called *qibla* and points toward the city of Mecca and, specifically, to the Kha'bah. Modern *qibla* pointers utilize "Google Maps."

Al-Biruni

Of particular importance in the development of the *qibla* was the role of the Uzbekistan-born astronomer, Ahmad Abu al-Rayhan al-Biruni (973–1050), who went on to author the *Kitab Tahdid al-Amakin* (The Demarcation of the Coordinates of Cities). He also determined the motion of the solar apogee, corrected Ptolemy's findings, and determined for the first time that the motion is not identical to that of precession or, change in the orientation of an astronomical body's rotational axis, but comes very close to it. He also employed mathematical techniques unknown to his predecessors, especially involving analysis of instantaneous motion and acceleration.[2] In particular, as Park (2012: 73) has highlighted, a sketch map produced by al-Biruni advanced upon other Arab schools (as with the Balkhi school maps with their focus upon Islamic regions). Al-Biruni produced a map offering an oceanic expanse in the southern hemisphere corresponding to the Indian Ocean and thus replacing the *terra incognita* of the Balkhi school. In turn, the Indian Ocean is connected with the Atlantic Ocean and with southern Africa depicted as a rectangular, not triangular, shape.

In order to pray, Muslims must face in the proper direction. Believers have to know with some precision the longitude and latitude of Mecca. Once determined, the values are applied to a spherical triangle, and the angle from the local meridian to the required direction of Mecca can be determined. Obviously, the problem of measuring coordinates admits of more than one method of solution and al-Biruni did his share in supplying the various methods in this book (Muslim Heritage 2014). As discussed in chapter 5, the question of determining direction as applied to maritime navigation would be taken up by Portuguese astronomers in the service of the kingdom.

Muslim Navigational Priority

As opposed to geographic accounts, virtually no Arab sea chart of the China trade survives from this era. Nevertheless, the Arabs did have navigational maps, as with Indian Ocean routes. Marco Polo makes more than one reference to such. At the time Santarém was researching in Paris collections, besides the al-Idrisi manuscript, there were seventy other Arab map geographies. Writing in the tenth century, al-Masudi (896–956) observed the existence of Arab mappemondes colored for clarity; "They represented the world with celestial spheres, astres, continents, seas, inhabited land, deserts, occupied regions, peoples, great towns" (Santarém 1852: 377).

As F. Jamal Ragep (2010: 121) points out, although Ptolemy's mathematical model was widely respected in the Islamic world, still there was ambivalence toward him when it came to the practical issue of astronomical observation as in the testing and confirmation of results. One example of Abbasid astronomy which fits this description is the project mounted by Caliph al-Ma'mun (r. 813–33) to measure the Earth's circumference, an activity carried on by al-Biruni in the eleventh century. Muslim Arab mapping also made strides under al-Ma'mun. According to Turkish historian of Islamic science, Fuat Sezgin (2004), the Caliph al-Ma'mun map discovered in the Topkapi Palace Museum in Istanbul illustrates just how far Muslim cartographers departed from earlier worldviews. Displaying large parts of the Eurasian and African continents with recognizable coastlines and major seas, it reveals the world as it was known to the captains of Arab sailing dhows. As he declaimed, "The maps of the Greeks and Romans reveal a good knowledge of closed seas like the Mediterranean but the Muslims knew the continents are islands." Known to Columbus for his corrections to Ptolemy, al-Ma'mun's lasting impact cannot be ignored.

Born 1431 in Dhofar, present-day Oman, Ibn Majid, author of *Book of Profitable Things Concerning the First Principles and Rules of Navigation* provides and offers rare information bearing upon Arab sailing and navigational lore at a time when the first Portuguese were also entering the Indian Ocean. Contemporaneous with Vasco da Gama, Ibn Majid should not be confused with the pilot hired by the Portuguese at Melindi. Offering meticulous instructions on how to sail the Indian Ocean using twenty-eight lunar mansions and star positions, compass rhumb lines, the monsoon, and other seasonal winds, in Park's (2012: 188) estimation Ibn Majid produced the best navigational work of its time, and a work that later Muslim navigators would consult.

From the ninth to the tenth centuries the Arabs were already using a celestial navigation device called the *kamal* to determine latitude. As such, the invention of the *kamal* allowed for the earliest known latitude sailing and was thus the earliest step toward the use of quantitative methods in navigation. A handheld rectangular wooden devise through which a knotted thread was attached, the *kamal* consisted of twelve square shaped wooden boards used to measure latitude. It was adopted by Indian navigators soon after (Ives 1961). Known to Chinese as *suoxing ban*, the *kamal* undoubtedly served Chinese navigators, as with those of the Zheng He voyages (Park 2012: 174).

More or less serving as a summary to the above, as Park (2012: 185–186) points out, having adopted the compass from the Chinese and combining it with their own astronomical methods for sea navigation and, following generations of acquired knowledge, Arab navigational lore reached a new level of sophistication. The two key navigational elements employed by the Arabs were "star altitude measurements and compass bearings." Arab navigational knowledge and its heritage was not only confined to the Indian Ocean but spread westward to other cultures including those of the Mediterranean and, as demonstrated below, to eastward-sailing Arabs, Persian and Indians traversing the Bay of Bengal, the Straits of Malacca, via coastal Vietnam en route to China.

ARAB/MUSLIM GEOGRAPHIES OF ASIA

More generally, Arab pilots began reaching the region known today as Southeast Asia from the seventh century in search of spices. By the ninth century, Arab traders had also voyaged to the Caspian area. But the lure east might also have been direct trade with China, especially linking up with the source of fine silks, porcelain, and other valued manufactures and commodities. As explained below, the first compilation of traveler's tales based upon these voyages emerged from the mid-ninth century, although still remaining in manuscript form until a later age. While, as discussed, Muslim/Arab navigational priority was both sophisticated in terms of its mathematical foundations and workable at a practical level at the point when the Europeans were entering the Indian Ocean, much more reserve can be attached to Arab/Muslim descriptive accounts of China and other destinations.

Some authors are less sanguine as to the veracity of Arab geographies. According to Arab navigation specialist Gerald Tibbetts (1992: 258–259), Arab navigational texts circulating before and after the advent of Islam were "to an extent representative of all Indian Ocean sailing." This he compares to the way that navigation in the Mediterranean area was knowledge shared by local sailors. Notwithstanding various historical references to charts (Polo; Barros, Varthema), "In reality, the texts do not mention charts at all." In any case, stellar navigation served the Arabs well even in the absence of ways to calculate longitude. He concludes that Arabs did not have the technical traditions to either construct navigation charts based on a matrix of latitude and longitude or even to use such a chart even if available.

According to Suárez (1999: 51), many of the places mentioned in these early Arab geographies cannot be identified with certainty and with many assuming a mythological status: "Some of the fanciful places may derive from the Alexander Romances, or perhaps from attempts to reconcile Ptolemaic geography with the Qur'an." Even by 1000, the Arab geographers continued to recycle three-century-old lore without updating. Santarém (1852: 334) is even more dismissive offering a wealth of examples. While Arab geographical knowledge was superior to that of Europe with regard to the east African coast, he asserted, they were basically ignorant of the West African coast, ignoring the Portuguese discoveries. Sicily-based al-Idrisi, as mentioned, was not immune from this tendency, even though dedicated to collecting traveler's information. As discussed below, neither did Ibn Battuta's fourteenth-century firsthand account feed into Arab geographies. More specifically, the Arab embrace of the Ptolemaic view created great confusion, especially the notion of an enclosed Indian Ocean, which led to the mixing of information pertaining to Southeast Asia (al-Zabaj) with the east Africa coast (al-Zanj). Still these criticisms could be culturally bound.

Ibn Battuta

Departing Tangier in June 1325, Ibn Battuta would return twenty-four years later having visited south and central Asia including China, the Swahili coast and west Africa. Among the Asian destinations to which he journeyed along the maritime route to China were the Malabar coast of India, the Maldives, Ceylon, Samudra Pasai in northern Sumatra, Malacca, central Vietnam and Zeitun (Quanzhou), Cabton (Canton), Hangzhou, and Beijing. His *Rihla* or *Journey* was dictated to Ibn Djozai (Juzay) and embellished by copyists, leading some Orientalists to question details. Though partly fictional, it nevertheless offered a state-of-the-art rendition of the world of the mid-fourteenth century. For centuries his book was obscure, even within the Muslim world, but in the early nineteenth century extracts were published based on manuscripts discovered in North Africa containing abridged versions of Ibn Juzay's Arabic text. Yet, by this age, it cannot be said that Ibn Battuta's Afro-Eurasian journeys influenced European images, at least not in the way of the Renaudot translations, as discussed below. Europeans of the sixteenth century found that Arab geographical knowledge of Asia had stagnated by three centuries, although Persian and Turkish geographies were rising. By then, Arab geography was incorporating European sources (Suárez 1999: 51).

Abu'l-Fida (Abduseda)

Between 1315 and 1329, Abu'l-Fida (Abduseda) (1273–1331) produced his *Tarikhu'al-Mukhtasar fi Akhbar al-Bashar* (*The Concise History of Humanity or Chronicles*), long considered in Europe as the finest Muslim history/geography. Drawing upon both Ptolemy and al-Idrisi, Abu'l-Fida offers a long introduction on various geographical matters followed by twenty-eight sections dealing in tabular form with major towns of the world, referenced according to longitude, latitude, climate, spelling, and other observations borrowed from earlier authors.[3]

There is no doubt as well that Abu'l-Fida was advanced from within the Muslim traditions, but even relative to Europe. As his geographical work *Taqwim al-Buldan* states, "if a person goes from a given point to the east, another to the west and the third remains on the spot, and if the two passengers one day return back to the starting point, the one who has gone in the eastern direction to make the round trip of the globe will count one day more, and the one gone in the west one day less (judged by the risings and settings of the Sun) in exactly the same space of time! For the one who goes in the west marches in the same direction as the Sun, and thus his day is always a bit longer than the day of the one who goes in the east, with the result that when the round trip of the Earth is completed, that makes the difference of a whole day" (Reinaud 1848–83: II/I, 3–5; Mohammed 1968: 7–9). Such direct understandings were only revealed to Europe by the survivors of the Magellan circumnavigation.

Abd-ur-Rassaq

A fifteenth-century Persian geographer born in Herat, Abd-ur Rassaq (b. 1413) was the ambassador of Shah Rukh, ruler of Persia under the Timurid dynasty to Calicut in India, from January 1442 to January 1445. His mission to India was chronicled in his book entitled *Matla-us-Sadain wa Majma-ul-Bahrain* (*The Rise of the Two Auspicious Constellations and the Confluence of the Two Oceans*). Offering an overview of the history of this region from 1304 to 1470, Abd-ur-Razzaq's book described the life and events in Calicut in south India under the Zamorin as well as the southern Hindu state of Vijayanegara. Additionally, he described shipping activities in the Indian Ocean during the period of his embassy. Abd-ur-Razzaq's *Matla-us-Sadain wa Majma-ul-Bahrain* also included a detailed account of diplomatic relations between Shah Rukh's state and Ming China. He also identified Tenasserim (Burma) and Shahr-I naw (Siam). His major legacy was thus to describe the sea journey from the Gulf to Calicut as well as to bequeath an account of his visit to the little-known Deccan region of India (see Suárez 1999: 51; Muzaffar Alam and Subrahmanyam 2007).

To be sure, as Park (2012: 82) declaims, the sheer variety of Muslim maps and treatises written in Arabic, Persian, and Turkish reveals a process of variation in Islamic geographical knowledge about the world based upon different traditions and sources. Such diversity matches the experience of Islam from around the tenth century onward as it expanded eastward in its encounters with different cultures and peoples. It is certainly true that the spread of Islam from the Arabian heartland to, in one direction, the Maghreb and, in another, to China and the Indonesian archipelago was also accompanied by the circulation of ideas. As the European translations from Arabic point up, and indeed the reception granted the Marco Polo legend, Europe actually lagged far behind the Islamic world in its knowledge of China where, as indicated, Muslims under the Khanate were actually embedded in government office alongside commercial activities.

Ninth-Century Arab Merchant Accounts of Asia as Translated by Renaudot

More generally, Arab accounts of Asia including China were unknown in Europe hence the sensational attributed to Polo's *Travels* in its first reception, just as the Polo narrative mirrors, adumbrates upon, and sometimes corrects typical Arab travel geographies. In fact, Europe would await the early eighteenth century before authenticated Arab travel accounts entered translation in European languages. Particularly notable were two ninth-century travel descriptions to China via India translated by French Orientalist Eusèbe Renaudot and published in 1718. Both are said to be taken down from statements made by a merchant named Sulaiman, the second a continuation of the first, by Hasan Ibn Yazid and Abu Zaid al Hasan (Abu Zayd). The voyages described therein date from 851 to 877 CE, corresponding to the Tang dynasty (618–907 CE) with its capital at Chiang'an, present-day Xian in what is now Shaanxi Province. Even so, according to the English translation of this work appearing in 1733, the travels seem to have been mostly performed by one person only; the latter portion being chiefly a commentary upon the former, and appear to have been the work of one Abu Zaid Zeid (Zayd) and to have been written about 915 CE. The manuscript employed by Renaudot belonged to the library of the Count de Seignelay, and appears to have been written in the year 1173 CE. Predating Marco Polo by 100 years, we wish to know the veracity of these accounts and how, if so, they informed general Arab understandings of India, Southeast Asia, and China?

As Renaudot (1718: iv) acknowledged, the importance of the manuscript was that it contains the earliest (foreign traveler) account of China entering, in this case, Arabic language. He also allows that there are many curious and fabulous passages in these travels, although far less than in many other Arab accounts. Moreover, they write of actual historical events as with the civil wars in China and India prior to the Muslim conquest. Nevertheless, there are lacunae. Notably, no mention is made of the magnetic compass. Yet they convey information respecting customs and events that are nowhere else to be found. He also argues that the general veracity of these accounts was subsequently confirmed by Polo and, in succeeding ages validated by such Jesuit accounts as those of Martino Martini (1654) and by the Portuguese voyages. As Renaudot (1718: xxxiv) also acknowledges, there is no exactitude in Muslim renderings of place names and, as Martini was the first to point out, in China place names frequently change. Still, Renaudot does impose names in the text, such as with Maldives, Ceylon, Canton, though he is also cautious.

Renaudot's account was challenged, especially as he neglected to explain the provenance of the Arabic manuscripts and also fell foul of Jesuit Sinologists, who contested certain of his negative views on governance. Yet, he was subsequently redeemed. In time, his merchant accounts would enter mainstream versions of Chinese history as received in Europe, thus becoming image-forming texts for cartographers, as with French mapmaker d'Anville, as discussed in chapter 7, but not of course image-making in the age of discoveries, as with Polo.

CHINESE ASTRONOMY AND THE MUSLIM WORLD EXCHANGE

Unlike ancient Egypt, Babylonia, and Phoenicia, the Chinese astronomical tradition is striking for its continuity over millennium. Although China was willing to learn new astronomical techniques, as with borrowings from the Arabs and Persians and, in a later period, from Europe itself, it is important to acknowledge that, unknown to Ptolemy and unknown to even late medieval Europe, the Chinese astronomical tradition was hegemonic within the vast Chinese empire wherever the tradition held sway, as in Korea, Japan, and Vietnam. In distinction to the Alexandrian tradition of philosophical mathematics, in China astronomical and calendrical calculations were ceremonially integral to the reproduction of the imperial system and with the emperor known as the "son of heaven" (see Needham 1959; Waley-Cohen 1999: 107–112).

More generally, modern archaeology confirms a wide range of star charts across East Asia, mostly ancestral to those of China ranging back to neolithic times and reaching increasing sophistication in Tang dynasty times, as with the Dunhuang star chart discussed below. In China, star charts were still made through the Qing era, in Japan down through the Edo period, and in Korea down to the late Choson period. Richard F. Stephenson (1994: 520) traces stargazing in China back to Qin and Han dynasties (221 BCE–AD 220), noting that a tradition of painting star charts on ceilings of royal tombs was common from Han times onward.

Also revealing of the Chinese civilizational footprint over East Asia is a star chart discovered in 1998 on the ceiling of a tumulus in Asuka town in Japan's Nara Prefecture (the *Kitora Kofun*), dating from the late seventh to the early eighth centuries (Asuka period). As announced in 2015 by the Japanese Agency for Cultural Affairs, the Asuka or *Kitora Kofun* star chart was, in turn, based on much older celestial observations made in China. Featuring sixty-eight constellations in which the stars are depicted using gold discs, the movement of celestial objects is also represented in the form of three concentric circles with another circle depicting the movement of the Sun. The Polar Star is depicted at the center. Designations for the horizon, Equator, and ecliptic circles, as well as recognizable patterns of stars are also discernible.

Although subject to scientific debate, and with many questions as to how this civilization transfer to Japan occurred, analysis of digital images reveal that the most likely observation sites were those located on the 34th parallel north, including the ancient cites of Chang'an and Luoyang in western Henan (Tsukamoto 2015).

If this analysis is correct, then the Asuka chart would be the oldest existing star map of its kind in the world although it was not actually designed for use in divination and was based upon or ancestral to an actual chart never recovered. According to Steve Renshaw and Saori Ihara (2002), the chart in *Kitora Kofun* utilizes observational data for plotting star coordinates that would have been in use in China around 70 BCE or late Qin-Han Dynasty (dating consistent with estimations made by Japanese historical astronomers). Moreover, the system was probably used by an astronomer in the Kokuryo kingdom based on Pyongyang who adapted it for that location's latitude. However, such adaptation was apparently not made when the chart was brought to Japan.

The earliest scroll star chart to survive is the Dunhuang star map, dating from Tang dynasty. Now kept at the British Library in London, the star atlas was found among the 40,000 other manuscripts discovered in 1900 at the Buddhist Mogao or Dunhuang cave complex in China's northwestern Gansu Province on the edge of the Gobi Desert. Once a frontier garrison on the land silk road, the site lies at 40° N latitude. A total of more than 1,300 stars are distributed in 257 different Chinese constellations. According to J. M. Bonnet-Bidaud et al. (2009), the Dunhuang chart is not only a complete atlas of the northern sky, but represents "a remarkable early achievement in terms of mathematical projection methods in China."

Unquestionably, the Chinese astronomical tradition was both observational and mathematical. As with contemporaneous Phoenecian-Sumerian and Grecian counterparts, the Chinese developed sophisticated star charts based upon continuous celestial observation matched by the creation of instruments undergoing significant refinement over time. The Chinese understood the Earth's sphericity and were aware of a heliocentric universe. Nevertheless, the Middle Kingdom's obsession with the cult of the emperor as the son of heaven, as with prediction of eclipses, the sighting of comets, and elaborate ceremonials surrounding calendrical calculations, deflected attention away from earthly concerns. Moreover, Chinese astronomy did not easily translate into the mathematical plotting of terrestrial coordinates as with the Ptolemaic template. Unlike the Mediterranean world, maritime navigation was not yet a priority.

But how did Chinese astronomical prowess translate into general cartographic knowledge or, more specifically, into knowledge of the world, its oceans and continents? By 1500, as Park (2012: 2; 15) has underscored, "the Chinese learned significantly more about the Islamic world, and the Muslims of West Asia and North Africa knew a great deal more about China, than either society had known about each other in the eighth century." In 770, she continues, "maps of China did not exist in the Islamic world, nor did the Chinese possess maps of the Islamic world." To illustrate the jump in knowledge in China during this period, Park offers a Korean "world" map of 1402 (embracing the China-Southeast Asia-India area also suggesting Africa), in turn based upon a Mongol map. We have already seen that, thanks to their assimilation of Ptolemy and their navigations, Muslim geographers had developed a sense of the coastlines linking the Islamic world and China much earlier than Chinese cartographers.

The Persian-Muslim Exchange with China

More generally, we should be cognizant of major geopolitical shifts across Eurasia not excluding the westward spread of Islam through the Mediterranean. As Park (2012: 7) summarizes in her study of the Islamic world exchanges with China, by continuing its eastward push through central Asia, the Abbasids came into political and military collision with the Tang dynasty of China, a reference to the Battle of Talas (751 CE), leading to a Muslim advance and a Tang retreat from the Transoxiana region (as it was known to the ancient Greeks). By the eighth century, with the central Asian silk roads disrupted and going into decline, trade across the maritime silk roads surged.

As Park (2012: 8) determined from a scrutiny of Arabic and Persian texts, when the Mongols ruled both China and Persia under the Il-khanate in the thirteenth and fourteenth centuries, their mutual geographic knowledge expanded greatly. As she explains, both overland and maritime societies experienced integration under the Mongols who conquered most of Eurasia and reopened trade routes that had been closed for centuries. During this period, Chinese-Muslim relations especially developed via the maritime route, culminating in the spectacular advance into the Indian Ocean littoral of the early Ming-era Zheng He voyages. As explained below, commercial, geographic, cartographic, scientific, and many other knowledge exchanges resulted from this broadly understood *Pax Mongolica.*

We have alluded to the exchanges engendered across the Mongol empire from the Mongol-Yuan center in Khanbalik to the outlier breakaway House of Hulagu in Persia under Genghis Khan's grandson, Hulagu Khan (r. 1256–65). More generally, the original contributions of Arab and especially Persian mathematicians and astronomers under the Mongol Khanate cannot be underestimated. Incredibly, their knowledge traveled in two directions, west toward Venice and Europe, and east toward Khanbalik. This book can hardly do justice to such an abstruse subject. Two individuals are outstanding in this sense, both overseeing the construction of celestial observatories, namely, Nasiruddin Tusi and Ulugh Beg.

Facing down the Mongol invasions that swept his homeland in Khorasan, Persia, Nasiruddin Tusi (1201–74) entered the service of the Khanate, constructing an observatory leading to the production of a major astronomical treatise, completed in 1272. Along with a treatise on the astrolabe, critiquing Ptolemy's *Almagest* and revising his heliocentric model, he offered a standard only eclipsed by Copernicus or, perhaps, even borrowed (Ragep 2007), indeed another example of the Eastern origins of Western civilization. According to George Saliba (2006: 273), given Copernicus's ten-year sojourn in northern Italy, it is a plausible setting in which he collaborated with Arabic translators, learning of Tusi's critical theorem on the concept of linear motion. As Park (2012: 128) points out, Tusi's astronomical table also contained an accurate description of the Chinese calendar, proving that Chinese scholars were also involved in its production.

On his part, Uzbek astronomer and Timurid ruler, Ulugh Beg (1394–1449) is known for the production of star tables, a star map, and latitude tables, superior in its time. He also constructed an observatory at Samarakand, destroyed in 1449, albeit recovered in 1908 with original drawings of Beg's star charts kept in the Arabic and Persian manuscript collection of St. John's College of Oxford University.

As now known, Safavid Persia (1501–1722 CE) produced sophisticated world maps with localities properly marked on a coordinate grid. Such comes to light with the recent (in the 1990s) discovery of two maps (A and B), actually instruments, ca. 1650–1850, taking the form of circular brass plates, also indicating the distance and *qibla* direction to Mecca. According to Islamic map specialist David A. King (1997: 62), seventeenth-century mathematical cartography as represented by Safavid world maps undoubtedly had its origins in Baghdad in the seventh century.

Astronomical Exchanges under the Mongol Khanate

As observed by the first-arriving Jesuits in China in the late sixteenth century, the Mongol rulers appointed Muslims (Persian speakers particularly) and other non-Chinese to important government posts, and even created special offices for Muslims in recognition of their unique astronomical and geographical knowledge. As Park (2012: 99) embellishes, notable was the Islamic Astronomical Bureau Observatory (*Huihui Sitiantai*) under its first director, Jamal-al-Din (Zhamaluding), established in 1271 in the then Mongol capital of Xanadu (Shangdu), parallel with the Chinese Astronomical Bureau. The city is also described by Marco Polo believed to have visited about 1275. As Matteo Ricci remarked in his time, there was nothing like it in Europe (Duteil 1994: 285). In fact, as discussed in chapter 6, the first-arriving Jesuits both witnessed this observatory (and met face to face with the leading Muslim court astronomer whose position and learning they would soon challenge). In 1267, working under Kublai Khan, Jamal-al-Din constructed seven astronomical instruments including an astrolabe, a colored terrestrial globe with grids, plausibly representing the longitude and latitudinal coordinates as used by Muslim cartographers (essentially a Ptolemaic inheritance). Among his contributions to Chinese astronomy, Jamal-al-Din constructed a 10,000 years calendar (Wannian Li) which, as Park (2012: 100) describes, essentially represented a transfer to China of Islamic planetary tables modified from Ptolemy's *Almagest.* As Park surmises, Jamal-al-Din brought maps, although these are lost.

The Chinese astronomical tradition such as encountered by the Jesuits in the sixteenth century was strongly infused with Persian influence, and with China gaining knowledge of algebra, along with the technical know-how (Duteil 1994: 284). Notable was the effort by Guo Shoujing in 1279 in constructing astronomical instruments in Beijing as well as in reforming the calendar. It is also true that the advent of the Ming in 1368 meant a restoration of a more independent astronomy tradition in China, allowing the coexistence of the Muslim tradition at the same time. Notably, a Muslim astronomical observatory was established in 1368 in Nanjing the first Ming capital, although offering results of decreasing accuracy whether through loss of tradition or through neglect.

Navigational Exchanges

We should not be surprised that the great Ming-era Zheng He seafaring voyages incorporated the full spectrum of Arab navigational knowledge, just as the Arabs gained the compass from the Chinese. As Park (2012: 49–50) embellishes, Chinese not only knew the lodestone back to antiquity but from ca. 1086 commenced to use it aboard ships. Also, as noted, sea charts produced by the Zheng He voyages of 1405–33 did incorporate the major sea routes connecting China to the Islamic world. Chinese navigators used the compass and made celestial observations (see Figure 2.2).

Maritime exchanges between Arab, Persian, and Indian merchants reaching China by the maritime route obviously brought with them transfers of navigational and technical lore. By at least Yuan times, Arab ships arriving in Quanzhou (Zayton or Chinchew in classical works), their major Chinese port of call, were required to surrender their navigational logs. The southern Chinese port city then hosted a Muslim harbor-master, a mosque (the Qingjing or Ashab Mosque dated from 1009), along with a *qadi* or judge (see Steinhardt 2008). It makes sense that, in the careful preparations for these

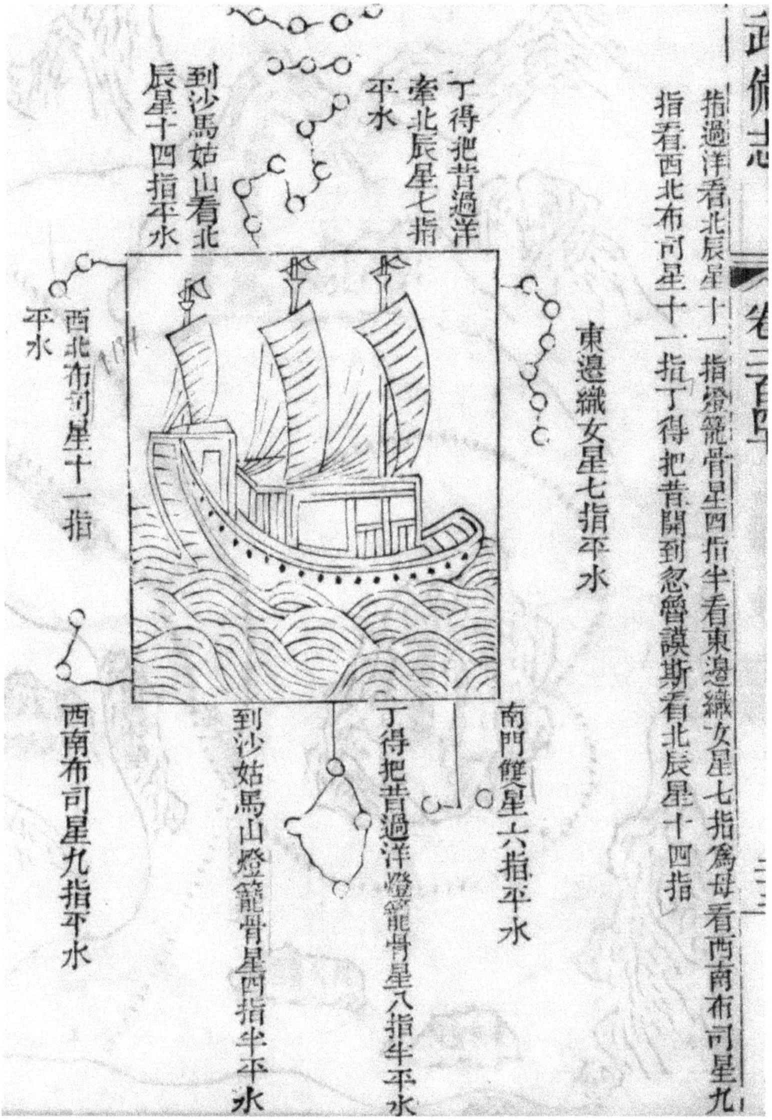

Figure 2.2 Wu bei zhi, Mao Yuanyi, 1594–1640. China: s.n; not before 1644. Wood block print. Early Qing Dynasty, prefaced Tianqi 1 (1621). Juan (chapter) 240 is one of 80 volumes in 8 cases (30 x 19 cm.) "Juan 240" shows Zheng He's sailing route from Nanjing, passing through maritime Southeast Asia, and the Indian Ocean all the way to the Persian Gulf. *Source*: Retrieved from Library of Congress, Geography and Map Division. Digital Id http://hdl.loc.gov/loc.gmd/g7821rm.gct00058.

massive Ming-era maritime expeditions, Arab navigators were brought on board. In turn, the Ming inherited the navigational skills of their Mongol-Yuan forbears.

As Park (2012: 58) points out, by the time they began to receive firsthand accounts of China derived from travelers, Muslim geographers had already established systematic frameworks to help them interpret the world, "because they assimilated the knowledge of the societies they conquered, like the Greeks and Persians." In other words, there was a major asymmetry between Muslim geographic description (and, I would say mapping as well) and its Chinese counterpart. Notably, Chinese writers first depicted their own empire (the Middle Kingdom), and with the outer world depicted as marginal. This is true and with the tradition also adopted in Japan continuing up until early modern times (Smith 1996: chap. 3). On the other hand, Muslim scholars conceived a larger world, as we have seen, especially as they inherited Greek and Ptolemaic geography before them. More than that, Muslim mapmakers, starting with al-Idrisi, learned map coordinates, albeit not replicated by practical navigators. Neither were map coordinates efficiently adopted or even accepted by Chinese until the age of Jesuit missions to Beijing, as discussed in a following chapter.

According to Park (2012: 173–174), Chinese navigators traversed the Indian Ocean using both a magnetic compass and through measurement of star altitudes to plot a ship's latitudinal position. This method required a celestial navigation device, as with the *kamal*, also known in Chinese as *suoxing ban*. Cartographers of the Zheng He voyages annotated each port city they identified with the height of such stars as the Pole Star or the Ursa Minor (Little Dipper) in addition to compass directions (Park 2012: 173–74). As discussed in a following chapter, Ming Chinese sea charts were not static but incorporated a corpus of geographical and navigational lore as confirmed by new map "discoveries" and textual analysis.

CONCLUSION

Offering a highly synoptic rendition of Muslim-Christian relations in the wake of the Arab conquest of the Iberian Peninsula, this chapter was necessarily selective. Especially we sought to understand through the medieval Arab-Muslim experience how adaptations or revisions to the Ptolemaic worldview were mediated, not excluding scientific contributions and new empirical knowledge of the world derived from Muslim merchant and scholarly contacts. The survival of Aristotelian philosophy in the Arab world and its revelation to late medieval Europe was also part of this story and the names of translators, interpreters, and princely patrons folds into the general narrative of European history and philosophy of science.

With Arab pilots reaching the region known today as Southeast Asia from the seventh century in search of spices, Muslim traders also gained a priority on practical navigation reaching beyond the Indian Ocean. It is certainly true, as Park (2012: 126) signals, that during the Mongol period (1260–1368), the level of Muslim knowledge achieved about the known world far surpassed anything Marco Polo provided. However, with respect to the Arab merchant accounts of India and China as translated by Eusèbe Renaudot, we observed that because of their late discovery and translation they never directly entered Renaissance thinking. With their knowledge traveling in two directions, west to Venice and east to Khanbalik, the original contributions of Arab and especially Persian mathematicians and astronomers under the Mongol Khanate cannot be underestimated.

Even so, while Arab and Persian prowess in navigational skills, celestial observation, measurement, and mathematics was demonstrated, cartographic empowerment was another matter. In Tolmacheva's (2015: 22) words, it was as if academic cartography remained "irrelevant to mariners' practical activity" (and this distance or gap, as argued, held for mariners of many traditions). Even if Persian and Turkish geographies were rising (Karamustafa 1992: 206), as Suárez (1999) contends, Arab geographical knowledge of the region actually stagnated by three centuries. Our conclusion must then heed Visconde de Santarém's (1852: 325) opinion that "the Arabs not only adopted the ideas of the ancients but also the fundamental basis of the Greek cosmographic system." Moreover, as Anthiaume (1916: 419) demonstrated, even Abbasid geographers corrected the false notion of an enclosed Indian Ocean, "the Arabs did not perfect the Ptolemaic view of the India region."

Notwithstanding Arabo-Persian borrowings through the Mongol exchange and empirical knowledge gathered by emissaries and others in the train of the Ming-era Zheng He missions, Chinese map renditions of the known world remained highly Sino-centric and culturally bound. In this light, claims by proponents of a Chinese discovery of America and other lands, matched by new Chinese map discoveries, leave much unexplained. Even so, we are not arguing a time bound and static China. Rather, the Mongol exchange and, as expounded in a subsequent chapter, the epoch of Catholic missionaries first arriving in Macau on the southern coast of China, accelerated change in a range of areas, not including a new sense of geography of space and empire.

NOTES

1. All modern translations of Benjamin's *Itinerary* derive from Editio Princeps, printed by Elizier ben Gershon of Constantinople in 1443 and the Ferrara edition of 1556, printed by Abraham Usque. Then followed an edition by A. Asher in 1840–41. In turn, Marcus Nathan Adler worked from Hebrew mss copies to produce *The Itinerary of Benjamin of Tudela: Travels in the Middle Ages* (1907).

2. Al-Birumi's work became better known via Edward Sachau's translation from the Arabic in 1887. The Arabic text was edited by Bolghakov and Ahmad (1962), and with a separate English translation achieved by Jamil Ali (1967).

3. With parts of the work published and translated in Europe in 1650, it also became an image-forming text, as with German scholar Johann Jakob Reiske's *Abulfedae annales Moslemici* (1754); and Reinaud's *Geographie d'Aboulfede* (1848–83) (which I have consulted). Renditions of Ab'ul Fida's maps were published, as with Pierre Duval's *Regionum ultra Oxum tabula geographica ex abulfeda ismaele, principe Hamah*, anno 1345 in 1665, and by Jean-Baptiste Bourguignon d' Anville such as profiled in chapter 7.

Chapter 3

Ptolemaic Revival and Cartographic Adjustments in a Transitional Age

Surprised at the apparent cosmopolitanism and geographical reach of Islam across the known world, such as described in the previous chapter, we wonder how the Ptolemy revival in late medieval Europe translated into new understandings of the Eurasian landmass? Notwithstanding the image of the European Middle Ages as "dark," it has long been recognized that important innovations in mathematics, astronomy, cartography, and other disciplines were apparent in Europe even prior to the classical revival and the Gutenberg revolution. As Lucas (1960: chaps. 9–10) long ago advised, the 1,500 years separating ancient Rome and the medieval era were not stagnant, especially in consideration of urbanization, the growth of markets and trade, business and finance, the rise of universities, the Aristotelian revival, along with science and developments in technology. More recently, Felipe Fernández-Armesto (1987) has drawn attention to the importance of medieval ideas to the age of discovery. These were not only Latin European innovations, but included Arab, Jewish, and other contributions. We have drawn attention to the role of such European centers of learning as Toledo in Spain, important as with the perfection of astronomical tables enabling calculation of eclipses and the positions of the planets for any given time based on Ptolemaic theory. The Majorcan school of cartography was a landmark in development of European knowledge of the Mediterranean and beyond. Neither can we ignore the special role of Venice in mediating east-west trade in the eastern Mediterranean, but also serving as a crucible of late-medieval-early Renaissance learning. Such an understanding jells with Fernand Braudel's (1992) portrayal of "wheels of commerce" linking up Mediterranean ports with land communication routes, and emergence of such "world cities" as Venice, Genoa, Antwerp, Amsterdam, and London.

In line with broad evolutionary trends taking place in late medieval Europe, this chapter commences with an investigation into the transitional status of astronomical and cartographic knowledge especially with respect to its classical lineages. Second, the chapter describes the primacy of Venice in connecting up Latin Europe with the Eurasian silk and spice roads during a period that might be described as the "Mongol Peace." Third, and in some detail, the chapter investigates the impact of the Ptolemy revival upon navigation and the Columbian discoveries. Finally, the chapter adverts to cartographic adjustments made to Ptolemy by key northern European map and globe makers during this transitional period.

MEDIEVAL EUROPEAN MAPPING

Long before the classical revival of the early Renaissance also touching cartography, a tradition existed in medieval Europe, in Arabic culture, and indeed in the Sinic world, of producing *mappamundi* (from the Latin) or *mappaemundi* (the plural form) or just *mappemonde* (the French form). The genre can be traced back to late antiquity, as with the five global (Ptolemaic) zonal-climata maps of the Macrobius (also known as Theodosius), a Roman provincial from the early fifth century whose works and images reappeared in the twelfth century.

According to Woodward (1987: 286; 290; 315), the primary purpose of *mappaemundi* in Europe was didactic, namely to "instruct the faithful about significant events in Christian history rather than to record their precise locations." Scale could be highly elastic to exaggerate areas of interest, as with Jerusalem, and with standard Roman sources often incorporated. The Hereford Map, ca. 1290, found in Hereford cathedral is but one example. A number of these medieval *mappaemundi* were circular, schematic, as with the T–O maps (*orbis terrarum*, orb or circle of the lands), with the letter T inside an O. Typically, they depicted a tripartite division of the world between Europe, Africa, and Asia, surrounded by a Mare Oceanus, as with that of Isodore of Seville (560–636). Some were quadripartite, allowing an antipodes, but they were still not Ptolemaic (see Figure 3.1).

Still, as Woodward (1987: 318) remarks, it is not easy to generalize as to *mappaemundi* over a 1,000-year-long period embracing monastic and nautical versions, especially as more scientific and Ptolemaic elements gained acceptance. It is also true that over time a trend toward realism appeared in *mappaemundi* with new concepts and techniques entering the picture. By the thirteenth and fourteenth centuries, the term "*mappaemundi*" came to apply to any world map. Nevertheless, given the Christocentric assumptions of the medieval European *mappaemundi*, with Jerusalem centered on T–O maps

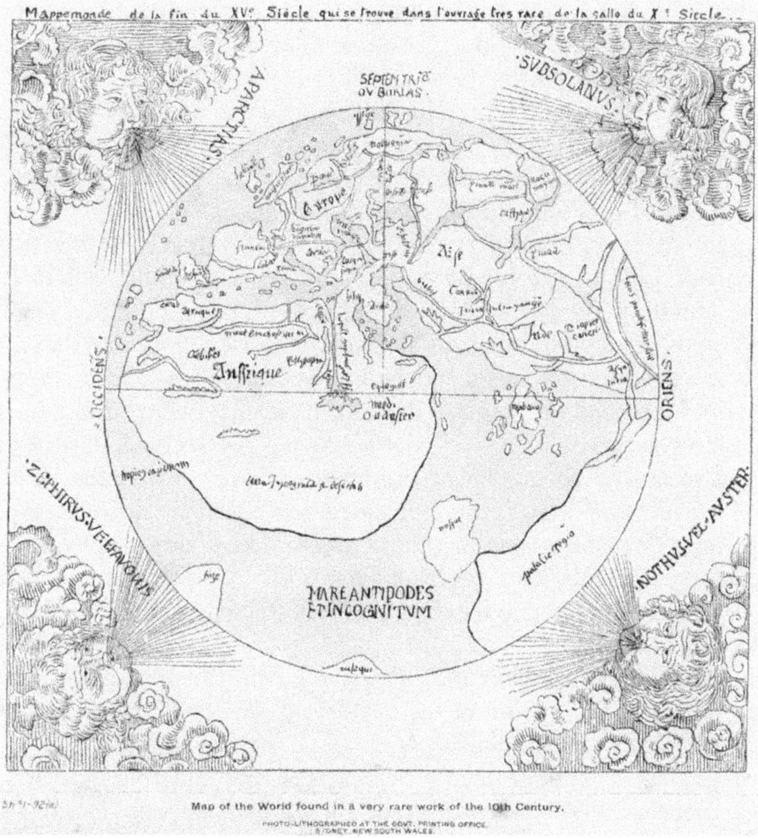

Figure 3.1 Mappemonde de la fin du XVe siécle qui se trouve dans l'ouvrage très rare de la Salle du X e siécle: Map of the World Found in a Very Rare Work of the Tenth [*sic*] Century. *Source*: State Library of New South Wales.

and with Asia representing the site of the Garden of Eden, we cannot ignore the weight of biblical conventions reaching through to the age of European discoveries.

Portolan Maps and the New Nautical Age

Typically portolan (*portolano* in Italian; *portulan* in French) charts were characterized by rhumb lines radiating from the center in the direction of wind and used by pilots to lay courses from one harbor to another (*Encyclopedia Britannica* 2014). More accurately these were wind rose lines as portolan maps lacked any map projection such as required to indicate true rhumb lines. As such they offered sailing directions in the way of a seachart.

Technically, a rhumb line or loxodrome is an arc crossing all meridians of longitude at the same angle and measured relative to true or magnetic north. The emergence of the portolan genre in Europe might thus be associated with the "nautical" phase, an important transition away from the purely monastic renditions of early medieval Europe. The earliest dated navigational chart extant was produced in Genoa by Petrus Vesconte in 1311 and is said to mark the beginning of professional cartography.

As Anthiaume (1916: 7–14) explains of an earlier style of written portolan existing alongside its graphic counterpart, the concern going back to Homer was the length of the journey and the direction taken. With the adaptation of the compass, so compass maps were drafted; Magellan used these. From 1300 when the compass was a standard, a wind rose was added, along with rhumb lines. Dubbed, in French, *carte-portulan*, they were written on sheep or goatskin parchment. They were not graduated and without a mathematical base and so lacked geographical coordinates or lines of latitude and longitude. However, they were favored by navigators for the ease of use in preference to lines of parallels and meridians entering later cartography. From the fourteenth century, the map portolan came to be crossed with a network of lines. A rose was situated in the central part of the map, with secondary roses ranged around it.

Originally in fourteenth-century maps, the central rose was placed upon Greece. Then, in the middle of the century, it was displaced to Sicily, and then northwest toward Sardinia suggesting an increasing interest in Western Europe and the Atlantic. Ptolemy misplaced Constantinople by 2° latitude. He also placed Gibraltar and Alexandria upon the same latitude, ignoring a real difference of 5°. The east-west length of the Mediterranean was exaggerated by 20° in Ptolemaic maps, but in portolans the error is around 1°. By the thirteenth century, the map, the compass, and star observation were standards on ships. As to origin, no Greek antecedents survive, and the origin is accredited to the Italians and with Sicilians, Catalans, and Majorcans all involved, or "Catalan on the Italian model" in Anthiaume's (1916: 7–14) description.

Nevertheless, as Anthiaume (1916: 25) stresses, up until the "Ptolemaic renaissance" the portolan maps were simply not adequate for navigators to undertake ocean voyages, especially as they failed to take into account the curvature of the earth. In the Middle Ages, maps did not deliberately bear projections with the exception of wind roses and were otherwise cylindrical (with meridians represented by straight parallels) or flat. In practice, by illustrating direct magnetic bearings between places, these maps actually anticipated the cylindrical map projection presented by Gerardus Mercator in 1569, as discussed below.

According to Anthiaume (1916: 26–27), from around 1300 Genoese navigators modified the cylindrical map and created the *carte plate carré* (square chart) or equidistant cylindrical projection. In this projection, meridians and parallels were strictly equidistant (constant spacing). Degrees of longitude and latitude were uniformly equal to the equatorial degree and were *carré*, hence the name. Some, as with Gaspar (2007), have rejected outright the notion that the Portuguese adopted the "square chart" method to nautical charts in the fifteenth or sixteenth centuries. Such a projection was not suitable for marine navigation, was not conformal as angles were not preserved, rhumb lines were not straight, longitude could not be reckoned, and they were not corrected for magnetic declination. Rather, the nautical charts of this epoch were planimetric, constructed as if the Earth was flat. Still, I would argue, there was evolution to a geometric form, just as the Ptolemaic mathematical grid model of the Earth was known.

As highlighted, the development of portolan maps was integral to the evolution of mapping in the Mediterranean world. Such description may appear irrelevant to a discussion on the European framing of Asia. Not so, however, if we consider the maps of Portuguese navigator Francisco Rodrigues (ca. 1512) of the sea route from Malacca to the Moluccas (the "spice islands") along with the first European representations of the China coast. In fact, the first nautical-navigational maps of Asia, whether Portuguese, Spanish, or Dutch were versions of portolan maps and there are many examples (see Winter 1949) (see Figure 3.2).

The Medici Atlas of 1351

With time, medieval maps began to incorporate African and Asian voyages as with those of the Franciscan travelers, Marco Polo, and also the Genoese and Venetian voyages. Besides Asia, the Italians showed great interest in Africa (which they knew to be a continent). Departing Genoa in May 1291, a company of Genoese under Tedisio Doria in conjunction with the two merchant-navigators, the Vilvaldi brothers, Vandino (sometimes Vadino or Guido) and Ugolino Vivaldi sailed out the Straits of Gibraltar in two galleys and headed south down the coast of Africa, plausibly in search of a route to India via the "ocean sea" (Lucas 1960: 380). Other Genoese voyages followed up, as with that by Lanceloto Malocello in 1321 who sojourned on one of the Canary Islands.

Such knowledge entered the Portolano Laurenziano-Gaddiano (Laurentian Sea Atlas or Medici Atlas), an anonymous fourteenth-century set of maps, probably composed by a Genoese cartographer and explicitly dated 1351. The Medici Atlas is likely the first to benefit from the 1341 mapping expedition to the Canary Islands sponsored by King Afonso IV of Portugal and commanded by the Florentine Angiolino del Tegghia de Corbizzi and the Genoese Nicoloso da Recco. It shows that Africa has a southern end, and that

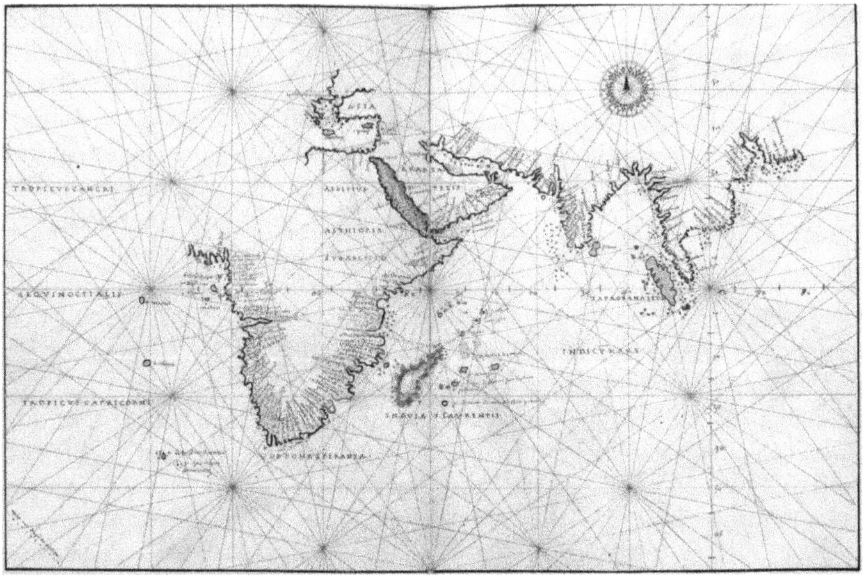

Figure 3.2 Portolan Atlas of Nine Charts, a World Map, and so on. Dedicated to Hieronymus Ruffault, Abbot of St. Vaast] Agnese, Battista, Active Sixteenth Century [ca. 1544]. Here We See the Indian Ocean Realm-sea-silk Road to China Represented in Familiar Mediterranean Portolan Style. *Source*: Retrieved from Library of Congress, Geography and Map Division. Digital Id http://hdl.loc.gov/loc.gmd/g3200m.gct00001.

the Atlantic and Indian Oceans are connected to each other below the African continent. If the original date 1351 is true, that would make it the first (extant) map to incorporate the travel reports of Marco Polo and Ibn Batuta. It displays Asia up to India, marking such places as the Delhi Sultanate with reasonable accuracy. The Atlas also shows the Caspian as a closed sea (unusual for maps of that time). Still, this map remains an enigma, especially for its precocious renderings of the coast of Africa including the Gulf of Guinea. As described below, the near contemporaneous Majorcan school of cartography considerably advanced upon this genre of map, especially with respect to Asia.

The Mandeville "Map" of the World

It is sometimes forgotten that the synthetic *Travels of Sir John Mandeville* or *Livre de Merveilles* as in the original French was actually a seminal text in the late Middle Ages finding wide audience among a reading public. While its fame may have gained from its Christocentric renderings of a first-person pilgrimage to the Holy Land, as much its rare culturally relative introduction to the virtually unknown religious doctrines and mores of the "Saracen" enemy, a section of the text devoted to stellar navigation and the possibilities of circumnavigation attracted more serious attention.

Claiming to be an account of a voyage east as far as China undertaken between 1332 and 1356, the *Travels* first began to circulate in Europe between 1356 and 1366. By 1400, versions of the book in manuscript were available in every major European language, Latin included. Starting with the "Pynson" edition of 1496, printed in England through to 1500, print editions appeared across Europe in German, French, Latin, Italian, among other languages, far exceeding even copies of Polo's *Divisement du Monde*. *Travels* was also extracted entering various collections.

Notwithstanding its late medieval reception, textual analysis has revealed to modern critics that Mandeville is informed by an array of texts, including that of Odoric de Poderone, the Italian Franciscan missionary traveler, suggesting a grand synthetic work with added fantastical elements as with a land of Amazons somewhere beyond Persia. But, by amassing information and injecting his own persona into the narrative as an albeit faux participant observer, Mandeville's account nevertheless did strike verisimilitude. Importantly, in C. W. R. D. Moseley's (1983: 31) words, "Much of Mandeville's material then flows indirectly through these conduits, as well as directly, into the sum of European knowledge in the Renaissance."

In framing his travels, Mandeville's work embraces a complex spatial topography consonant with medieval *mappaemondi* of the T–O version placing Asia to the top (labeled east), and with Europe and Africa to the bottom (labeled west) separated by the Mediterranean and with both separated from Asia by the Don and Nile rivers. Jerusalem is at the center (see Moseley 1983: 192–193). The geography of the text is also matched by discussions on Cathay under the Great Khan and the 5,000 islands of further India, especially Java. But what is striking is the inclusion of a long passage on the Antipodes and the possibility of circumnavigation. In discussing Lamory, a kingdom on Sumatra Island straddling the Equator, he reveals that the Pole Star such as guided navigators cannot be seen in these southerly latitudes, and so seamen steer according to the Antarctic Star, not visible to "us" (in the north).[1] "It can be seen from this that the world is quite round; for the parts of the firmament which can be seen in one country cannot be seen in another." He thus conjectured that the two unmoving stars divided the whole firmament into two equal parts (two spheres). "Each part of the earth and the sea has its opposition which always balances it" (Moseley 1983: 129). Half the firmament contains 180° and so, theoretically, a man could travel around the world in one direction and arrive home. More than that, he offers an example. Men living under the Arctic Pole stand foot to foot with those under the Antipodes. It was not possible to fall off the Earth. Noting a convention of ascribing the circumference of the world at 20,425 miles, he reckoned that it was considerably larger at 31,005 miles (somewhat larger than modern reckoning).

What is actually novel—even striking—in this rereading of Ptolemy and medieval *mappaemundi* is the possibility of reaching any place on the globe by sailing or moving in one direction. Such knowledge was undoubtedly inferred in a range of esoteric medieval texts including cosmographies and even Arab lore, but it is unlikely that the possibilities of circumnavigation had ever been so explicitly circulated in popular form. This is important, especially as it is known that Columbus perused a copy of Mandeville before undergoing his voyages believing that he could sail west to China without interruption. It is also understood that Infante Dom Henrique of Portugal, better known as Henry the Navigator, read Mandeville seeking information or confirmation that he could strike a diplomatic partnership with the Mongols and the legendary Prester John as a way of undercutting the menace of Islam once the discovery voyages were launched (Moseley 1983: 16). The *Travels* was also used as a source for the Abraham Cresques's Catalan Atlas, as discussed below.

It is noteworthy, as Moseley (1983: 31) asserts, that maps inspired by Mandeville's *Travels* "represent the picture of the world the explorers took with them and the basis upon which their backers financed them." This is an allusion to the German House of Fuggers known to have financed Henry the Navigator's voyages. More than that, according to Moseley (1983: 321), Mandeville's "confident and unusual insistence that the world was everywhere traversable radically altered the esteem in which the book was held and the uses to which it could be put." Nevertheless, by around 1600 the scientific revolution had caught up and with Mandeville's *Travels* having run full course as far as the credulous were concerned though, as indicated, his "cosmology" was precocious.

The Catalan Atlas of 1375

A product of the Majorcan school of cartography, the Catalan Atlas has rightly been hailed as a landmark in European knowledge of the Mediterranean and beyond. In this section, we will also explain how the Catalan Atlas both revealed and represented Asia, obviously a major leap in cartographic development. For Woodward (1987: 314–315, 322), the Atlas represents "the finest example of a *mappaemundi* in its final transitional stage." "It is transitional in the sense that it included information derived from portolan charts, from Ptolemy's *Geography,* and from the new discoveries in Asia." While acknowledging the general late medieval consensus that the world is a sphere, the Atlas was nevertheless composed at a time when Ptolemaic world atlases were not known (see Guerreiro 1993: 456).

Of the Catalan Atlas, according to Lach (1965: 66), "no map produced before the sixteenth century gives so comprehensive picture of Asia." Executed in Catalan language in 1375 by Abraham Cresques who worked in the court of Peter of Aragon, ruler of the Mediterranean island of Majorca, it was also a work of great labor and richness. As such, it was a gift to Emperor Charles VI (1380–1422), the French monarch who wished to better know the world by the intermediary of Guillaume de Courcy. From a Jewish family, Cresques was eventually forced to convert to Christianity in 1391 taking the name Jacobus (Jaume).[2]

In 1137 the kingdom of Aragon incorporated Catalonia, a trading partner of North Africa where mapmaking was a strong tradition, and with the Balearic Islands annexed in 1229. These circumstances fostered the rise of a highly important and flourishing cartographic school owing in particular to the contributions of Arab and Jewish scholars. What characterized this school was its lavish, vividly colored illustrations of cities, significant geographical features, portraits of foreign rulers, rich toponomy, and other features. Even before the first map was produced in Majorca, the Catalan scholar Ramon Llull included the sea chart among navigational instruments placing it on par with the compass. In 1354, King Peter of Aragon decreed that two charts should be carried on all the ships of the realm. The portolan chart drafted in 1339 by Angelino Dulcert (Dalorto) (BnF, Res Ge B 696) was the first map known for certain to have been produced in Palma, on Majorca. It is thus one of the "prototypes" of Catalan maps drawn in the fourteenth and fifteenth centuries.

The first folio of the Atlas opens with a cosmographical diagram; the Earth is personified by an astronomer holding an astrolabe situated in the center of the Universe. It is surrounded by concentric circles representing the four elements, the seven planets (the spheres and their personifications), the signs of the zodiac, the positions and phases of the Moon. The diagram is framed by the personifications of the four seasons (BnF, Esp 30). The four remaining leaves make up the actual map, in turn divided into two principal parts. The map displays illustrations of numerous cities, and with their political allegiances symbolized by flags. Christian cities are marked with a cross, Muslim or oriental towns by domes. Seas and oceans are symbolized by wavy blue vertical lines, as with the Atlantic, the Mediterranean, the Baltic, the Black Sea, the Persian Gulf, the Caspian Sea, and the Indian Ocean (BnF Gallica online date: December 19, 2011).

The date 1375 appears in several places. Eastern Asia including the ocean as far as India is represented. The Caspian Mountains are guarded by Alexander the Great. China; Catayo, the Island of Taprobane and Jana Island, India; Asia from India to the Caspian Sea; the Persian Gulf; Arabia and Mecca; the Near Orient and the eastern Mediterranean up to the Black Sea; Palestine, and the Red Sea are all represented. Closer to home, the western Mediterranean and the Atlantic coast, Scandinavia; Bohême mountain; Corsica and Sardinia are also represented, as are the British Isles, Thule (a metaphor for a distant place, as with Iceland); and the Canaries. All maps portray south above, although the maritime charts have no special orientation (BnF 2011).

Numerous flag symbols are used to represent churches and religious monuments along with representations of the Saint-Sépulcre of Jerusalem, the monastery of Saint Catherine in Sinai; Mecca, the tower of Babel to the south of Baghdad, Kings, and other personages. In Africa are found representations of a Berber on camel; the black king Mussa Melly of Mali in 1307 (ruling from Timbuktu until 1337); the king of Organa (a Turkish title), a king in the Nubia of Prester John, and the sultan of Egypt. In Asia the following are found: representations of the sultan of Turkey, the Queen of Saba (Sheba), a Muslim pilgrim praying toward Mecca, the king of Tabriz (Rey del Tauris), a Mongol prince to the north of the Caspian (identified with Janï Beg, king of the Golden Hoard, died in 1357), a caravan (men, camels, horses) proceeding east (plausibly Venetian merchants as with Marco Polo), and with the Magi kings traveling west, a person guiding an elephant, the king of Delhi (lo rey Delli), the king of Colombo in India (identified with the ruler of Vijayanagara), the Chabech king, identified with Kebek Khan, who reigned between 1309 and 1326, King Etienne looking toward Butifilis (Motupalli) where Saint Thomas was buried (and with Motupalli an ancient trading port in the present-day Indian state of Andhra Pradesh), a queen on the island of Women, the King of Taprobane, Holubeim, and "the grandest prince of all the Tartars," Alexander. The figure of the Last Judgment appears, as does Antichrist. Pearl fishers, Aragonese (?) ships offshore Africa, tents in northwest Africa; elephants in Africa and on Taprobane, dromedaries in Africa and Asia, all add an ethnographic touch (BnF 2011).

According to the Bibliothèque nationale de France (BnF) description, realized for a Christian king (the king of France) the Atlas is enriched by knowledgeable anecdotes as well as scholarly references. The representation of the world is written as sacred history in the fashion of the *mappaemonde* of the thirteenth century. The Christian order of the world is represented from the opening with its recall of cosmography (the calendar and page of commentary), as well as allusions to the Apocalypse as with the representations of the prince of Gog and Magog and the Last Judgment, Antichrist). The Orient positioned at the end of the world is also a sign of its temporal end. Opposing it, the western limit of the known world is situated in the Fortunate Islands, designated as the mythical place of earthly paradise. Better known as the Canaries, lying 100 kilometers off the coast of Morocco, they were also designated by Ptolemy as his prime meridian. The circularity of the *mappaemonde*, reveals a beginning and an end, an Orient and an Occident, the Terrestrial Paradise and Last Judgment end up meeting each other (BnF 2011).

In the BnF description, the Catalan Atlas is a work halfway between a marine chart, replete with lines of direction (or rhumb lines), and a *mappaemonde*. The map also offers one of the first displays of a wind rose on a nautical map. The reader must take north as the base and read from left to right, from the Far East to the Atlantic. The Atlas shares many features with the Majorca school for the western part. On the other hand, the description of Asia is "highly innovative." The interior is replete with details of persons and animals in a space which is the least known of the epoch, namely sub-Saharan Africa and the Far East. The geographic space, dotted with indications of people, flora and fauna, and riches, according to information furnished by traveler accounts of the epoch, is also studded with biblical references, saints, and allusions to evangelization (BnF 2011).

According to the BnF source, despite its innovation and despite its "linear reading," the Catalan Atlas did not depart from the concept of curvature of the Earth. The sacred journey from the beginning to the end of time is part of the history and geography as measured by the milestones of biblical places and oriental networks in the settlements of Christianity. The diplomacy of the time was entirely focused on the desire to outflank the Muslims who monopolized the spice trade and, more or less, this is a conventional understanding in the European world order of that time. As with the Atlas, so with the historical reality, the East was a space for missionization before its opening for commercial conquest. There is no contradiction in juxtaposing different moments in the history of the countries and peoples of the same image. Medieval art until the late fifteenth century willingly fell into this mode. The map is enriched and an encyclopedic knowledge of the world (BnF 2011).

Asia in the Catalan Map

Asia including China is also represented in the Catalan Map. For the Far East, the names of towns collected from various sources appear to be randomly distributed. To the south of Catayo (Cathay), three circular bays are associated with three great ports, Zayton (identified with Quanzhou), Cansay (frequently Quinsay, modern Hangzhou in eastern China), and Cincolam (Canton/Guangzhou), well known to Arab navigators though not Marco Polo. Certain of the islands facing Quinsay represent the Chusan (Zhoushan) archipelago and further south. The large island of Caynam (Hainan) is shown. Kao-Li is the name of Korea (as identified by Sir Henry Yule). There is no allusion to Japan in the map (BnF 2011). As the BnF analyst surmises, it is probable that Marco Polo, Odoric de Pordenone, as well as Mandeville were the inspirational sources for the Asian part of the Catalan Atlas, adding to the habitual (and classical) medieval cartography from Pliny onward (BnF 2011).

As Lach (1965: 66) remarks of the Atlas, although the peninsular shape of India is hinted before in previous maps, it "here emerges for the first time in the history of cartography and it is truer than that found on some other maps prepared a century later." Southeast Asia is less well developed, indeed distorted by a misplaced mythical Taprobane. The *mappaemonde*, as suggested below, shows the semi-circular coastline of China, marking the cities immortalized by Polo and successors (although not the Arab voyages). The chief divisions of the Mongol empire and the placements of lakes are mostly accurate. Notably, Cresques did not follow Polo in dividing China into Cathay and Mangi. Inland waterways and mountain ranges are not highlighted with any precision. Still, Lach allows "no better portrayal of the Eastern Hemisphere was available in Europe until the production of the world maps of the sixteenth century."

From Portolan to Latitude Charts

As recent research has highlighted, a long period of stasis in Iberian nautical chart making was scarcely even interrupted with the Ptolemaic revival. The first nautical charts used to explore the African coast and to navigate in the Atlantic in the late Middle Ages and (early) Renaissance were identical to the portolan charts of the Mediterranean, only differing in their geographic limits. They were constructed using the navigational information collected at sea with places represented by compass courses and distances between them. More generally, they conformed to the Majorcan model (Gaspar 2013: 68). There was also a credible crossover from Majorca to Portugal with the arrival in Lisbon of Abraham Cresques's son Jafuda, also known as Jaume of Majorca (or a namesake), even prior to the Henrican period of Portuguese mapmaking (Marques 1987: 72–74).

According to Gaspar (2013: 68), very little is known about the technical evolution of nautical cartography during the sixteenth century and remains a subject of scholarly inquiry. While, as observed in chapter 2, the Arabs had prior knowledge of astronomical navigation, as Gaspar points out, the Iberians only adopted the method in the second half of the fifteenth century. Revolutionary in its full consequences, this innovation was made possible by the simplification of observational instruments already perfected on land, as with devices to observe the Pole Star and the Sun to determine latitude. With time, the practice of navigating by compass courses and estimated distance gave way to the "set point" method where the observed latitude became the prevailing element of navigational information. Dias, da Gama, and Cabral all adopted astronomical navigation, although the "latitude chart" as a nautical-navigational guide was also slow in evolving, especially pending the resolution of accurate means to determine longitude and magnetic declination. As Gaspar (2013: 69) reveals, these charts did not conform to projections "in the sense that they were constructed plotting directly on the plane the observed latitudes, courses and distances, as if the Earth were flat." In any case, as Gaspar (2007: 67–68) points out, the latitude chart "would serve as a valuable tool for more than two centuries, well before the appearance of the revolutionary Mercator projection of 1569." As such it represented a major breakthrough in the nautical cartography of the Renaissance, marking the beginning of the evolution from "maps based on routes, to Ptolemy's system of geographical coordinates." As discussed in chapter 4, the earliest extant chart to incorporate astronomically observed latitudes is the Cantino planisphere, although no graphical scale of latitude is explicitly shown.

THE "MONGOL PEACE" AND THE VENETIAN (POLO) LEGACY

It is sometimes neglected in historical writing that, 200 years prior to the arrival of the first Jesuit missions to China, the Holy Roman Empire backed by the Franciscan Pope Nicholas IV (1288–92), initiated a series of missions to the Mongol-Yuan court of Khanbalik (identified by modern archaeology as located in Beijing). As explained by Lach (1965: 32–33), this was part of a geopolitical push by Western Christendom to seek out new allies in the struggle against the Muslim "other."

It was not entirely coincidental that the Reconquista only gathered momentum in the wake of the Mongol destruction in 1258 of the Baghdad-centered Abbasid Caliphate. Inspiration for the missions also stemmed from the loss in 1291 of Acre (modern-day Haifa), the last Crusader stronghold in the Holy Land, to the Egyptian Mamluks (Spooner 2013: 17). Catholic Europe's search for strategic allies thus led to a fifty-year outreach to the Mongol court, also involving the reciprocal dispatch of envoys by the Great Khan to Avignon (1336; 1353). Lach (1965: 46–47) calls this the "Mongol peace (1240–1350)." Other elements were also involved, as with connecting up with the presumed kingdom of Prester John founded upon a medieval belief that an isolated Christian kingdom existed in, variously, central Asia or eastern Africa, given credence by secondhand knowledge of scattered Nestorian Christian communities in India and China.

Prominent among the Papal missions was the voyage of the Italian Franciscan Giovanni Montecorvino conducted between 1292 and 1305. Having arrived in the Mongol-Yuan capital via Persia and India, he established the Roman Catholic Patriarchy of Khanbalik, surviving down until the Ming ouster of the Mongols (and the Christian communities) in 1368. The Flemish Franciscan William of Rubruck who undertook a mission to the Mongols (1253–54) was received at the court of the Great Khan at Karakorum. The oldest European account of the Mongols is that of John of Plano Carpini, written in the 1240s. In October 1338, the Italian Franciscan Giovanni di Marignolli or Jean de Marignol was summoned to Avignon by Pope Benedict XII (1334–42), named ambassador legate to the Great Khan as successor to Montecorvino, and arrived in Khanbalik in 1342. Decades earlier (1316–30), Odoric de Pordenone had taken the sea route to China via Sumatra and Champa, the Hinduized kingdom occupying central coastal Vietnam, and was received at the court of the Great Khan in Khanbalik prior to returning to Europe overland. He also bequeathed an account (Dawson 1955; Lach 1965: 31–33).[3]

One upshot of the Franciscan voyages was that medieval Europe misleadingly came to know northern Mongolia-China as Cathay, named after the semi-nomadic Khitai or Khitan people who conquered the area from the tenth century. In any case, with the eclipse of the Mongol-Yuan Dynasty (1271–1368) by the upstart Ming, the overland connection to Europe was ruptured and with Europe ignorant of internal changes in China practically until the Portuguese established a sea route. As discussed in a following chapter, even Columbus believed that, in crossing the Atlantic, he would make contact with the Great Khan or Cathay such as introduced by medieval traders and missions, Marco Polo included.

Venice and the "Near East"

From its origins as an outlying colony of the Byzantine empire, Venice gradually carved out its own cultural and political dominance. Unlike Spain or Sicily, Venice was never part of any Islamic Caliphate. From at least the ninth century, Venice's political and economic viability depended upon commercial links with the East. Undoubtedly, as Lucas (1960: 39) points out, the power and position of Venice rested upon its trading monopoly, and with its key industry, shipbuilding. For Braudel (1984: 93), Venice was simply the font of the capitalist revolution associated with the beginnings of the European world economy appearing around 1200, and with financial-commercial centers developing in the fourteenth century. In large part this epochal transformation was stimulated by the Crusades. By the late thirteenth century, Venice was the most prosperous city in all of Europe, a veritable thalassocracy or "trading-post empire forming a long capitalist antenna; an empire 'on the Phoenician model', to use a more ancient parallel" (Braudel 1984: 119). For Jack Goody (2010: 24; 28–30), high levels of literacy made possible the diffusion of new technologies including the new print revolution at the heart of the Italian Renaissance. On its part, Constantinople served as the occidental terminal of trade flowing from India and beyond, including the distant "spice islands." With Constantinople as the economic "bridgehead" between Europe and the East, Venetians, Genoese, and Pisans began competing with Byzantine merchants (Lucas 1960: 138).

After the First Crusader War (1096–99), Venice lost no time in establishing trade relations with the Kingdom of Jerusalem based upon the model of their trading privileges with Constantinople since 1082. Over time Venice established a network of trading posts dubbed "colonies," such as at Acre, Alexandria, in the Sea of Azov at Tana, Trebizond (Trabzon) on the Black Sea, and at such inland cities as Damascus and Teheran (Howard 2007b: 73). With Constantinople falling to the Crusaders in 1204, Venice gained considerable privileges. Although hampered well into the fourteenth century by a papal ban on trading with the Muslim enemy (at least down until 1344), Venice also gained its exemption (Christ 2012: 19).

However, the recapture of Constantinople by the Byzantine Emperor in 1261 gave trade preferences and a territorial base to Venice's rival, Genoa. Still, Venice retained its Greek colonies and with Venetian shipping reentering the Black Sea trade at a time when the Mongols reopened the silk route through Central Asia. Importantly, in 1291, the Genoese defeated a Moroccan fleet controlling the Straits of Gibraltar, thus opening the way for European commerce from the Mediterranean to the Atlantic. Thereafter, Venetian galleys used this route to trade with London and Bruges, and with Venetian ships also turning south down the African coast, epochal actually, as demonstrated below.

As Adam Smith (1999: 138) observed in his late-eighteenth-century classic, the Venetians mostly purchased "spiceries, and other East India goods" from Egypt under the domination of the Mamluks "the enemies of the Turks, of whom the Venetians were the enemies; and this union of interest, assisted by the money of Venice, formed such a connection as gave the Venetians almost a monopoly of the trade." In Deborah Howard's (2007b: 79) interpretation, as land-based powers who controlled the eastern Mediterranean from 1250 to 1517, the Mamluks depended upon Venetian naval strength and goodwill to protect the Mediterranean littoral. Although Muslim diplomats kept up a presence in Venice, it was Venetian merchants who dominated the sale of Oriental goods at the Rialto marketplace.

From its trading post at Tana (in part shared by Genoese), at least until conquered by Tamerlane (Timurlane) (r. 1370–1405) in 1392 and the Turks in 1471, Venice tapped the overland trade from India and Persia via Tabriz at a point far closer to its source than European rivals. But with the Mongols under Tamerlane extending their sway over Persia and Mesopotamia, trade activity on the northern land route as well as a middle route via Hormuz,

Mesopotamia, through Aleppo to the Mediterranean, was effectively blocked. In John Darwin's (2007: 35) description, as the last of the great Mongol-Turkic "world conquerors," Tamerlane "understood the point of commercial as well as military strategy," destroying the towns of the north Caspian Sea while enforcing his monopoly over the trans-Asian trade that passed to the south. Such geopolitical shifts playing out across the terrestrial silk roads also met with reaction. As Geog Christ (2012: 22–25) explains, the sea route via Aden actually became dynamized with the "Karimi" or pepper fleets merchants sailing direct to the Red Sea area for transshipment via the Nile to Cairo. Neither was the trade stable with such Red Sea ports as Aydhab, Tor, Ainuna outside Aqaba, shifting according to political contingencies.

Venetian agents were especially trained and skilled in international relations (Lucas 1960: 340). As Abbé Raynal (1784: 105) explains, perceiving "that her commerce and, consequently her power was being transferred to the Portuguese," Venice worked through its emissaries and through local Arab merchant contacts, seeking to persuade them that it was in their interest to make common cause against a nation that sought to ruin them both. The idea was taken up by the (Mamluk) Sultan of Egypt who saw his trade declining before his eyes. To this end, Venice and Egypt entered a trading confederacy, and with Venice helping Egypt to fit out a small fleet of ships launched in 1508 to preempt Portuguese supremacy over the Red Sea route, as with its control over Socotra Island (present-day Yemen). The coffee trade was also an issue and with Venetian merchants connecting up with the trade at its source.

The Fra Mauro World Map

No better expression of the Venetian priority in conceptualizing geographic knowledge of the known world can be found than in the Fra Mauro world map composed by the Venetian-born monk of that name. Commissioned by King Afonso V of Portugal in 1457, it was finished on April 24, 1459, and, in some accounts, delivered up to Prince Henry the Navigator, Afonso's uncle, encouraging further funding of exploration trips. Because of the meticulous craftsmanship and cost, the map was undoubtedly a collective effort. The result was a work covering over five square meters, extremely detailed, and containing many thousands of descriptive texts and illustrations. The single surviving copy of the original is a version commissioned by the Republic of Venice (and with an 1804 copy of the lost Portuguese version held in the British Library).

As Evelyn Edson (2007:140) summarizes, in the fifteenth-century sea charts, *mappaemundi*, and Ptolemaic atlases all coexisted, but it was Fra Mauro who attempted to integrate the three traditions into a single map. True to Ptolemaic convention this was a three-continental construction. Fra Mauro drew from Venetian merchant accounts including Polo and, especially, as profiled below, the early-fifteenth-century merchant-traveler to Southeast Asia, Niccolò de' Conti, as well as Arab and even oral sources. Notwithstanding its argumentative texts, Venetian idiom, and other unresolved questions, according to Edson (2007: 366), the Fra Mauro map still lies firmly within the *mappaemundi* tradition.

But, we may well ask, was it post-Ptolemaic? According to Piero Falchetta (2006: 584–85), who issued a critical edition of the map, Fra Mauro also departed from a number of the Alexandrian's conventions. For instance, following Arab conventions, he placed south at the top of the map. In a textual note he explained that in order to accommodate new data, he was not persuaded to incorporate Ptolemy's meridians or parallels or degrees "because I would have to omit many provinces which Ptolemy never mentioned at the extremes of south and north latitude said by him to be *terra incognita*" (Edson 2007: 152). Rhumb lines are also absent. Biblical imagery is abundant.

Among the *terra incognita* filled in on the Fra Mauro map was a part of Japan appearing below the island of Java, with the legend "Isola de Cimpagu" (a first on a map). Ayutthaya, Ava, Cathay, the Yangtze (Quian River) all figure as toponyms. Notwithstanding mistakes in the Southeast Asia region, the Fra Mauro map offers up a fluid rendition of the world known to Venice, especially its precocious knowledge of the Indian Ocean and the "spice" trade linking east and west. We don't know how the Crown in Portugal reacted to the imagery portrayed by Fra Mauro, but the open sea route to the East must have appeared as tantalizing in this era prior to the discovery of the Americas. Importantly, in a major revision to the Ptolemaic stereotype, the map depicts Asia, Africa, and Europe in a way that does away with the land bridge by offering an encircling ocean around Africa leading to an island-studded Indian Ocean connecting up with the coast of China. Based on de' Conti, the map misleadingly situates Taprobane as Sumatra. It also records the presence of Chinese naval junks off the coast of Africa, a probable allusion to Zheng He derived from Arab sources (Suárez 1999: 49; 79; 87). It is undoubtedly true, as Dalché (2007: 317) sums up, that the Venetian cartographer's attitude toward Ptolemy was far from a straightforward rejection. For Fra Mauro, "Ptolemy could represent either a rigid corpus of knowledge or an opening towards innovation." Indeed, we might add, this was a paradox confronted by many among the scholarly elites in Europe across these centuries.

It is important then to assess the de' Conti contribution to the Fra Mauro Map, at least as a source. Following a sojourn in Damascus, between 1414 and 1429, de' Conti visited India and various points in both mainland and maritime Southeast Asia. Returning to Venice via Cairo in 1438, he joined the Council of Florence instantly gaining an influential audience (Christ

2012: 12). Duly questioned by the papal secretary, Poggio Bacciolini, the latter produced a Latin manuscript account of the Venetian's travels. According to Lach (1965: 61–62), de' Conti's rendition of an open ocean route along with his description of India as true to natural and not imposed geographical boundaries was entirely outside of Ptolemaic conventions, whatever Poggio believed.

According to Durand and Curtis (2013: 27), information provided by de' Conti fed into the anonymous "Genoese map" of 1447–57 adding a peninsula at variance with the Ptolemaic stereotype as well as the islands of Java and possibly Borneo, such as mentioned by Marco Polo 150 years earlier. Not only did de' Conti influence Fra Mauro, but he actually met and helped the monk in the preparation of his world map. The Portuguese who attended the Council of Florence took note, and so did Columbus in his time. With de' Conti's sea voyages chronologically overlapping with the Ming-era expeditions of Admiral Zheng He, his accounts can also be measured against published Chinese accounts of the Indian Ocean trade. In Lach's (1965: 63) estimation, thanks to Poggio, de' Conti's account "undoubtedly" influenced contemporary thought on geography.

The Marco Polo Rendition and Legacy

Not surprisingly, Venetian traveler accounts of the ancient silk roads across central Asia to China during the *Pax Mongolica* achieved a special priority in Europe even if, as stressed in chapter 2, such knowledge had long circulated in the Arab world. Seminal, in this sense, was the account of Marco Polo (1254–1324) who followed in the footsteps of his father and uncle who preceded him in the 1260s to the court of Kublai Khan at Khanbalik (Lach 1965: 34–35). Departing Venice on his fabled peregrination in 1273, and returning to his native land twenty-four years later, Polo was imprisoned at the hands of the rival Genoa Republic. In these circumstances, his travels were dictated in 1298 to fellow inmate, Rusticello da Pisa, first entering manuscript form in Italian as *Il Milione*, in French as *Livres des merveilles du monde*, in English as *The Travels of Marco Polo*, alongside some 150 surviving manuscript copies, many of them variant in detail and certain illuminated before the age of print (Larner 1999: Appendix I). By 1500 the work had been translated into Latin, Italian, German, and Spanish (Wallis 1992). As Jonathan Spence (1998: 1, 12) points out, later versions of Polo (after 1320) had many new sources to draw upon allowing copyists to add embellishments rather than going back to original drafts. All in all, he finds a combination of "verifiable fact, random information posing as statistics, exaggeration, make-believe, gullible acceptance . . . and . . . . outright fabrication." But neither does Spence, as with Hans Vogel (2013), doubt that the Polo family made the voyage (see

Wood 1995, 1996). At least until the arrival of the Jesuit missions in China 200 years later, the Venetian offered to readers the first authentic account of the inner workings of the Mongol-Khan government, besides a wealth of information, much of it subsequently confirmed.

Polo and Revisions to Ptolemy

Invariably, given the number of editions of Polo's *Travels* circulating in late medieval-Renaissance Europe, his ideas began to translate into new imaginings and conceptions of Asian geography. Polo's *Travels* not only influenced the 1453 Fra Mauro map but, as Larner (1999: Appendix III) examples, a range of fifteenth-century world maps were so inspired. Readers of the French version of Polo would equally have gained insights into the relationship between the Mongols and the Christians (Spooner 2013: 17). It is not irrelevant that illuminated texts of the period, certain combining Polo and Odoric, drew attention to diplomacy, protocol, mutual interest in keeping the Muslim enemy at hands distance, and a generally pacific encounter from cultural-religious opposites at opposite ends of the Eurasian continent. Polo would also be a game-changing text when it came to revisions of Ptolemy at the hands of late Renaissance globe and mapmakers. As Wallis (1992) points out, Polo's travels (1271–95) to China and back with a stay in that country of seventeen years transformed Europe's ideas of the Orient and, demonstrably, revealed a navigable route to China via maritime Asia. As advanced by Larner (1999), Polo's influence on later geographical and literary treatises as well as his indirect importance in advancing European expansion cannot be underestimated. As discussed below, Columbus also looked to Marco Polo as a late medieval authority.

COLUMBUS AND HIS PTOLEMAIC CONCEPTION OF ASIA

As alluded in chapter 1, the Ptolemy-revival template sowed much confusion in the minds of Columbus and his contemporaries. Especially at issue was the actual circumference of the globe and the number of miles (or leagues) to a degree, as well as the size of the continental landmass of Europe, Africa, and Asia, in relation to the whole circumference. Constructed on a framework of latitude and longitude, the Ptolemy-revival map projections revealed the extent of the known world in relation to the whole. Typically, they displayed a Eurasian landmass extending through 180° of longitude from a prime meridian in the west (variously the Canary Islands or Cape Verde) to a location in the "Far East." As Wallis (1992: 17) points out, Ptolemy produced a great overestimation, as the true distance is only 100°. In fact, Ptolemy had taken the circumference of the Earth to be a quarter too small. In correcting Ptolemy, as Wallis continues, Columbus made an even greater error. He preferred the estimate of Marinus of Tyre, who calculated the Eurasian ecumene to cover 225° in east-west extent. Columbus then made a further adjustment, pointing out that Marinus and Ptolemy did not know the full extent of eastern lands. As discussed below, it appears that Columbus misread the discrepancy between the measurement of Arabic and Italian miles.

Besides mariner's lore and, possibly, information gleaned from prior voyages down the African coast, Columbus's ideas about the measure of the Earth and the extent of Eurasia combined with the evidence of a navigable ocean were a major driver behind his search for a western route to Asia. It is also true that Florentine adventurer and Columbus rival Amerigo Vespucci made numerous allusions to Ptolemy in his writings (Fernández-Armesto 2007: 18). But unlike Pedro Álvares Cabral's voyage of 1500 to Brazil, and unlike the Vespucci (voyages of 1497? 1490–1500, 1501–1502) coasting along the south American coast, until his death Columbus believed he had reached the Indies or the Orient. He did not conceive that he had discovered a new continent or New World but still firmly believed in a three-continent depiction of the globe. It is now common knowledge that Polo's Cathay and Cipangu (Japan) were key elements in his planning. The supposed wealth of these countries, the site of spices and gold was the lure to convince his backers. As mentioned in chapter 1, Strabo and Pliny together would also be indirect influences upon Columbus (Wallis 1992: 18).

Columbus and his Conception of Asia

The Genoa-born Christopher Columbus (1450–1506) is known to have had an early seafaring career that took him to Lisbon and into the Atlantic, trading along the coast of West Africa as far south as the coast of Guinea. From the Canary Islands he also gained invaluable knowledge of the trade winds, indispensable for his successful cross-Atlantic navigation. Here we are mindful, as Fernández-Armesto (1987) argues, that Columbus's voyage was a new step in a centuries-old process of European expansion from the Mediterranean to the Atlantic around the rim of Africa, and to the Canaries. Having presented his sailing plans to King John II of Portugal in 1485, and again in 1488, he failed to win support for his venture. Turning to Spain, he eventually gained success in January 1492 at a meeting in Cordoba with the Catholic sovereigns, Ferdinand (of Aragon) and Isabella (Queen of Castile), at the historic juncture when Grenada, the last Muslim stronghold on the Iberian Peninsula, was conquered.[4]

As recounted by Wallis (1992: 20), in its meetings at Salamanca in 1486–87 and again at Santa Fe near Grenada in 1491, the Spanish Junta debated the question of the proportions of land and sea on the terrestrial sphere. Two divergent theories were current concerning the size of the ecumene, both derived from writings of Aristotle. The discouraging experience of the long navigations down the African coast probably explains why King John II had called Columbus to Lisbon in 1488. Columbus was present in Lisbon in December when Bartolomeu Dias arrived home with his report of discovering, naming, and rounding the Cape of Good Hope (not incidentally exposing the Ptolemaic notion of a southern land bridge enclosing the Indian Ocean and blocking access to an open sea). Dias's success in entering the western approach to the Indian Ocean explains why the Portuguese monarch abandoned further consideration of Columbus's project. As Wallis continues, the Admiral's command as granted on April 30, 1492, by Ferdinand and Isabella was, however, "conveniently vague:" "Whereas you, Cristobal Colon, are setting forth . . . to discover and acquire certain islands and mainland in the Ocean Sea."

As known to Columbus' son, Hernando, as well as to Spanish historian Bartolomé de las Casas (1484–1566), Columbus was greatly influenced by his reading of the medieval theologian Pierre d'Ailly's *Imago mundi* (1410). As much conjectured, *Imago mundi* taught Columbus the possibility of reaching the East (Indies) from the West (see Nunn 1935). As Nicolás Wey Gómez (2008, chap. 1) interprets, the theologian also drew attention to the character of "torrid" as opposed to temperate and polar zones, hence Columbus's inclination to strike tropical latitudes as opposed to a more northerly course which would have led to the Great Khan (Marco Polo's East). In so doing, the "Admiral" potentially tapped a lucrative trade in spices, dark-skinned slaves, gold and future colonies, and with the Portuguese experience along Africa's coast a model at hand.

Nevertheless, as Wallis (1992: 18) points out, Columbus owned the 1485 abridged Antwerp edition of *Il Millone* reading and annotating this volume in the period when his project was well advanced. Among the hundred annotations inscribed by Columbus in his copy were remarks on gold, silk, and other trade goods, monsoon fleet sailings, as well as the names of such trading ports as Hangzhou and Yangzhou and with Khanbalik identified (Spence 1998: 17–18). Columbus had learned about Polo from Paolo dal Pozzo Toscanelli, the Florentine physician and cosmographer, to whom he wrote in 1481. As Columbus's most important contemporary authority, Toscanelli's sources also included de' Conti's travels. In 1474, King Afonso V of Portugal (r. 1443–77) dubbed the "African" owing to his conquests of parts of northern Africa, also consulted Toscanelli over the shortest route to the lands of spices in Asia.

Toscanelli's answering letter was sent to Canon Fernão Martins of Lisbon who acted as intermediary. In response to Columbus's inquiry, Toscanelli sent a copy of the earlier letter to Martins, and Columbus transcribed this letter on a page of his copy of Aeneas Silvius Piccolomini's *Historia* (Venice, 1477).

It is well recorded that, on his first voyage across the Atlantic in 1492, arriving in Cuba and Hispaniola, Columbus believed that he had landed in India. But, between the voyage of 1492 (discovery), 1493, 1498, and 1502, he also modified his conceptions of geography. As the lead American researcher on these questions, George E. Nunn (1952: 15) explains, on the third voyage coasting Venezuela and sighting the Orinco River, he conceded the discovery of a new continent but one attached to China. An Italian version of Columbus's letter of July 7, 1503, under the heading "*Como Colombo navigò per ponente et trovò l'Asia, et navigò per la Costa da India de l'equinotial ver il Polo Arctico et arrivò nel Regnio del Cataio*," is also indicative of Columbus's mistaken belief that India and Cataio (Cathay) could be reached directly by sailing west across an open sea.

According to Nunn (1952: 15), armed with a concept of the geography of eastern Asia (similar to that of the Behaim Globe, as discussed below), Columbus planned to follow the Polo route around the Malay Peninsula into the Indian Ocean. As revealed, he carried a letter of March 14, 1502 from Isabella and Ferdinand addressed to the Portuguese commander in India. He was also granted the right to carry Arab interpreters. But matters worked out differently on his fourth voyage and he changed his thinking as to the relation of the New World and Asia (or, at least reverted back to his earlier thinking). Now he believed that the Mondo Novo (South America) was a peninsula of Asia. "Columbus' concepts of old world geography were those of Ptolemy [not Marinus of Tyre] with the new combination of eastern Asia derived from Marco Polo," albeit modified as to longitude with his degree measure of 56 2/3 miles (Nunn 1952: 18).

Undeterred, Columbus persisted in his plan to reach the Great Khan. The island of Cuba (also called Juana) raised his hopes. As he wrote on October 24, 1492, "I believe . . . that it is the island of Cipango of which marvelous things are told." On the second voyage 1493–96, he became convinced that Cuba was part of a mainland, and took sworn statements from his men to that effect, believing it to be the Golden Chersonese or Malay Peninsula. In his fourth voyage 1502–04 at a higher altitude, Columbus sought a strait between Cuba, seen as a promontory of China, and the newly discovered continent. From Honduras and Darien he followed the coast of Central America which he identified with Polo's Ciamba or Champa, the strategically located Hindu kingdom straddling the central coast of Vietnam (until its eclipse in the sixteenth to seventeenth centuries) (Sauer 1966: 142; Wallis 1992: 21).

Plausibly, Columbus was also swayed by biblical belief. Sometime after his second voyage, he received a letter written to him by the Aragonese cosmographer and Royal counselor Jaume Ferrer de Blanes, sent from Burgos and dated August 5, 1495. As such, this letter adverted to a land in the *Sinus Magnus* where the apostle St. Thomas reputedly died, actually on the Malabar or eastern coast of India. As the letter closed, "and I say that within the equatorial regions there are great and precious things, such as fine stones and gold and spices and drugs" (Thatcher 1903: 368). Later-arriving Portuguese in Brazil replayed this fable as well. As Suárez (1999: 97) comments, the mixing of fundamental Christian belief with the lure of riches found among people of black or tawny complexion also fed into the Columbian vision.

As Wallis (1992: 22) submits, Columbus was hoping to find a navigable passage into Ptolemy's Great Gulf, the *Sinus Magnus*, a theme to which we shall return in a subsequent chapter. Parenthetically, survivors of the Magellan voyage were similarly attracted at the point when they headed west from Palawan Island in the southern Philippines to Brunei on Borneo, until they (or, at least, their Malay pilot) figured a way round by a southerly course, before boldly entering the Indian Ocean via Timor in the Lesser Sunda Islands. Columbus, at the isthmus of Panama, believed he was only ten (or nineteen) days sailing from the Ganges, whereas it actually took Magellan ninety-eight days to cross the South Pacific (Skelton 1969: 1). The natives he encountered informed him that he was on an isthmus between two seas, but a chain of mountains impeded his progress. While at least correct as to the general latitude of the now-lost Hindu kingdom of Ciamba (Champa) on the ancient eastward-sailing maritime route to China, famously Columbus fell into the trap set by Ptolemaic-revival atlases of exaggerating the eastward extent of the Eurasian landmass, thus deluding him as to a quick westward-sailing passage to the Indies.

Even prior to the printing of Latin texts on astronomy, as mentioned in the following chapters, we should not ignore the influence of Arab conceptions of geography upon Europe on the cusp of the age of discovery. As Wallis (1992: 18) has pointed out, in support of the view that the distance between Western Europe and Asia was small, Columbus turned to the opinion of the Arab or Persian geographer Alfraganus (al-Farghani) of the late ninth to tenth centuries, who gave the length of a degree to be 56–2/3 Italian miles. Moving from Baghdad to Cairo, Alfraganus's textbook on astronomy, *Kitāb fī Jawāmiʿ ʿIlm al-Nujūm (Elements of Astronomy)*, a capitulation and revision of Ptolemy's *Almagest*, was translated into Latin in the twelfth century by John of Seville in 1137 under the title *Differentia Scientie Astrorum.* Although manuscript or print editions of *Alfraganus* remained current at the time of Columbus (and with printed editions of the first translation appeared in 1493, 1537, 1546, 1556, 1564, and 1618), it is known that he sourced the figure from d'Ailly's *Imago Mundi.* It is believed that Alfraganus's *Elements* was responsible for spreading knowledge of Ptolemaic astronomy in Europe, at least—as explained in chapter 4—until this role was taken over in the thirteenth century by Sacrobosco's *Sphere.*[5]

The Columbus disambiguation of a new continent from Asia was obliquely acknowledged by his son, Fernando (Hernando) Colón at the Junta de Badajoz-Elvas of April 1524 bringing together specialist cartographers and others from Portugal and Spain with a view to determining the location of the antemeridian of Tordesillas (a reference to the Treaty signed at Tordesillas on June 7, 1494, dividing the newly discovered lands outside Europe between the Portuguese Empire and the Crown of Castile, along a meridian 370 leagues west of the Cape Verde islands, off the west coast of Africa). As Hernando explained, a wide variation obtained between different calculations of the length of a degree and the circumference of the Earth. To this end, he cited Aristotle, Strabo, Ambrosius, Theodosius, Aurelius Theodosius Macrobius, Eratosthenes, Marinus, and Ptolemy. He also referenced three Arab astronomers, namely, Tebit ibn Qurrah (836–901), Almeon, a name associated with the Abbasid caliph al-Mamun (r. 813–833), Alfraganus also attached to the Abbasid court, along with Cardinal Pedro de Aliaco [Pierre d'Ailly], and Fray Juan de Pecan, author of a treatise on the sphere. Columbus (who may have been cognizant with all these texts, directly or indirectly) is cited as having adopted the measure of "fifty-six and two-thirds miles" or "fourteen leagues and two-thirds of a mile" to a degree, and a circumference of 5,100 leagues (Blair and Robertson, 1903–07, vol. I, no. 1: 203).

The Alessandro Zorzi/Bartolemeo Columbus Maplets

Reaching back to Ferraro (1875), Wieser (1893), Bigalow (1935), Almagià (1944–45), Nunn (1952), and Laurencich-Minelli (1985), map scholars have widely debated maps attributed to Columbus's brother, Bartholomew, who had accompanied the Admiral on this final voyage. From Jamaica on July 7, 1503, during his fourth voyage, Columbus wrote a letter to King Ferdinand, reporting on his exploration of Nicaragua and Panama. A copy of the letter was brought to Rome by Bartholomew in 1506. Alessandro Zorzi, a Venetian who gathered accounts of explorers and travelers, was in Rome when Bartholomew arrived. Zorzi, assisted by Bartholomew, embellished an Italian translation of Columbus's letter with three sketch maps dated to 1503–06/1516–22 (the Zorzi or Alberico codex as found in the Biblioteca Nazionale Centrale in Florence). They appear as marginal illustrations embedded in text and together comprise an equatorial zone.[6]

As Suárez (1999: 91; 98) relates, some contemporaries of Columbus held that the Southeast Asia subcontinent area (as with Ptolemy's Great

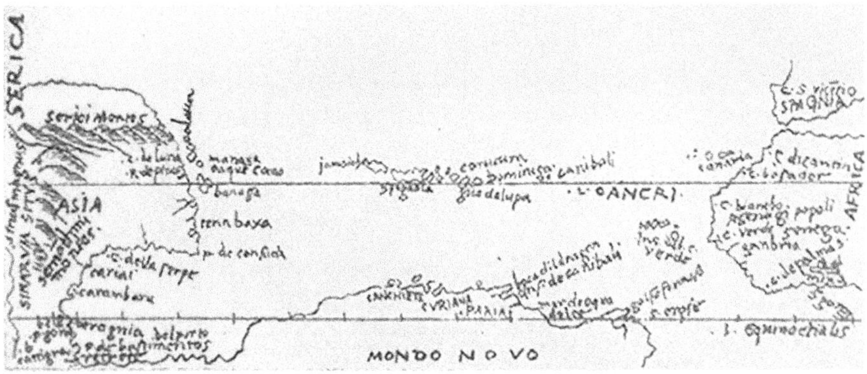

Figure 3.3 Bartholomew Columbus/Alessandro Zorzi Maplet of the West Indies (1506) Illustrating the Columbian Conception of an American-Asian Continental Conflation. *Source*: Biblioteca Nazionale Centrale, Florence. This version sourced to Wieser (1893).

Promontory) was in fact America. The Zorzi/Bartolemeo Columbus sketch map (ca. 1506/22), such as first appeared in print in Wieser (1893) offers convincing evidence. As revealed, the Caribbean islands (Jamaica, Hispaniola, the coast of Venezuela, etc.) are pictured offshore a depiction of the China coast (Champa). The South American continent is merged with the Ptolemaic land bridge or its vestiges, connected up with China. *India Extra Gangem* is clearly marked shading into *Senarum Situs* (China) and Asia. As indicated, Cattigara floats in a *Magnus Sina* east of the *Aurea Chersonesus*, south not north of an equinox. With the *Magnus Sina* practically a China Sea it lies between 170° and 180° E., a major distortion of the reality (see Figures 3.3 and 3.4).

The Behaim or Nuremberg Globe

Although less accessible than map reproductions, the production in Europe of terrestrial (as opposed to celestial) globes was also an index of cartographic progress. Columbus was also alert. The earliest surviving terrestrial globe (the Behaim or Nuremberg Globe) created in 1492 or on the eve of the first Columbus voyage was a production of the Nuremberg-born Martin Behaim (1459–1507), a commercial agent on behalf of the Hanseatic League. A supposed student of cosmographer Johannes Müller von Königsberg (Regiomontanus), Behaim moved to Lisbon where he worked on navigational instruments and tables and, supposedly, participated in voyages down the African coast. He may well have sojourned in the Azores. Falling in with navigators and explorers at the court of King John II, he gained a reputation

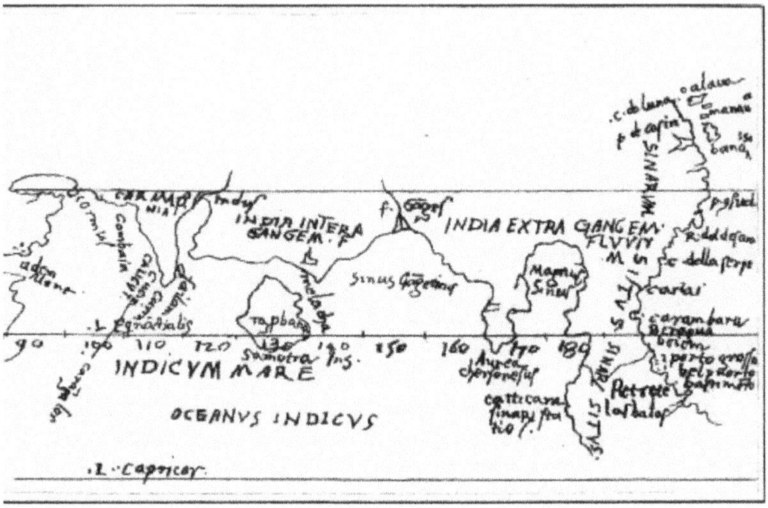

Figure 3.4 Bartholomew Columbus/Alessandro Zorzi Maplet of Asia (1506) Illustrating the Columbian Conception of the Indian Ocean-China Coast. *Source*: Biblioteca Nazionale Centrale, Florence. This version sourced to Wieser (1893).

for his mathematical and practical navigational knowledge, also apparently making the acquaintance of Columbus.

Essentially a representation of a Eurasian continent surrounded by islands and ocean, Lach (1965: 270) is dismissive declaring that Behaim adds nothing on the eastern hemisphere that is not derived from Ptolemy, Mandeville, or Polo. Focusing upon the Behaim representation of the Indian Ocean, Inácio Guerreiro (1993: 456) points to incorporation of knowledge derived from the Dias voyage of 1488. Although lacking longitudinal and latitudinal markings described by Ptolemy, the legacy is evident. Such understanding leads this author to suggest that the Behaim Globe represents a "Luso-Ptolemaic" synthesis. As suggested above, Behaim modified Ptolemy by illustrating a communicable Indian and Atlantic Oceans. With Cipangu at the eastern extremity of the eastern sea (Oceanus Orientalis Indic), the disproportionate extension of the Asian continent eastward on the globe offers a major distortion. There is no Pacific Ocean, setting the trap for Columbus. Tartaria, Cathay, and Mangi (as denoted by Polo) and even India melt into a poorly delineated China. By placing Cipangu (Japan) on the eastern extremity of the "Indian Ocean" within sailing distance of the Canaries, Behaim, and Columbus were privy to a common (mis)understanding. As discussed in chapter 5, the scribe aboard the surviving Magellan navigation revealed that Magellan had perused Behaim's Globe with reference to a gap in the Antarctic Circle, vital knowledge for passage though the straits named after the expedition.

The Place of Marinus of Tyre in the Columbus Concepts

Whether or not Columbus placed great store upon Marinus of Tyre's reckoning of longitude as relating to the easternmost extent of the Eurasian continent has also been the subject of some discussion. As Nunn (1937: 27) pointed out, Marinus's eastern Asia was 45° closer to Western Europe by way of the Atlantic than Ptolemy. Ptolemy placed the extreme east of the known world at the 180th meridian, as opposed to Marinus who placed it at 225° east of the Fortunate Islands (Canaries). After Ptolemy, no geographer placed the extremity of the known world beyond 180° until sometime in the fifteenth century. The breakthrough, he asserts, came with the Behaim Globe now extending the Eurasia continent to the 240th meridian. Marinus, by contrast, knew nothing about this extended 60° realm. As Nunn (1937: 35) concludes, "The Behaim globe was drawn on the Ptolemy measure, and therefore it was the Ptolemy concept which indirectly affected the Columbus concepts. Apart from the longitude of Cattigara, Columbus and Marinus of Tyre had nothing in common concerning the geographical concepts of eastern Asia."

Tellingly, Columbus made three references to Marinus in relation to his fourth voyage. An inscription along the west coast of Africa on sheet three, reads, "According to Marinus and Columbus, from Cabo de São Vicente (Cape St. Vincent) [in Portugal] to Cattigara is 225 degrees, or a difference in time of fifteen hours. According to Ptolemy to the extremity (of the known world) at Cattigara there are 180 degrees, which are equivalent to twelve hours" (Nunn 1937: 28–29). The argument becomes very technical, as with Nunn (1937: 30–33) who deems it necessary to examine three different mileage circumferences of the globe, namely Behaim's, that of Columbus, and present-day, to illustrate the concepts of Columbus. It is also important in this discussion that Columbus's degree value supposed the Earth's circumference 2,000 miles smaller than Behaim, placing Cipangu within 750 leagues of the Canary Islands.

According to Nunn (1952: 22), who sought to correct other interpretations (Almagià and Bigelow), the maplets are not in entirety the work of Bartholomew but were marginal drawings by Zorzi on the July 7 letter where significant statements by Columbus are found. Bagrow (2010: 108) reckons that, dated about 1506, the Zorzi map was actually drawn twenty years later to illustrate Columbus's own concepts as to the western discovery which "he never conceded to be a new continent." As such, according to Nunn (1952: 22), they represent an interpretation of the Columbus brothers' concept of world geography based upon Columbus's 56 2/3 concept of a mile value for an equatorial degree (and that measure is acknowledged in Fernando's petition at Badajoz in 1524 as taken up in chapter 4). This concept made the transatlantic discoveries part of Asia. The maplets are significant, he concludes, "because they are contemporary illustrations of one of the theories of world geography held while Christopher Columbus, Balboa, Magellan, Cortes, Pizarro, De Soto, Coronado, Cabot and Cartier made their explorations." In other words, the maplets offer perfect evidence of the Columbus thesis that the transatlantic discoveries constituted part of Asia. One of the maplets shows the New World ("Mondo Novo") as a big continent connected to Asia.

As the author of the anonymous website, "Cartographic Images," summarizes, and I agree, "Columbus thus owed his success to two fundamental errors, his underestimate of the size of the earth and his overestimate of the portion already known." Despite this dismissal of Ptolemy's reckoning, Bartholomew Columbus drew the map according to Ptolemy's figures, making the distance at the Equator from Western Europe to eastern Asia 180°.[7]

Amerigo Vespucci

Writing of his 1499 voyage, of exploration, Amerigo Vespucci also stated that he had hoped to sail westward from Spain across the Western Ocean (the Atlantic) around the Cape of Cattigara mentioned by Ptolemy into the *Sinus Magnus*. Ptolemy understood Cattigara to be the most eastern port reached by shipping trading from the Greco-Roman world to the lands of the Far East. Specifically, as Clements R. Markham (1894: 55n) established, Vespucci's second voyage from Lisbon in 1503–04 (which he misleadingly termed his fourth voyage) was undertaken for the discovery of Malacca which he falsely believed to be south, rather than just north of the Equator and "33 degrees from the Antarctic Pole." Vespucci failed to find the Cape of Cattigara on his 1499 voyage; he sailed along the coast of Venezuela but not far enough to resolve the question of whether there was a sea passage beyond leading to Ptolemy's *Sinus Magnus*. In any case, unlike Columbus, Vespucci announced in letters that he had indeed discovered a "Mundus Novus" (New World) or, as given new meaning by Martin Waldseemüller in his map of 1507, "the fourth part of the world."

CARTOGRAPHIC ADJUSTMENTS TO PTOLEMY

The framing of continents and regions through text cannot be disassociated from cartographic representations, just as they were frequently married. Neither can we neglect the outpouring of scholarly production from both south and north of the Alps. The Gutenberg revolution was at the heart of this progression. Indubitably, challenges to the strict Ptolemaic blueprint emerged in the light of both Renaissance science and the evidence produced by the discoveries. Some have termed this the "Waldseemüller revolution," after mapmaker, Alsace-based Martin Waldseemüller (1470–1520)—and that is just—but other names (and schools) such as Martin Behaim, Johannes Ruysch, Sebastian Münster, Lorenz Fries, Gerardus Mercator, Abraham Ortelius, Jodocus Hondius, Johannes Janssonius, Willem Blaue, and family equally merit of attention for their bold innovations. Notably, all emerged from the Low Countries, part of the bourgeois revolution emerging in northwestern Europe, in turn fed by the discoveries, and the ascendancy of print. Many of these names were active in the decades crossing the fifteenth and sixteenth centuries and the "revolution" insofar as it occurred was also realized within these short years (though consistent with my thesis, it would take a century or more to smooth out the rough edges and to fill in the blanks on the world map). The point that should be taken is that the cartographic revolution actually preceded the text renditions of the New World or, at least the authors of the great travel collections lagged the mapmakers by at least a generation in many cases.

Whether "revolutionary" or not, Waldseemüller stands out for his understanding that the methods developed by Ptolemy could not portray the shape of the world accurately. Nevertheless, according to Wallis (1992), as with other Renaissance cartographers, Waldseemüller "paid homage" to Ptolemy with maps in the "ancient form," adding "new maps" to his atlas. He also used the innovation of a quadratic plane projection to create "a representation of the world more proper to our time." As such he was the first to recognize the necessity of moving forward and questioning the conclusions of ancient authorities. He therefore led people to recognize geography as an "evolving science."

Accordingly, it can be said that in his Atlas, Waldseemüller sought both to celebrate and to revise Ptolemy's *Geographia.* Such can be seen with his *Tabula Nova* as well as in *Tabula Moderna Indiae* from the 1513 edition of his *Geographia.* Centered on the Indian Ocean, *Tabula* is described by Suárez (1999: 109) as the first "modern" map of Southeast Asia in the way of drawing upon the Portuguese discoveries, yet its Ptolemaic features are obvious. As Durand and Curtis (2013: 30–31) observe, Waldseemüller's *India Extra Gangem* (1513) bears a resemblance to the Cantino world map smuggled out of Portugal. Here, Waldseemüller vastly exaggerates the Golden or Malay Peninsula compared to the Indian subcontinent. Malaqua (Malacca) is indicated and so is an island called Timonia (Tioman). Located off the east coast of the peninsula, Tioman was a landmark port of call for sailing vessels passing the Singapore Straits and coasting southern Vietnam, entering a more extensive *Sinus Magnus*, approximating the modern South China Sea. *Tabula Moderna Indiae* from the 1513 edition of Waldseemüller's *Geographia* also displays an enormous distortion to the Malay Peninsula, a veritable "dragon's tail" or false peninsula, as discussed below. Certain information is also incorporated from the Portuguese discoveries. Notably, the map is the first to depict *Sinus Magnus* as the South China Sea. Yet, it would be Waldseemüller's rendition of the Americas, as discussed in the following chapter, that would secure his image as "revolutionary."

Another noted mapmaker from northern Europe (Germany) was Sebastian Münster (1488–1552). A mathematician, geographer, and professor at Basel University, in 1540 he published a Latin version of Ptolemy and, fifteen years later, his landmark *Cosmographia* appeared, the first German language description of the world going through repeated editions.[8] His *Neuw India / mit vilen anstossenden lendern / besunder Scychia / Parchia / Arabia / Persia* and so on offers a Ptolemaic template but adding knowledge from the Portuguese discoveries. Goa and Malacca and the Moluccas are shown. Cathay is derived from Polo. Japan is absent but appears on his Pacific Ocean map. As explained in chapter 4, both Waldseemüller and Munster would produce bold new interpretations of the globe incorporating the Americas as distinct continents separated from Asia by a Pacific Ocean.

The Flemish school of mapmakers also left their mark in both reviving and revising the Ptolemaic edifice. As Gaspar (2013) points out, the major problems inherent in the "latitude maps," especially errors in the directions and distances between places when measured across huge space, was corrected by Mercator in 1569. Initially an engraver, Mercator (birth-name Gerard de Kremer) (1512–94) went on to create the map projection that bears his name. By combining his skills as engraver and mathematician, in 1569 he presented his cylindrical map projection in the form of a large world map printed on eighteen sheets.

Given its enduring influence, we detour to examine the Mercator projection. According to Crato (2014), Mercator knew the results of Pedro Nunes's work leading him to create a map that would directly serve navigators, especially as flat maps carry their own distortions, especially so when the areas involved are on a global scale, such as confronted by the first Portuguese navigators. This result was the Mercator projection (still in common use, and notorious for the way that it distorts the size and shape of continents especially the polar regions). A "conformal map," it preserves the direction between any two points on the globe. Basically the Mercator map was based upon a net of lines of equal latitude, parallel to the Equator and perpendicular to the meridians and with the meridians all parallel among themselves. As the latitudes approached the poles, Mercator progressively increased distances between the parallels. According to his principle rhumb lines would appear as straight lines. As Gaspar (2013) acknowledges, notwithstanding errors introduced into Mercator's 1569 production resulting from inaccurate rhumb tables, he was well before his time in an era when longitude could not be accurately measured at sea. According to Durand and Curtis (2013: 38), this innovation coupled with the collection of data enabled Mercator to produce a relatively accurate map even for such lesser known regions as Southeast Asia. Toward the end of his life, Mercator created a Latin edition of Ptolemy's *Geographia* (1584) said to be the most faithful interpretation ever published.

A broad consensus holds that, together with Mercator, Abraham Ortelius (1527–98) was one of the two greatest figures in modern cartography. Born in Antwerp and starting his cartographic career in 1547, he traveled extensively through Europe also making contact with Mercator. In later life he was appointed Royal Cartographer to King Phillip II of Spain. Notably, his *Theatrum Orbis Terrarum* (Atlas of the Whole World) originally printed in Antwerp on May 20, 1570, is considered to be the world's first modern atlas and a turning point in the history of cartography. According to R. V. Tooley (1979: 29), *Theatruim* was also "the first uniformly sized, systematic collection of maps of the countries of the world based only on contemporary knowledge since the days of Ptolemy, and in that sense may be called the first modern atlas." Comprising both text and maps, *Theatrum* presented maps in a uniform format. Ortelius thus produced the first major printed work of any kind to include scholarly citations of authorities (i.e., by naming the original mapmakers) and, in Tooley's description, was the first to introduce the concept of footnoting to Western scholarship.

Ortelius's interest in history is seen in the addition of the *Nomenclator Ptolemaicus*, with its own title page. This was a list of place names, areas, islands, and their modern equivalents, taken from Ptolemy's *Geographia.* Ortelius further included a massive appendix (*the Parergon*), consisting of a detailed classical atlas in way of appealing to Renaissance Europe's fascination with the ancient world. In fact, this collection of maps of the ancient world was so significant that it became the model for all historical atlases published throughout the seventeenth century (see van den Broecke 2011). Also included in *Theatrum* were two maps of Asia; *Indae Orientalis*, a map corresponding to Southeast Asia; and a more general map of Asia, *Asiae Nova Descriptio*. Although different renditions, both were published without change in the *Theatrum* from 1570 to the last atlas of 1612.

It is important for our study to acknowledge that the major cartographers of the day replicated the faux "Southeast Asian-sub-continent of America" of Columbus' belief. The 1531 map by French mathematician Oronce Finé offered a third Ptolemaic peninsula, comprising Mexico, Central and South America. It combined the Indian and Pacific Oceans and also displayed parts of the Moluccas lying off the coast of Mexico. Suárez (1999: 93–94), who has made a major contribution to the "three India" depiction of Ptolemy's single elongated Golden Peninsula, explains that this is actually a relic of his land bridge enclosing the Indian Ocean, and as depicted in a range of world maps from the fifteenth and early sixteenth centuries. Each depiction reaches below the Equator and each trends toward the southwest (unlike the true Malay Peninsula). He links this misconception with biblical notions of three Oriental kings or Prester John's "three Indias." He believes the prototype of this kind of map is that by Martellus (Henricus Martellus Germanus) of ca. 1489, offering a false peninsula subcontinent (or "dragon's tail") in addition to a Golden Peninsula. Bagrow (1990: 107) finds it convincing that Columbus may have perused a world map graduated in longitude produced by Martellus who moved to Florence from his native Nuremberg from 1480 to 1496. Versions of the false peninsula carried through to the Cantino planisphere of 1502 as well as to the Waldseemüller world map of 1513. He has Malacca and Cattigara marked on the western coast of the great peninsula and just to the south is the inscription: *hic occisus est S. thomas* (here St. Thomas was killed). Waldseemüller had confused Malacca with Mylapore in India. The Alsace-born Lorenz Fries (1522) produced another version, warping Mangi (Mongolia-China) into a huge peninsula, intersected toward the south by an almost enclosed *Sinus Magnus*, separated by an enveloping ocean.

The Dutch advance on intelligence on maritime Asia as well as cartography was given a major assist by Jan Hughen van Linschoten whose map, published in 1596, was drawn primarily from Portuguese sources styled after portolan charts of the sixteenth century. As such, his vision or cartographic intelligence in combining maritime with mainland Southeast Asia and with China, Japan, and Korea was precocious for the age. His rendition of Japan, for example, is grafted from the chart by the Portuguese, Lázaro Luiz of 1563, to whom is attributed the first shape and nomenclature related to Honshu, Kyushu, Shikoku, and southern islands. This cartographic knowledge also appeared in Fernão Vaz Dourado's map of 1568 and, in turn, that by Bartolomeu Lasso. A Portuguese working under Spanish royal patronage, Lasso's nautical charts and geographic intelligence leaked out to Dutch mapmakers. We reserve a discussion of Portuguese mapmaking to chapter 5.

CONCLUSION

As explained, Ptolemy's instructions on how to render three-dimensional geographic space on a two-dimensional surface were not surpassed until Renaissance times or after. Thanks to the contributions of Arab mathematics and geography, his legacy endured 1,500 years. Contrary to myth, the notion of a spherical Earth dating back to Aristotle received no major challenge in medieval Europe. Nevertheless, as described, the late medieval age was a transitional period in terms of geographic awareness, even prior to the

Columbian voyages. This we have seen with the Medici Atlas, the Mandeville "map," the Catalan map, and the Behaim Globe with its still far from complete revision of Ptolemy.

Venice, we have seen, was ahead of the curve in its diplomatic prowess as well as "political risk" analysis in dealing with the Oriental "other." It was also ahead in assembling geographic knowledge. This we have seen with the de' Conti travel account and the spectacular—for its age—Fra Mauro Map. In sum, Venetian merchants filled the niche of middlemen in the East-West trade, especially in silks and spices sourced from India and distant Sumatra, just as the Venetian republic would be energized by seaborne commerce and links with the broader Mediterranean world. Venetian fleets also entered the Atlantic but, for a combination of factors, in part locational and also political, it was the Iberian nations who, in the wake of the *Reconquista*, turned the tables on historical enemies in the western Mediterranean (Ceuta). Ahead of Arab navigators, they also pioneered the navigation of the western coast of Africa, leading to the now classic voyages of Columbus, Vespucci, Cabral, Dias, da Gama, and, eventually, the Magellan circumnavigation. These early navigators were well served by the development of portolan maps, not only as part of a Mediterranean tradition but, as demonstrated, a model which would come to embrace representations of Asia and regions within Asia.

Needless to say, the Columbian voyages touched off major adjustments and reconsiderations as to the configuration of the continents. As demonstrated, bold new interpretations of Ptolemy did emerge in this "early modern" period especially on the part of map and globe makers in the Low Countries. Nevertheless, with a rekindled interest in mapping the New World, as per the Columbian-era discoveries, scholarly elites still looked back to classical conceptions as a guide and point of reference. Consequently, as discussed, it would take a small revolution in thinking to modify Ptolemy's geography in the light of newly discovered lands (and advances in astronomy). The rising humanistic and scientific revolution associated with Galileo Galilei and the Copernican revolution was not irrelevant to innovations upon the Ptolemaic canon and will be discussed in chapter 5.

NOTES

1. Lamori (*La-mu-li*), located just north of Samudra-Pasai in northeast Sumatra, sent envoys to the Mongol court and, as recorded in Chinese accounts (Ma Yuan), the Ming admiral Zheng. He stopped over at *La-mu-li* on each of his first six voyages (1405–22). It was also the first port polity in maritime Southeast Asia to embrace Islam and therefore was well connected with the eastern Indian Ocean trade in pepper and other tropical goods.

2. A copy is held in the Bibliothèque Nationale de France (BNF). As described, the mss comprises a dozen leafs fixed on wood, and is richly illustrated with a "cosmographic description" attributed to Cresques, or to his atelier.

3. Some seventy mss versions of his travels are known in Latin, French, and Italian. The major mss is the translation by Jean de Long held in the BNF dated to 1350. (Manuscrit latin 2584, folios 118 recto à 127 verso.) Ramusio included a version in his second volume (1574).

4. The Columbus historiography is obviously long and detailed and begins with the first reception and interpretation of his letters. The controversy over Columbus, reaching a new peak during the 400th anniversary of the "discovery" is well covered in a dedicated issue of *History Today* 42, (May 1992).

5. The website of ViaLibri, an antiquarian bookseller, offers a precise description of the 1546 Paris edition of this book https://www.vialibri.net/552display_i/year_1493_0_407297.html

6. See the privately hosted website showing the three maplets along with thick description at http://cartographic-images.net/Cartographic_Images/304_Sketch_maps.html. Also see Macías (2017).

7. See the private website "Cartographic Images." From which I have adapted the anonymous text. *Sketch maps of the equatorial belt of the world* DATE: 1503–06/1516–22 AUTHOR: Bartolommeo Columbus and Alessandro Zorzi DESCRIPTION: http://cartographic-images.net/Cartographic_Images/304_Sketch_maps.html

8. Sebastiano Münster, *La Cosmographie universelle contenant la situation de toutes les parties du monde, avec les proprietez & appartenances.* Henry Pierre marchant-libraire, Basle 1552. This work including maps can be consulted at http://www.istitutodatini.it/biblio/images/it/lazzer/munster/htm/carte.htm Fondazione Istituto Internazionale di Storia Economica "F. Datini." A detailed biography of Münster appears on the website of the commercial house Paulus Swaen http://www.swaen.com/sebastian-munster.php

Chapter 4

Revisions to the Ptolemaic Template in the Age of Discovery (1300–1500s)

The impacts of the Ptolemaic revival upon European conceptions of geography in Renaissance Europe were described in the previous chapter. Still, we need to arrive at an understanding of the process whereby revisions to the Ptolemaic template led to its eventual disaggregation and abandonment. Allowing that Iberian "expansion" lacked the mature capitalist character of the later-arriving European trading companies, as a process it nevertheless called down profound changes in the territorialization of discrete regions of the globe, such as revealed by the progressive European "discovery" of the Indian Ocean area, the coast of China, the vastness of the Pacific Ocean and, eventually, the realization of the full extent of the Eurasian landmass. Obviously where conquest, control, and missionizing was a goal, as in the Philippines, Goa, and Malacca, the changes were profound. Where the Portuguese and later-arriving Europeans established trading posts, certain of them on sufferance, as in Macau or Nagasaki in Japan, then the local impact was more subtle. With the new navigational and cartographic information arriving back in Lisbon and Madrid, eventually leaking to Calvinist rivals in northern Europe, European Jesuits, themselves products of the Catholic Counter-Reformation, fed a virtually uninterrupted stream of knowledge home, joined by traveler's accounts and collections, much of it entering print. By around 1530–90, Europe was inundated with a torrent of new information, spatial and territorial included, challenging local verities, revising the Poloean narrative and foreshadowing even further revisions to classical understandings of the Eurasian landmass, *Sinae*/China included. As noted, the European Renaissance played a key role in this development and with Venice a crucible where this new knowledge was assimilated and reproduced. But the impact would also be two-way, as discussed in the following chapters, with Qing China and Tokugawa Japan, especially, reacting to the European encroachment in

Asian seas (New Worlds for the Iberians) by firming up their own boundaries and setting down their own meta-geographical renditions of imperial region and space.

Specifically, this chapter sets down the main lines of the Iberian encounter with Asia especially with respect to the construction of regions and how knowledge of this meeting across civilizations was received back in Europe. A first section is concerned with Portuguese and Spanish navigational advances in the age of discovery. A second section develops the themes of Iberian "expansion" with specific reference to the Columbian and Magellan "revolutions," as they may be interpreted in terms of introducing radically new understandings of the oceans and continents. A final section focuses upon Iberian mapmaking in the wake of Columbus/Magellan.

PORTUGUESE AND SPANISH NAVIGATIONAL ADVANCES IN THE AGE OF DISCOVERY

As alluded, it was the Portuguese monarch Henry the Navigator (r. 1394–1460) who literally inaugurated the European "Age of Discovery," notwithstanding the actions of Venetians, Genoese, and others. Motivated by religious as much as commercial zeal, he devised the formula to outflank the Ottoman Turks from around the coast of Africa. Son of King John I, as described by Raynal (1784: 30), Prince Henry "availed himself of the little knowledge which was preferred among the Arabs," and, reputedly, established a training center for young courtiers at Sagres in the Algarve (from the Arabic Al-Gharb, meaning west), a place associated with the sacred promontory as referenced by Ptolemy via Marinus. Reputedly, "He had a considerable share in the construction of the astrolabe, and was the first who was sensible of the advantages that might be drawn from the compass, which, though already known in Europe, had never been applied to the purposes of navigation."

Allowing that the existence of the Sagres school is not documented fact, it is certainly true that knowledge of the compass, the lateen sail, and navigational lore gained from the Muslims was a necessary condition in this first wave of European expansion, alongside mastery of shipboard cannon crucial to naval supremacy. Madeira, identified by the Sicily-born Greek historian, Diodorus Siculus, and Plato, as Atlantis of the ancients (Raynal 1784: 30), was officially discovered by Portuguese mariners in 1418–19, and the Azores in 1427, although sea charts from as far back as 1370 showed faithful representations of those islands. Between the 1440s and 1460s, the Cape Verde archipelago had been discovered off the coast of Guinea. The Congo region had been reached by 1483 and, in 1487–88, Bartolomeu Dias had rounded the

Cape of Good Hope. With the voyages of Vasco da Gama to Calicut in India in 1497–99, and Pedro Alvares Cabral to Brazil in 1500, the Portuguese had established regular links between Europe, the Atlantic Ocean, and the Indian Ocean.

Portuguese scholars, such as Alfredo Pinheìro Marques (1987: 39; 74–82), have acknowledged the importance of Genoese and other Mediterranean collaborators in forging these early voyages. The influence of Catalan cartography upon pioneering Portuguese mapmaking as the Atlantic voyages commenced is also acknowledged. Such is registered with the Carta Penela, a crystallization of the state of knowledge of the discoveries as of 1443. Another register was the instance of capture of Muslim slaves in the early Henrican period (1435–45), such as described in the Zurara chronicle (ca. 1453), that of Gomes Eanes de Zurara relating to the discovery and conquest of Guinea. In this narrative, there is no doubting the primacy of Portuguese mapping of the discoveries in a European context. Nevertheless, the lack of physical evidence of such early (fifteenth century) maps (aside from three or four examples) remains a matter of speculation, whether owing to a policy of *sigilo* (secrecy), natural deterioration, or loss owing to the Lisbon earthquake of 1755.

Technical Empowerments

In an age when seaborne measurement of longitude was imprecise, just how did Iberian sailors navigate in the sixteenth century? Equally, how did the effort to calculate coordinates translate onto maps of the time? As W. G. M. Randles (1984: 143) summarizes, four different navigation methods or instruments were used in the sixteenth century with varying success: first, the hourglass; second, observation of the eclipse; third, the lunar distance method; and fourth, the magnetic variation of the compass method. The two astronomical methods led to error even when measured on land and were simply unreliable at sea. Moreover, they were also vitiated by the lack of reliable tables and, as discussed below, with those of Abraham Zacuto (1496) and Regiomontanus (1574) found wanting, and with João de Castro's magnetic variation method disproven.

The organized character of Iberian "expansion" into new worlds should not be ignored. State sponsorship became the hallmark of European trading and colonial endeavor but the venture could not have succeeded without the acquisition of key maritime and navigational technologies or empowerments. Notable was Portuguese and Spanish mastery of the nautical astrolabe, cross-staff, and quadrant. Typically a bronze-cast instrument, the astrolabe, as with the cross-staff and quadrant, were inventions or adaptations linked with the appearance in fifteenth-century Europe of stellar navigation, necessary for sailing beyond sight of coastlines. Especially, instruments were crafted to determine latitude through the observation of the Sun during the day and stars at night.

As known since antiquity, the altitude of heavenly bodies corresponds with latitude. Accordingly, measuring the altitude of stars and planets was vital for navigation. Perfecting the measuring method also became a vital task. A number of pioneers entered the picture. Knowledge of longitude, as discussed below, would take centuries to perfect. By the mid-fifteenth century, Portuguese pilots were already adept at estimating the latitude of their location. Writing in the early eighteenth century, Manoel Pimentel (1712: 15) names the astrolabe, the *anel graduado* or nautical ring, the *balhestilha* (cross-staff or backstaff), and the quadrant as the key instruments, among other innovations, complimented by the *nonius* (a process of subdividing the scales).

As Pimentel (1712: 17) declaims, the astrolabe was the single-most important instrument used by the Portuguese facilitating the discoveries. Notwithstanding its Greek ancestry, he also claims that the nautical astrolabe was a Portuguese invention dating back to the time of King John II, namely as devised by Masters Rodrigues and Joseph (both Jewish mathematicians) along with Martin of Bohemia (Martin Behaim). In practice, the marine astrolabe was used to measure the altitude of the Sun (at noon) or certain stars, especially the North or Pole Star (at night). By measuring the angular altitude of the North Star, the navigator could calculate his latitude (with the North Star vertically above the observer at its zenith equaling 90° north, and with the North Star appearing on the horizon at the Equator, corresponding to 0°). Because the North Star does not precisely correspond with the North Celestial Pole (magnetic north), compensation had to be made according to certain rules. Where in southern latitudes the North Star was no longer visible, early navigators searched for an equivalent South Star. Although the Southern Cross would become a navigational guide, as with Cabral in the wake of his voyage to Brazil, more generally measurement of the Sun at noon became the standard.

This was done by observing the Sun's maximum daily height, and comparing it with the declination of the Sun, known for each day of the year according to the almanac first drawn up by Zacuto. In practice, the nautical astrolabes used to measure the altitude of the Sun differed from those used to measure the altitude of the stars. Sighting devices were different. The pilots would hang the astrolabe at waist height, allowing the light of the Sun to pass through an aperture and project onto a floor (or ship's deck). The altitude of the Sun was then measured using a graduated circular disc. According to Crato (2014), the procedure was given the name of "weighing the Sun" and so, by weighing the Sun, the Portuguese navigators also "plowed" the oceans.

The most common method, as with crossing the Atlantic from east to west or even the Magellan circumnavigation, was that of the marine sand or hourglass to measure time at sea converted to the distance run (sometimes calculated into longitude). Attested back to antiquity, the method was revived in the Middle Ages. It was also used in combination with the "chip log" a knotted string cast overboard to determine the ship's speed (in knots or nautical miles per hour). Known in Portuguese as a *barquinha*, or *barca*, its invention has been attributed to the Portuguese Bartolomeu Crescêncio, from the end of the fifteenth century.[1] As explained in the following chapter, accurate determination of longitude would not make progress until the eighteenth century, leaving highly inaccurate forms of "dead reckoning" subject to vagaries of winds and currents as the primary form of navigational guide.

New Printed Astronomy Texts

Portugal was not isolated from European knowledge of astronomy entering print. It is a statement of fact that the early navigators empowered themselves with the best-known selection of maps, almanacs, and instruments. In other words, without scientific advances and exchanges of knowledge among scholars and their courtly patrons, key navigational feats could not have been accomplished. Here, we mention four individuals whose scientific studies fed directly into Portuguese seafaring ventures, Johannes de Sacrobosco, Abraham Zacuto, his student, Pedro Nunes, and, in turn, his student, João de Castro.

Johannes de Sacrobosco

The first printed astronomical book deemed fundamental text of medieval and post-medieval astronomy owes to Johannes de Sacrobosco (1195–ca. 1256), the English-born and University of Paris-educated astronomer. First composed in 1220, Sacrobosco's *De sphaera mundi* (1472) offered a synthesis of Ptolemy and his Arabic commentators. Numerous reprints and commentaries followed, not only in Venice, but from other European cities. As described by Woodward (1987: 306), "it was a textbook for beginners in cosmography, fully illustrated with world maps and diagrams." From the middle of the thirteenth century, *De sphaera* became the most authoritative astronomical textbook of its time, both in manuscript and print form, and with Sacrobosco's Earth-centered diagrams of the solar system in use even after the Copernican revolution.

Entering Portuguese translation ca. 1509–16, Sabrobosco's *treatise* also came to influence astronomer Pedro Nunes and navigator João de Castro, foundational figures in Portugal's pioneering oceanic navigational feats.

According to Luis Filipe Barreto (1985), just as Sacrobosco was highly influential in implanting Aristotelian ideas in Renaissance Portugal, so his work entered a "discursive space" or part of an intertextual dialogue between scientists and philosophers. On his part, João de Castro "creatively appropriated" Ptolemaic and Aristotelean logic not slavish to Sabrobosco but as an innovator.

Abraham Zacuto

Fleeing persecution in Spain for Lisbon in 1492, the Salamanca-based Jewish scholar and mathematician Abraham Zacuto (b. 1452) went on to make a major contribution to letters and science in Portugal, before joining an exodus of Jews from Portugal in 1496 with himself moving on to Damascus. Drawing upon the works of earlier astronomers, including the Majorcan School, in 1478 Zacuto produced an astronomical treatise written in Hebrew. This comprised sixty-five detailed astronomical tables with the radix or base set in the year 1473 and with the meridian at Salamanca, literally charting the positions of the Sun, Moon, and five planets. In turn, Zacuto's Portuguese disciple Joseph Vizinus adapted it into a Latin translation under the title *Tabulae tabularum Celestium motuum sive Almanach perpetuum* (Book of Tables on the Celestial Motions or the Perpetual Almanac) (Buchman 2015: 8–10; Albuquerque 1986: preface). Published in Leira in 1496, Zacuto's work became an important element of the print revolution occurring in Portugal, especially at the hands of Jewish immigrants from Germany, some of them publishing in Hebrew (see Gunn 2003: 21).

Zacuto's *Almanach perpetuum* (or *Biur Luhoth*) immediately helped to revolutionize ocean navigation. Prior to the *Almanach*, navigators seeking to determine their position in the high seas had to correct for "compass error" (the deviation of the magnetic north from the true north) by recourse to the quadrant and the Pole Star. But this proved less useful as they approached the Equator and the Pole Star began to disappear into the horizon. Zacuto's *Almanach* supplied the first accurate table of solar declination, allowing navigators to use the Sun instead. Zacuto's tables in conjunction with the new metal nautical astrolabe allowed navigators to take accurate readings anywhere. Zacuto also developed a new type of astrolabe, specialized for practical determination of latitude while at sea, in contrast to earlier multipurpose devices intended for use ashore (Buchman 2015: 8–10). Nominated Royal Astronomer and Historian by King John II, he held this position until the early reign of Manuel I (r. 1495–1521).

Zacuto's tables would continue to be used by Portuguese ships thereafter to reach such far destinations as Brazil and India. In Portugal in 1493 at a time when preparations for Vasco da Gama's voyage to India were underway, it is understood that King Manuel I of Portugal (r. 1495–1521) held frequent conversations with Zacuto on meteorological conditions relating to a proposed India voyage. Notably, in 1497, da Gama took Zacuto's tables and the astrolabe with him on his pioneering trip to India. According to Luís de Albuquerque (1986: 36), Zacuto met with Columbus in Salamanca and supplied him with the tables and with Columbus—famously or not—predicting a timely eclipse based upon Zacuto in order to awe the natives he encountered.

The Pedro Nunes Advance

Portuguese navigation also took a practical turn, as with the Sagres school. One of the trailblazing Portuguese astronomers of this age was Pedro Nunes (1502–78), also of Jewish descent. Joining Coimbra University in 1537, he was nominated principal cosmographer to John III (r. 1521–57). In 1557, he published *Libro de Algebra,* and *O Tratado da Sphera,* a work containing Sacrobosco's *Tractatus de Sphaera*, the Austrian astronomer Georg von Peuerbach's presentation of Ptolemaic astronomy, *Theoricae nouae planetarum*, and Book I of Ptolemy's *Geographia.* Nunes was also author of *De arte atque ratione navifgandi librij duo* (1573). As this work attests, he invented the "nautical ring" similar to an astrolabe minus the central pin in order to permit greater precision in measuring the altitude of the Sun. Nevertheless, his most enduring publication was *De Crepuscullis*, published in Lisbon in 1542. Here, he resolved "a very difficult problem of spherical geometry," namely "the precise determination of twilight for a given latitude and given day of the year" (Instituto Camões 2014).

The question of direction and its measurement, such as pioneered by Arab mathematicians, was developed further by Nunes with respect to maritime navigation. Especially, he was concerned with determining the direction a ship should follow in order to reach its destination across the open ocean. Such led to his discovery of the loxodromic curve or rhumb line which, as alluded, cut meridians of longitude at oblique angles and which spiral toward the poles (as opposed to a great circle route). In doing so, he may have found suggestions in Ptolemy's *Geographia* which he translated. Randles (1990: 123), however, is skeptical, believing that Nunes's discovery derived from practical navigational experience "and was in no way a theoretical postulate," although may have seen a justification in Ptolemy as well.

As Crato (2014) explains, by generalizing the *qibla* problem, such as discussed in chapter 2, Nunes turned it into "a dynamic problem." Establishing the difference between the orthodrome or great circle route and the loxodrome or rhumb line route (or path of constant compass bearing crossing meridians at the same angle) was one of Nunes's greatest feats. As acknowledged by

Randles (1990: 129), Nunes anticipated Gerardus Mercator's famous projection as presented in 1569 or, stated another way, Mercator's revolution in cartography was based upon the foundation laid by Nunes.

João de Castro's Navigational Priority

As a student of Nunes, João de Castro (1500–48) also borrowed his teacher's navigational methods and instruments for his voyage from Lisbon to Goa in 1538. Drafted in 1545, and comprising a 106-leaf manuscript, his *roteiro* found its way to Evora in 1578, only entering into print version in the mid-nineteenth century. Nunes was of the opinion that magnetic variation could serve navigators in determining longitude. Conducting numerous navigational measurements during his voyage, de Castro was obliged to adjust for magnetic variation. From observation in Bombay, he observed that the deviation of the ship's compass needle was affected by the magnetic effect of iron objects. This led to some tergiversation as to the method of measuring longitude and the relation between geographic meridians and those called "magnetic meridians." But Columbus had already observed this phenomenon on his first voyage, and it was confirmed by other navigators. De Castro's "scientific" approach to navigation also caused him to comment when coasting Tenerife that "these islands are poorly marked on our maps." Of the Canaries, he also remarked, Ptolemy, Pliny, and Pomponius Mela were "very confusing" (Corvo 1882: 381–84; see Appendix).

As de Castro wrote with confidence in his *Da Geografia por modo de diálogo* of ca. 1535:

> . . . having found the means of putting each one of the lands and seas of this world in its correct place, all the ancient navigations become very simple, many new seas and lands were discovered, all trading became easier, a new world was discovered, and today it is as easy to travel right around the world as it was in ancient times to navigate from Italy to Africa, and finally, with great ease one can now navigate and communicate with the whole world.
>
> This, the true and perfect geography, consists primarily of demarcating lands by the correlation of each with the sky, giving its proper length and width. In this manner it is possible to place on a brief chart and painting the entire world, and any part, province, kingdom, or district of it with much accuracy. (Barreto n.d.: 3)

This of course was a brilliant summation. It was also post-Ptolemaic in the sense of announcing the method of correcting millennium-long conventions though empirical observation, although the details of accuracy would take centuries to work out.

THE IBERIAN PUSH INTO NEW WORLDS

The Iberian push or "expansion" into New Worlds has obviously spawned a vast literature, some of it technical some of it philosophical or, combining both. Few understood better than Adam Smith the importance of the great discoveries. As he put it, "The discovery of America, and that of the passage to the East Indies by the Cape of Good Hope, are the two greatest and most important events recorded in the history of mankind" (Smith 1776: 209). The broad raison d'être of Portugal's push south down the Atlantic has engaged historians over the centuries. There are a number of tangents, as with a religious need to link up with the legendary Christian Kingdom of Prester John (now identified with Ethiopia) and, on another level, to strike political alliances against the Turks as with the Mongols, though it was Safavid Persia which became a privileged ally for Portugal since first making direct contact at Hormuz in the 1510s (*Encyclopedia Iranica* 2014).

Another literature looks back, as with Maria Portuondo (2009), to wonder as to the questions raised for Renaissance thinkers and proto-scientists by the new discoveries, namely as to their locations in relation to Europe, their inhabitants, cultures, and so on. More than that, in the wake of Columbus, Cortes, Pizarro, and other conquistadors, Spanish cosmographers in particular were charged with turning their observations of the New World into a body of knowledge that could be used for governing their expansive imperial domain, not only in the Americas but across the Pacific Ocean to the Philippines. Having discussed technical empowerments wielded by the Iberians favoring their oceanic expansions, discoveries, and conquests, we also seek to know how navigational and cartographic knowledge of the New World was produced in the Iberian nations, who were the patrons, and for what uses?

The 1494 Treaty of Tordesillas

Alongside nautical charts and the measurement of latitude and longitude, Spanish and Portuguese seafarers were obliged to take cognizance of "lines of demarcation" separating spheres of influence in the New World but also impacting upon conflicting claims in the western Pacific. This was a political divisioning of the globe created by a Papal Bull with the mutual consent of the two sixteenth-century maritime superpowers. As Christian nations, they needed Papal validation of their discoveries as they sailed west across the Atlantic and south down the coast of Africa (in the case of the Portuguese). Starting with the Treaty of Alcáçovas of 1479, Portugal renounced all claims to the Canary Islands which Spain had begun to colonize in 1402 and, in return, Spain under Ferdinand and Isabella renounced designs upon the Guinea coast, the Azores, the Madeiras and Cape Verde Islands, or other islands yet discovered. This led to the first papal bull. But when, in 1493, Columbus arrived in Lisbon announcing his discoveries, Portugal entered into dispute with Spain over ownership. Having appealed to the Spanish-born Pope Alexander VI (1492–1503), Spain was supported by four bulls issued in short succession in 1493, setting down a temporary divisioning between the two claimants at 38° west longitude, essentially granting everything in the New World to the Spanish. By default, territories to the east would fall to Portugal, effectively limiting Portugal's discoveries to the coast of Africa at a time when it was also scouting the maritime route to India.

Seeking to avert conflict, Spain and Portugal sought to arrive at a compromise independent of the papacy. The result was the 1494 Treaty of Tordesillas, wherein a line was drawn from the Arctic to the Antarctic Poles at a distance of 370 leagues (approximately 1,770 kilometers) west of the Cape Verde Islands (a line which actually cut through the American continent at 48°25'), effectively separating out the eastern extremity of the South American continent (the future Brazil) from those lands west becoming Spanish colonies). Although coming at a time when the measurement of longitude was yet to be perfected, as discussed in chapter 7, the Treaty set the tone for the early exploration of the New World and in the disputed Moluccas Islands on the other side of the globe. As a result, Portugal was given "rights" to most of maritime Asia but was confined to Brazil in the Americas, and with Spain given free rein in nearly all of North and South America and the Pacific. Effectively barred under the papal dispensation, England and France became preoccupied with seeking a northwest passage to Asia via a route north of what is now Canada or, equally vain, a northeast passage to Asia via Norway and Russia.

From West to East: The Portuguese Maritime Expansion

Aside from early Chinese, Arab, and other navigators who sailed these seas, by the early decades of the sixteenth century they were joined by eastward-sailing Portuguese and westwards-sailing Spanish. As alluded in the previous chapter, Portuguese navigators had progressively pushed southwards down the West African coast, mastering wind and tide patterns and navigational procedures. The stages of Portuguese voyages around the Cape of Good Hope; the Dias voyage of 1488, and the epochal push into the Indian Ocean trading zone by the da Gama voyage of 1497–99, is well traveled terrain in standard histories. No less epochal was the Portuguese conquest in 1511 of the Islamic emporium of Malacca (also a Ming tributary) situated astride the straits that bears its name leading east to the spice islands and, by rounding the southern tip of the Golden Peninsula of the ancients, leading to China itself. As we have labored, none of this could be taken for granted but implied high navigational skill, mastery of the Atlantic and Indian Ocean monsoon wind patterns, naval prowess, and much good luck given the risks, although local knowledge in the way of Arab and Malay pilots was also taken on board.

Prior to the permanent Portuguese settlement of Macau ca. 1557, the first European ships (the Tomé Pires mission) reached the Pearl River estuary in 1517 foreshadowing a major moment in the world historical expansion of the Iberian center into the Asia-Pacific periphery. Having pioneered the westward-sailing route across the Pacific Ocean in 1521 "discovering" the Marianas and the Philippines, Magellan's fleet also entered the South China Sea. Even earlier, eastward-sailing Portuguese had pushed beyond Malacca and explored the coastlines of Indochina through the voyages of Fernão Perez de Andrade in 1516 and Duarte Coelho in 1524.

Manoel Pimentel's *Arte de Navegar*

Although postdating the Columbus and Magellan voyages, the *Arte de Navegar* by Manoel Pimentel deserves special attention. According to Carlos Correia (2011) who made a comparison of both the 1691 and 1712 texts, the importance of the work arises from its concise presentation of the empirical and scientific knowledge developed over the time period between the two editions. As such, he situates Pimentel within the cultural and academic environment in Portugal between the fifteenth and seventeenth centuries. This was no accident, but a product of state investment. Notable in this respect was the establishment in 1590 of the Aula da Esfera (class of sphere) or nautical science course at the Jesuit-run Santo Antônio de Lisboa down until the suppression of the Jesuits in the middle of the eighteenth century.

In Pimentel's (1712: iii) own words, the work divides into two parts, one part is "scientific," and the other is "experimental." The first part is devoted to the "rules" of navigation including the uses of instruments and astronomical measurement techniques, magnetic variation, the making of marine charts, and mathematical science. The second part sets down the *roteiros* (saiing routes) and *derrotas* (or voyages), distances, coasts, bays, depths, currents, monsoons, and so on, based upon the experience and observation of the pilots. As Pimentel also acknowledges, the first printed versions of the *roteiros* in Portuguese owes to official cosmographer Manuel de Figueiredo, a hundred years earlier (in fact, published in 1608) and he has obviously drawn upon this earlier work.

It is intriguing, as revealed by Pimentel (1712: 423) that, from the early seventeenth century, the Portuguese were well aware of the southern Indian Ocean route from the Cape of Good Hope to Australia (Terra Nova), albeit with Timor as destination. As adverted, "*Mas devese advertir, que a distancia desde o Cão de Boa Esperencia até a Terra Nova não he tanta quanta mostrão os Cartas*" (But it should be noticed that the distance from the Cape of Good Hope until Terra Nova is not much revealed on the charts), setting aside unknown and adverse ocean currents. As alerted (and as well known to the Dutch from the 1610s), the reef-strewn coast of Terra Nova at 27° south was perilous, especially if approached at night. Riding the southerly or south-easterly winds at 22° latitude was optimum before turning north to Timor Island (Babao) at 11°5'. As an alternative to the Malacca Straits, Portuguese navigators setting out from Goa also coursed west of Sumatra, coasting south of Java at 7–8° latitude south, prior to entering the Java Sea via the Sunda or Bali Straits.

In reflecting upon the longitude issue, Pimentel (1712: 422), reckoned that the difference of longitude between Lisbon and Macau was 127°. However, in a knowledgeable aside, he remarked that Easter Day in Portuguese Macau was then commemorated one day variant from that of Spanish Manila, the reason being (citing the *Tratados da Esphera*),[2] that the westward-sailing Spanish following the setting sun added time (gaining fifteen hours from Cadiz to Manila), whereas the eastward-sailing Portuguese following the course of the Sun, counted nine hours, and registered time accordingly. This allusion to the so-called Circumnavigator's Paradox was first recorded on July 9, 1522, by the Magellan-Delcano circumnavigation when they registered a loss of a day upon arriving at Cape Verde, and with the Pope duly informed.[3]

West to East: Magellan in the Wake of Columbus

In the broader sweep of European expansion, read imperialism, when it comes to the conquest of the Pacific Ocean and its peoples, we cannot neglect the Spanish era in the wake of Columbus and Magellan. Understandings of the Columbus-Magellan voyages reached a new sophistication with the research and writings of George E. Nunn (1932), demonstrating that the Magellan voyage was actually a logical consequence of the final views of the Columbus brothers. More recently, Kevin Sheehan (2008: 75) has asserted that Balboa's vision was also a logical extension of the Columbus project. Working from an array of Spanish including archival sources, he posits a continuity from Columbus's mind-set through to Balboa, and even guiding the Spanish push across the western Pacific.

While the vast ocean later dubbed Pacific was of course known to China, although not its extremities, it was only vaguely apprehended in Europe via the Ptolemaic template, at least allowing *Sinae*'s proximity to an ocean. Having crossed the Isthmus of Panama on September 25, 1513, Vasco Núñez de Nunez Balboa (ca. 1475–1519) became the first European to observe the Pacific Ocean from the New World and, reaching the sea four days later, grandiloquently claimed all its waters from both sides of the Equator and from Pole to Pole for the Crown of Castille. He also dubbed it *El Mar del Sur*. But, in the words of Sheehan (2008: 75), "he saw not a new sea but rather one that had been know(n) since Antiquity." In his mind, "Balboa now saw before him the great *Sinus Magnus* separating what he presumed to be the region of Cathay to the north, and *Chersonoso Aurea*, and which contemporary geographies broadly defined as India to the West." As Sheehan continues, after struggling across terrain which he probably believed were mountains featured by Ptolemy at the easternmost boundary of ancient geographical knowledge, what Balboa saw fitted eastern shores of Ptolemy's enclosed Indian Ocean. While Balboa did not survive to express his thoughts on paper, it nevertheless makes sense, after Sheenan, to place this individual within the ideological framework of his peers.

If indeed, as Sheehan contends, that much of Spanish voyaging in the Pacific in the sixteenth century was, via Balboa, a logical extension of the Columbian venture, we still cannot ignore Portuguese-Spanish rivalry over the Moluccas (such as mediated at Tordesillas). As discussed below, even the Treaty of Zaragoza of April 1529 under which Spain resigned all claims to the Moluccas did not lessen Spanish interest in the western Pacific especially following the Spanish conquests of Mexico and Peru in the 1520s. Continuous tensions besetting the Spanish and Portuguese also played out in Macau during the period of the union of two crowns.

The Magellan Voyage

As summarized by Durand and Curtis (2013: 34), failing to convince his king of the feasibility of a westward-sailing route to the spice islands, the Portuguese Fernão de Magalhães (Ferdinand Magellan) turned to Spain. Backed by the Spanish Crown, Magellan departed Seville in August 1519, the year that Charles I was crowned Holy Roman Emperor as Charles V (r. 1519–56). Surviving storms and mutinies, literally eating rats and leather to survive, the expedition famously passed through the straits that bear Magellan's name and headed north-west across the ocean via the Marianas and the Philippines, reaching the spice islands in 1521 following a detour to the Brunei Sultanate on northeastern Borneo (and so entering what would later be known as the South China Sea).

As well known, Magellan lost his life in the islands that, within decades would be taken in possession by the Spanish crown and named the Philippines. Of the five vessels that departed Seville, the *Victoria* alone made the return journey with eighteen Europeans of the original group of 237. Among them were the Basque navigator, Juan Sebastián del Cano (Delcano); Francisco Albo, the pilot; and Antonio Pigafetta, the Italian scribe and diarist, the latter two of whom we owe an account of the voyage aside from interviews conducted with the survivors. Pigafetta also drew maps of the Pacific, measured longitude, and drew naive map images of Borneo, Timor, the Moluccas, and other locations, certain also profiled from local knowledge by Francisco Rodrigues a few years earlier. The voyage was also politically challenging for Spain as it thrust across the antemeridian set down by Tordesillas as Portuguese domain (Skelton 1969; Durand and Curtis 2013: 34).[4]

Importantly, methods for determining longitude were given special attention during preparations for the Magellan-Delcano circumnavigation (1519–22). An early candidate for ship astrologer, the Portuguese Rui Faleiro, was replaced by the Spaniard Andrès de San Martin. Nevertheless, Faleiro's *Regimento* or instructions on measuring longitude via three methods were taken on board. These were, respectively, (a) measuring the latitude of the Moon; (b) the lunar distance method; and (c) the magnetic variation method (Randles 1984: 147). In practice, the celestial methods were only used off the east coast of South America with one sighting tested against a solar eclipse. However, Pigafetta and Albo both recorded latitude and longitude measurements as with the Straits of Magellan (52° toward the Arctic Pole), enigmatically known to Magellan from having consulted the Behaim Globe. Pigafetta was also the author of a *Treatise on Navigation* and well understood the system of the world according to Ptolemy (Stanley 2010: 164–174).

As R. A. Skelton (1969: 51; 155) speculates, could Magellan have known of a break in the assumed encircling Antarctic landmass, or was there an earlier discovery known to Behaim? Besides keeping a daily log, Pigafetta recorded latitude and longitude for a number of western Pacific Ocean locations as with the still undetermined "Mazaua Island" in the Philippines, Brunei ("five and a quarter degrees towards the Arctic Pole and the longitude of one hundred and seventy-six and two-thirds degrees from the line of demarcation"), and Oecusse on Timor Island. More generally, the Magellan voyage calculated latitude with remarkable accuracy but with longitude inaccurate although relatively consistent (as with a 2° longitude differential between Mazaua and Oecusse which is roundly correct). Armed with a compass, marine charts, and a method of telling the time, Magellan displayed all before a friendly chieftain in the Philippines (Skelton 1969: 68).[5] Upon arriving in the Philippines, the expedition was undoubtedly served by Magellan's indentured Malacca Malay-speaking slave, and with other locals assisting the complex navigation through the archipelago. For mapmakers, the true size and scope of the planet had been revealed, and with the distance around the globe calculated.

According to Pigafetta, "It is supposed that all the circumference of the earth is divided into 360 degrees; and each degree is of seventeen leagues and a half, consequently the circumference of the earth is 6,300 leagues." Land leagues are of three miles and sea leagues of four miles (Stanley 2010: 164–174). The need to measure longitude was sharpened in 1524 when Portuguese and Spanish delegates came together to discuss the exact positioning of the antemeridian with respect to the Moluccas islands.

The World Mapping of the Pacific

Above all, by pioneering a route through the strait at the southern tip of the Americas, Magellan and his companions established the connectivity of the Atlantic and the Pacific Ocean, as he named it. Needless to say, his accomplishment facilitated a radical modification of theories represented by earlier cartography. As Rainer F. Buschmann et al. (2014: 10) state in a critical interrogation of the Spanish presence in the Pacific, "It was Magellan's paradigm-shattering circumnavigation of our planet during this period that illuminated the vastness of the Pacific Ocean and obliterated Ptolemy's centuries-old accepted 'truth' of a land-locked positioning of the Earth's Oceans." This is correct although the true measurement of the Eurasian landmass awaited the future. As noted below, the Spanish entry into the Pacific challenged the Iberian truce of Tordesillas.

Obviously mapping the Pacific after Balboa and Magellan represented a paradigm shift in conceptualizing the geography of the Pacific and the China seas. Yet one theoretical leap stands out, that taken by Waldseemüller, some thirteen years after Columbus's first voyage, prior to Balboa's sighting of the "other sea," and thus prior to Magellan. According to Buschmann et al. (2014: 17), Waldseemüller's separation of "America" from Asia in his *Universalis cosmographia secundum Ptholomaei traditionem et Americi Vespucii aliorumque lustrationes* (The Universal Cosmography according to the Tradition of Ptolemy and the Discoveries of Amerigo Vespucci and others) (1507) marked the conceptual birth of the Pacific. He is also the first to name America on a map and so, as mentioned, announced the "fourth part" of the world, alongside Europe, Asia, and Africa. The full title of the map is also suggestive of Waldseemüller's composite method, namely grafting new discoveries to the Ptolemaic template. As he explained in an auxiliary text, "In designing the sheets of our world map we have not followed Ptolemy in every respect, particularly as regards the new lands. . . . We have therefore followed, on the flat map, Ptolemy, except for the new lands and some other things, but on the solid globe, which accompanies the flat map, the description of Amerigo that is appended hereto."[6] Both Cattigara and Malacca are indicated.

In revealing a new ocean beyond the western Atlantic, the Waldseemüller map thus achieved a new cartographic priority even though fitting in with the Asia of Ptolemy and Marco Polo as exhibited on the 1492 Behaim Globe. Unlike the Iberian tradition of mapmaking yielded by direct knowledge, Waldseemüller and his collaborators working from Strasbourg in France drew upon inductive reasoning to conclude that the natives encountered by Columbus could not be the peoples of Asia as described by Polo and others. They obviously concluded that the long southward journeying of Vespucci down the east coast of the land now known as Brazil could not have been a Ptolemaic promontory. This enigma also attracted much specialist discussion, possibly causing Waldseemüller to shift back to the older paradigm in a subsequent map[7] (see Figure 4.1).

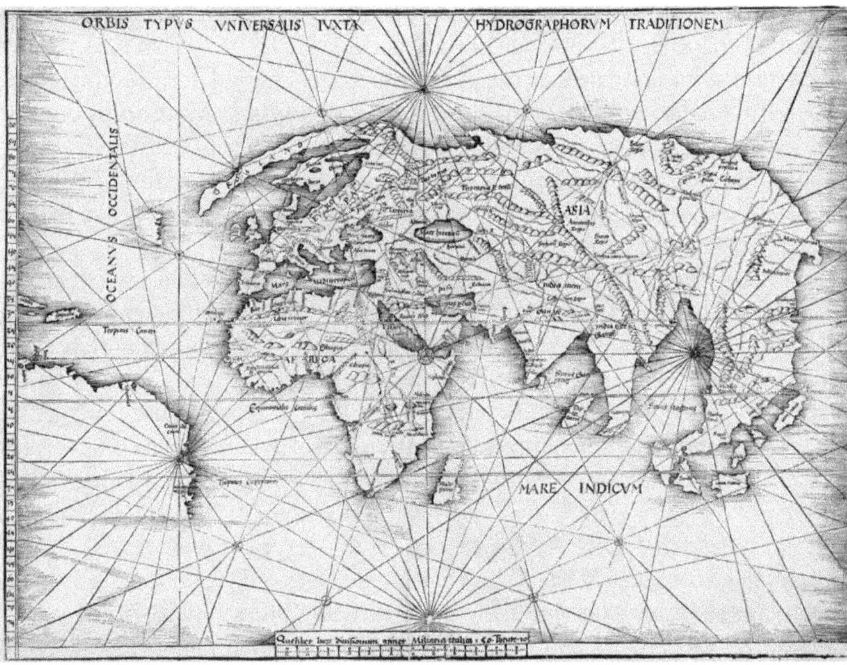

Figure 4.1 ***Orbis typus uniuersalis iuxta hydrographorum traditionem,*** **Martin Waldseemuller.** In Ptolemy's *Geographia* (Strasbourg, 1513). Although Waldseemuller had proposed naming the new continent "America" back in 1507. In this 1513 map, often referred to as the "Admiral's Map," he reportedly reflects Columbus's interpretation of the New World discoveries, with its focus upon the Caribbean islands and with no indication of a western coastline. In this version, the name "America" is excluded. *Source*: Map reproduction courtesy of the Norman B. Leventhal Map & Education Center at the Boston Public Library.

New Map Images after Magellan

As Sheehan (2008: 194) found, new outlooks and images flowing from Magellan quickly influenced leading cartographers, as with the Miller Atlas or world map of Portuguese cartographer Lopo Homem (1519), further discussed below. As such, Homem's chart continued to reveal Ptolemaic inheritance but was modified to accommodate the New World discoveries, Balboa and Magellan included. Now the new label Magellanica became synonymous on some maps with *Terra Australis* such as those of the Flemish school (Ortelius).[8]

Sebastian Münster's *Novae Insulae XXVI Nova Tabula,* published in 1540, was also reprinted in his *Cosmographie* (1552), as introduced in chapter 3. This is widely hailed as the first map of the continent of America, with north labelled Novus and south, Orbis. This map was also the first to name the Pacific Ocean (*Mare Pacificum*). In this map an enlarged Zipangu (Japan) lies just east of the North American continent surrounded by "Archiplegaus 7446" and with the landmass northeast of Japan labelled, variously, India, Quinsay, and Cathay. An image of the *Victoria* of the Magellan voyage reveals awareness of the circumnavigation. While Münster's revelation of America obviously advanced upon Ptolemy, the conceptualization still reveals a mind-set dating back to Columbus and Polo[9] (see Figure 4.2).

Going by the Latinized name, Ludovicus Georgius, Luiz Jorge de Barbuda was a Portuguese cartographer in the employ of Philip II of Spain. He is also regarded as the creator of the influential map of China that would in turn be published by Flemish mapmaker Abraham Ortelius in his *Theatrum Orbis Terrarum* (1584). This was *Chinae, olim Sinarum regionis, nova descriptio. Auctore Ludovico Georgio.* The basis for this map appears to be *China Regio Asiae* (1575), a simplified and miniaturized version reprinted in atlas editions of Ortelius's larger and more detailed work. Strikingly revisionist in its framing of China including the South and East China seas, *China Regio Asiae* was highly influential over the next decades at least until Jesuit productions offered closer firsthand descriptions. It might also be hailed as the first post-Ptolemaic rendition of China from a European perspective.

Not only does *China Regio Asiae* place China at the center, a device imitated by Matteo Ricci in producing his world map to please China, but identifies Japan, the Ryukyus, the coast of Vietnam, and the Malay Peninsula in spatial relationship. Such a production could not have been made without incorporation of knowledge derived from Portuguese voyages. Macau and the surrounding islands are depicted. Neither could the map have been produced without some intricate inside knowledge of Ming China. Importantly it names the two capitals and thirteen provinces of Ming Dynasty. Still, there are some Ptolemaic or classical influences here, as with large interior lakes feeding

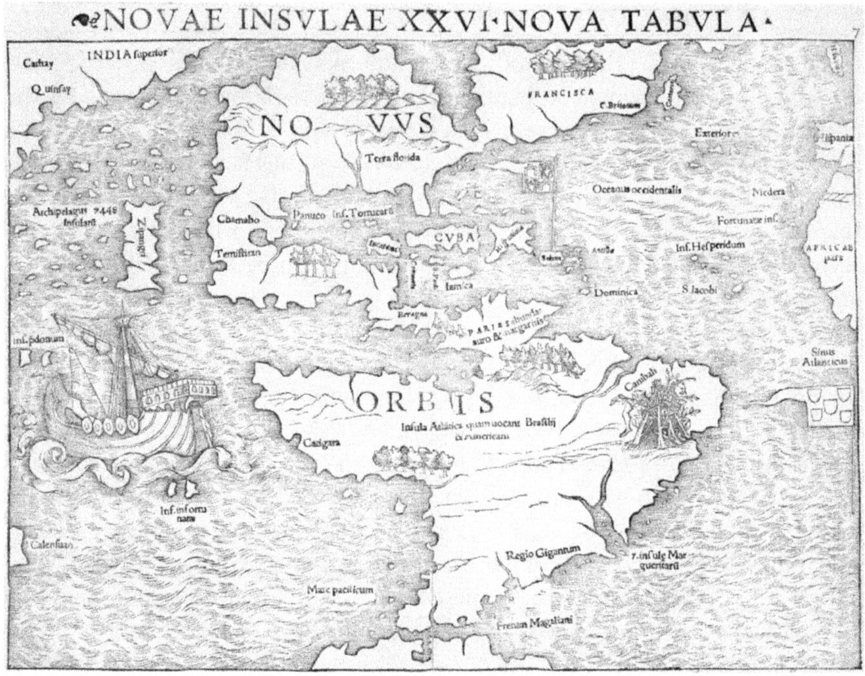

Figure 4.2 ***Novae insulae, nova tabula*** **(New Map of the New World), Sebastian Münster (ca. the 1540s).** First appearing in *Cosmographia*, this is the earliest map to show all of North and South America in a true continental form. It is also the first to name the Pacific Ocean (Mare Pacificum), also displaying Magellan's ship *Victoria*. Here we see the earliest appearance of Japan on a map, even though predating European contact and based solely on such legends as Marco Polo. *Source*: From the New York Public Library. https://digitalcollections.nypl.org/items/510d47df-f8e6-a3d9-e040-e00a18064a99.

river systems as well as an indication of the *Sinus Magnus*, marking the South and East China Sea zone, already known to the Spanish as the Pacific Ocean. The Great Wall is shown, as are Tartar "yurts" on the steppes of central Asia. Japan is represented in a curved projection reminiscent of Portuguese charts of the period (see Figure 4.3).

The Moluccas Longitude Issue

When, in 1511, Portugal conquered Malacca and explored eastward toward the Moluccas, then the Papal "line of demarcation" with respect to the other side of the world came to matter. In 1514, responding to a petition from King Manuel of Portugal, Pope Leo X (r. 1513–21) formally issued a bull extending the line of demarcation to the Pacific. With the Portuguese established

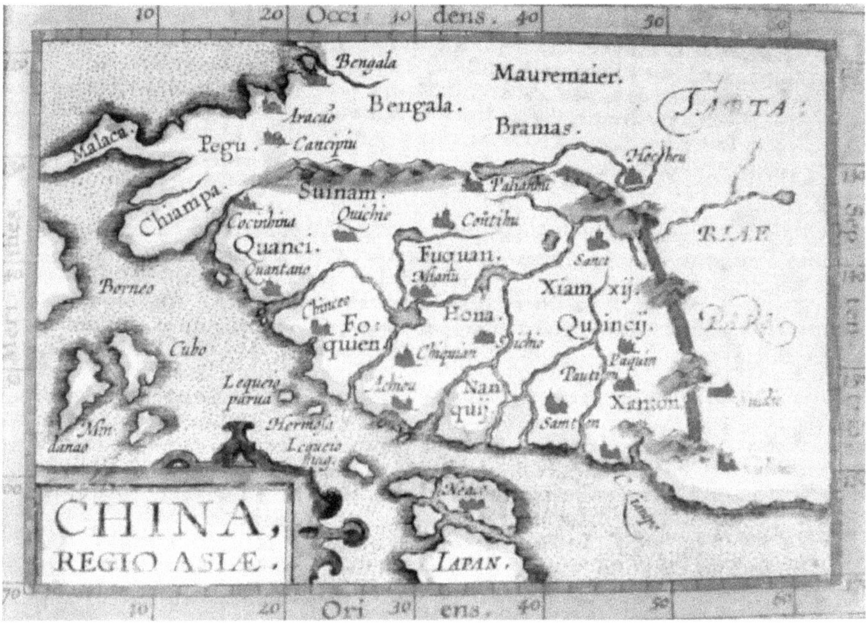

Figure 4.3 ***Chinae, Olim Sinarum Regionis, Nova Descriptio,*** **Abraham Ortelius.** *Theatrum Orbis Terrarum* (1575). An early post-Ptolemaic map of Asia placing China at the center in turn derivative of the map template created by the Portuguese cartographer, Luiz Jorge de Barbuda. It is also the first European map to illustrate the Great Wall of China. *Source*: Courtesy of Arquivo de Macau.

in the Moluccas at least ten years prior to the arrival of the westward-sailing Magellan navigation in November 1521, the ownership question over the "spice islands" became a matter of high diplomacy between Portugal and Spain for the next fifty years. In the meantime and in quick succession, Charles V of Spain sent three expeditions in Magellan's wake, namely under Loaisa (1525), Cabot (1526), and Saavedra (1527), to occupy and colonize the Moluccas. Only the Loaisa and Saavedra voyages reached the Moluccas and with the survivors surrendering to the Portuguese (Lach Vol. 1 Book 2, 1965: 639; Suárez 1999: 161–162).[10]

According to Suárez (1999: 162–163), Pope Leo X who had read the writings of the Italian historian in the service of Spain, Peter Martyr d'Anghiera came to understand something of the intractability of the issue. Disagreements stemmed not only from the actual measurement of longitude, but also entered a kind of "psychological" contestation with prominent historians and mapmakers entering the fray. Spain and Portugal literally fought over the "spice islands" of the Indonesian archipelago, such as instanced by Magellan's division of the islands between the two powers in his memoir dated

September 1519 prior to his departure (Randles 1984: 149). We should recall that Magellan's brief was to reach the spice islands, not to circumnavigate the globe.

The Lingering Weight of Ptolemy in the Wake of Magellan

Although practical Spanish navigation westward across the Pacific largely ignored Ptolemy, indeed added an ocean on top of a continent in the Americas, classical conceptions still spellbound cosmographers in Portugal and Spain. This was especially the case in connection with the "ownership" of the Moluccas such as entered the conference of the Junta of Badajoz-Elvas in April–May 1524 bringing together representatives of King John II of Portugal and Charles I of Spain.

In order to determine the exact location of the antemeridian of Tordesillas, both Spain and Portugal were represented by three-member teams of astronomers, cartographers, pilots, and mathematicians; Portuguese cartographer and cosmographer Lopo Homem was one. The Portuguese "conquistador" Diego Lopes de Sequeira was another of the participants, along with cartographer Diogo Ribeiro another Portuguese working for Spain rising to become head cosmographer at the Casa das India in Seville. In his petition of April 13, 1524, as Hernando Colón advised, owing to lack of agreement as to the circumference of the earth, "neither side can convince the other that he is trying to shift his ground; and therefore, I say, no sentence can at the present time be passed upon this case, except that it will be necessary to agree upon an expedition to compute the size of the degrees for the purpose of measuring the longitude and for definitely determining and marking the beginning and end of the said demarcation, and the lands falling in each part or hemisphere" (Blair and Robertson, 1903–09, vol. 1, no. 1: 205). In the event, the result was indecisive. As Colin Jack-Hinton (1964: 133; 152–154) explains, reaching back to the Papal Bull of 1493, the value of a league was not defined and even when the degree was adopted as a standard, its linear extent remained a mystery. As he argues, in the absence of real knowledge, classical concepts were referenced with Spain tending to Ptolemy (underestimating the width of the Pacific) and Portugal tending to Eratosthenes.

Intriguingly, the question of the location of the *Sinus Magnus* relative to Cattigara and both relative to *regio Sinarum* or China also entered the discussions. As noted in the joint submission of Fray Tomás Duran, the Venetian explorer Sebastian Cabot, and Juan Vespucci (the Italian-Spanish cosmographer), all in the service of Spain:

> Ptolemaeus locates this cape of Catigara at the point of the gulf Magnus, next to the gulf of the Ganges and the Cresonensus bay [*Chersonesus Aurea* or Golden

> Peninsula], which conforms wholly to the account now discovered, so that the description and figure of Ptolemaeus and the description and model found recently by those who came from the spice regions are alike and not only alike in appearance, but in name. That region is now called China; Ptolemæus styled it _regio Sinarum;_ the barbarians also compressing the _s_ say Sina instead of China; and the Portuguese themselves place China in this region.
> _Item_: in everything discovered by the Portuguese of which Ptolemaeus has any notice, the former conform in their navigation to the latter. They locate China north of the Moluccas in the gulf Magnus, just as Ptolemaeus locates it. (Blair and Robertson 1905–09, vol. 1, 210)

Accordingly, they argued, Ptolemy's Cape Cattigara actually corresponded to the Moluccas. Consequently, the Pacific line of demarcation fell 32° further west passing through the mouth of the Ganges (in line with the Treaty of Zaragosa). Therefore, Sumatra, Malacca, and the Moluccas fell within the Spanish realm (Blair and Robertson 1903–09, vol. I: 210). Obviously this was a major misrepresentation of the truth and the Spanish deception would also play out on maps. Still, longitude could not be determined accurately at this time and, in the event, no agreement was reached at Badajoz-Elvas. This may well have been the first and last international fora where Ptolemy actually gained an airing but it is also indicative of the staying power of the classical geographical template even in the wake of the Magellan-Delcano circumnavigation.

The process continued with the Treaty (or Capitulation) of Zaragoza (Saragossa) between Spain and Portugal signed on April 22, 1529, by King John III and Emperor Charles V in the Spanish city of that name. Under the Treaty, the two sides proclaimed a boundary near 142° E longitude (far to the east of the Philippines and bisecting the island of New Guinea), and with the new line solidifying Portugal's claim to the Indian Ocean and all of Asia while recognizing Spain's dominion over the Americas and most of the Pacific. Although effectively settling the "Moluccas issue," at least until the Dutch blustered in taking monopoly control, the two sides still continued to wage a sniping war.

Addressing the Moluccas longitude issue in 1596–1600, the Malacca-based cosmographer, Manuel Godinho de Erédia, also recalled how, in 1524, Pope Clemente VII had shifted the goal posts as it were in Spain's favor by modifying Pope Alexander VI's 1493 Bull. Accordingly, he advised his superiors that "the only way would be for more learned Cosmographers to go to the island of Molucca itself, and for the vertical and meridian thereof, observe some eclipse, solar or lunar, so that the difference in the hours of time at the position of Moluccas and at Cape Verde Islands respectively could completely and accurately determine the position of Portugal." Obviously alert to European science, Erédia also suggested that the same could also be determined by the "New Art of Navigation for East and West," namely "by the mechanical calculation of the time with a clock worked by wheels" (Mills 1997: 249). Erédia was far ahead of his time in appreciating the merit of determination of longitude by clockwork and, as taken up in subsequent chapters, it would be long centuries before that method was perfected.

IBERIAN MAPMAKING IN THE WAKE OF COLUMBUS/MAGELLAN

Commensurate with the Portuguese discoveries, Lisbon emerged as a preeminent center of mapmaking. From 1511, with the establishment in Lisbon of the Casa da India out of the earlier Casa da Índia e da Guiné and even earlier royal trading houses monopolizing the Africa trade, a general mapping office was established. As such, the Portuguese version paralleled that of the Casa de Contratación at Seville in 1503. In the closing decades of the sixteenth century, the Royal Council of the Indies (*Real y Supremo Consejo de Indias*), initially set up in 1524, took over many of the functions of the Casa, namely the collection, compilation, and production of cosmological knowledge.

As Portuondo (2009: 95–96) explains, with the establishment of the Casa, Seville became the locus of Spanish navigation and exploration. More or less through the sixteenth century it became a place where cosmographers and pilots launched experiments in mathematics, navigation, and in the production of descriptive rutters (the standard). As she writes, "By the latter half of the sixteenth century, the nautical charts prepared by the cartographers at the Casa de Contratación drawn according to Ptolemaic rules became the accepted visual representation of the newly discovered geography." Notably, as well, the Royal Mathematics Academy in Madrid, in part set up to prepare pilots and cosmographers versed in mathematics, placed Ptolemy's *Geographia* firmly on the curriculum as a guide to knowledge of the sphere and planetary theory (Portuondo 2009: 84).

By 1553 there were eighteen professional mapmakers employed at six different establishments in Lisbon making marine maps, certain introduced below. As Portuondo (2009: 81) explains, the union of two crowns eased some of the tensions surrounding the Treaty of Tordesillas/lines of demarcation, but it also offered an opportunity for the Spanish to take stock of Portuguese cartography and to adjust Spanish maps accordingly. The Spanish Casa de Contratación also made a failed bid to reconcile Portuguese and Spanish navigation charts. In the first half of the sixteenth century, several Portuguese mapmakers moved to Spain to collaborate with preparations on the Magellan voyage and were placed at the disposition of the Casa. After the return of the survivors of the expedition, an even greater movement of Portuguese cartographers to Spain occurred and with certain selling secrets of state along with maps to the benefit of Spain.

The two great Portuguese cartographers of this age were the father and son team of Pedro and Jorge Reinel. It is via these two individuals that a unification between Portuguese and Spanish mapmaking occurred between 1520 and 1523. As mentioned below, the French Dieppe school would also benefit from the *mappamundi* of Diogo Ribeiro (1529) and the map of Gaspar Viegas (1534), but it would be the Reinel maps which served as the prototype for the Dieppe school of the first half of the century. The French school was more practical and less scientifically inclined, at least ignoring the application of new science in cartography then developing north of the Alps (Anthiaume 1916: 35–37; 43).

We should pause to consider that mapmaking proceeded from small to big or local to global, meaning that originally all navigational maps as with Mediterranean portolan, Arab dhow or Chinese junk watches measured short routes, perhaps coastal. But with longer-ranging journeys and with distance and space compounded, a sense of scale had to be added to keep the map manageable. Compass directions and rhumb lines were added. But typically regional maps were not mathematical and could not be fitted into a larger graticule. Such a development awaited more sophisticated measurement of latitude and longitude. Over even longer time, hydrographic charts linked with sea charts produced increasing geometric ocean to ocean charts. Globe makers actually worked the other way round, that is, from global to local. The world was a sphere; coordinates were indicated and regions had to be fitted in, sometimes in highly creative ways, when it came to fitting the new discoveries alongside a more formal Ptolemaic template. As Berggren and Jones (2002: 58–59) state the matter, with regional geography there was no need of a mathematical method but in the world of cartography, "this element takes absolute precedence." My reading is that just as the great discoveries encompassed vast distances requiring cartographic and navigational understanding superseding local and even regional-focused portolans, so ancient Ptolemaic conceptions were again awakened, hence we should not be surprised to learn they entered the curriculum of the Casa de Contrataci6n.

Map Secrecy

For the most part, the general diffusion of cartographic knowledge in the "age of discovery" was blocked by the need for state secrecy on the part of all the key stakeholders, Portugal, Spain, joined by Holland, England, and France. Notably, Portugal hid evidence of its expeditions to the Moluccas in order to avoid condemnation from Spain also holding out pretensions over the islands and with each power seeking to make the most of the line of division laid down by the famous Papal bull. The Sunda Islands of which Timor Island is the most easterly, as visited by the first Portuguese expedition to the Moluccas in 1511–12, were incorporated into the map by Jorge Reinel of 1516–20 but of such vagueness to be useless to pilots. In his *Tratado de Marinharia* of 1514, João de Lisboa censored out his route to the Moluccas via Singapore. Andre Pires also skipped the route in his 1515 "regimento de navigacion" (Anthiaume 1916: 37–38). Writing of Spain, as Portuondo (2009: 95–96) explains, secrecy and the protection of crucial geographical knowledge was essential to guard against a range of potential enemies, not only the Dutch and English, but heretics and plotters. Censorship of travel accounts and books was also part of the project (formalized by royal decree in 1556). Neither were foreign vessels welcome in Portuguese ports just as the Portuguese posted agents in European ports in an attempt to invigilate its doctrine of *mar clausum* or closed sea to foreign intruders (Anthiaume 1916: 417).

Maps and Atlases

While the route to Calicut in India and especially information on the Moluccas was closely guarded on pain of death, the restriction did not apply to general atlases. As alluded in chapter 3, the anonymous "Cantino planisphere" or world map of 1502 represented the first "latitude map," suggesting its revolutionary character. In the words of Barreto (n.d.: 28), the "Cantino planisphere" offered "the first representation of Asia which did not follow the Ptolemaic tradition" (Barreto n.d.: 28). Such would include representations of the Bay of Bengal, the Malay Peninsula and the coast of China, as well as indications of place names (including "land of the Chins"), trade goods and latitude. To be sure, as Guerreiro (1993: 468) summarizes, although not completely liberated from the Ptolemaic force of tradition, Cantino certainly amended the Alexandrian by adding a New World. Major innovation is also true of Africa and the Indian Ocean littoral, including the triangular shape of India and the Bay of Bengal but not absolutely because the Malay Peninsula and the Indochina Peninsula are still run together and appear as an elongated Golden Chersonese. As revealed, the sea route to China was now open. Malacca is named and located as a thriving port and the coast of China is well delineated. Of course, such an advance was only possible through exposure to Arab, Persian, Gujarati, Malabar, Malay, Javanese, and Chinese knowledge, because at the time of its composition in Lisbon, the Portuguese had yet to proceed directly beyond the coast of India.

Drawn in Lisbon in September 1502, the Cantino planisphere was also subject to an early act of cartographic intelligence, having been spirited out of

Portugal by Alberto Cantino an agent of the Duke of Ferrara, hence the name. The Cantino planisphere is also outstanding for a number of features. As the earliest extant example of the latitude chart, it thus departs from the older tradition of portolan charts of the Mediterranean constructed on the basis of magnetic courses and estimated distances between places. Another innovation of the Cantino planisphere is the construction of the rhumb lines system with the western circle centered on Cape Verde, a central circle centered on India, and an eastern circle centered on *Oceanis Orientalist*, somewhere in the South China Sea. The depiction of the Equator, the tropics, and the Arctic Circle suggests that places are represented according to observed latitudes. Latitudes were incorporated only in the coasts of Africa, Brazil, and India, while Europe and the Caribbean Sea continued to be represented according to the portolan chart model. The rhumb line system was used in navigation as a reference, for reading and marking directions (courses) between places (Gaspar 2012: 181–182).

According to Barreto (n.d.: 5), such an evolution permitted Portuguese cartographers to advance over the use of portolan charts which typically represented coastal features in accordance with the dead reckoning employed in coastal navigation. The new Portuguese sailing charts that evolved in tandem with Atlantic explorations combined the traditional grid of magnetic courses with the introduction of scales of latitude. In Barreto's (n.d.: 5) estimation, the introduction of scales of latitude was "the great technical innovation made by the Portuguese in sixteenth century cartography." Nevertheless, other innovations would follow, such as in indicating horizontally how coastlines appeared as well as the practice of plumbing and recording depths.

Overall, as Gaspar (2013: 69) estimates, the Cantino planisphere set a new cartographic standard promptly adopted on many other world maps of the time, as with the Reinel chart of ca. 1504, albeit with a double latitude scale. But outside of Lisbon, important contributions were also made, as with the Goa-based Fernão Vaz Dourado along with Lázaro Luís, and Manuel Godinho de Erédia.

To illustrate this advance, the planisphere attributed to Jorge Reinel ca. 1519 is noteworthy. The Malay Peninsula, Sumatra, and the south coast of China are roughly drawn. On the planisphere attributed to Diogo Ribeiro ca. 1527, the Champa coast of Vietnam, Java, and the spice islands of the Moluccas make an appearance. In his planisphere version of 1529, China, which previously had been designated on European maps as Cathay or Upper India, is noted, along with Tamão (Tunmen), the trading post in the Pearl River estuary visited in 1513 by Jorge Alvares, the first-arriving Portuguese. Timor island makes an appearance in an anonymous Portuguese planisphere of 1545, noted also for the introduction of lines of latitude and longitude (Barreto n.d.; Barreto & Martins 1999).

Roteiros and Conquest Atlases

To recapitulate, *roteiro* or rutter were written sailing directions devised by early Genoese for coastal sailing on the Mediterranean. *Il compasso da navigare*, written ca. 1250 and published in Genoa in 1296 was exceptional as they were often memorized. They accompanied or formed the basis for portolan charts, later adopted by the Iberians for long-distance oceanic sailing. Portuguese *roteiro* comprise a special genre, as with the *roteiro* of Vasco da Gama; *roteiro* of the coasts of Portugal; the *roteiro* of the last voyage of Fernão Mendes Pinto from Goa to Japan in 1554–56; the *roteiro* of Portuguese voyages to the Red Sea; *roteiro* of Brazil and, as discussed below, the first *roteiro* of the coast of India from Goa to Diu, 1538–39, which is an appendix to the *Roteiro de Lisboa a Goa, 1538*, by Joâo de Castro (republished in 1882 from a manuscript discovered in Evora). As discussed below, the genre continued with Manoel Pimentel's state-of-the-art *Arte de Navegar* (1681; 1712).

It is easy to see how the early *roteiro* of the Indian Ocean region serviced future voyages of conquest and commerce, later morphing into cartographic representations. There are several map and graphic atlases depicting this space. One of the first, if not the first is Erédia's (1563–1623) *Plantas de praças das conquistas de Portugal: feytas por ordem de Ruy Lourenço de Tavora Vizo rey da India.*[11] Dated 1610, this album features twenty aquarelles including in west to east sequence the *fortalezas* of Mozambique Island, Muscat, Hormuz, Diu, Daman, Chaul, Goa; and to the south Mangalor, Cananor, Cranganore, Cochin; and across the Bay of Bengal Aceh and Malacca. An additional map named "Platasformas do Citio de Malacca," supplies rhumb lines departing from a single wind rose, a feature shared with the Goa map. The Malacca and Goa maps exclusively offer scale (in Spanish leagues). The Goa map exclusively displays coordinates. Notably, Goa is situated at a latitude of 15°45' north (close to the modern measurement). The atlas is closed with a small index listing of the maps.

Another rendition of this Portugalized world as of the late sixteenth century appeared in a series of *Cartas* and *Plantas* composed in Portugal. One of these is the *Livros das Plantas das Fortalezas, Cidades e Povoaçoes do Estado da India Oriental,* dated to 1635 by Visconde de Santarém and unfolding a west to east visual panorama or narrative of conquest and domination (Silveira 1991). Unlike general atlases, the *Livros das Plantas* depicts focused scenes, mostly *fortaleza* but certain with surroundings and certain showing regions or localities. Toponyms are Portuguese although Spanish is added to the map of Formosa. The codex also bears some annotations in English language. Undoubtedly the cartographer drew certain information from Spanish, English, and Dutch sources, though Portuguese sources predominate. More generally, the *cartas* are in the style of portolans with magnetic compass bearings radiating from one or two wind rose stars. *Legoas* or scale is added.

Portuguese toponyms are imposed upon coastal locations, rivers, and islands, especially, but less so with inland locations (Silveira 1991). Other versions of this genre exist as with the *Livro das Plantas de todas as fortalezas, cidades e povoaçoens do Estado da Índia Oriental* of António Bocarro (1635).

To further embellish, the trailblazing Portuguese cartography following in the wake of Cabral, da Gama, and others also began to change the Ptolemaic template, somewhat in some cases, radical in others. Notable was the *Atlas Nautique du Monde,* or the Miller Atlas, with Lopo Homem the presumed author (also allowing contributions from Pedro Reinel, his son Jorge, and artist Gregorio Lopes). Dating from 1519, it is understood that the map was commissioned by King Manuel of Portugal for Francis I (r. 1515–47) of France. Following the BnF description, it carries map depictions of the north-east Atlantic Ocean and northern Europe; the central Atlantic Ocean with the Azores; the northern Indian Ocean with Arabia and India; the south Indian Ocean with *insulinde* (East Indies) to the right and Madagascar to the left; *Magnus Sinus*; the China Sea with the Moluccas; and the southwest Atlantic with Brazil.

As Suárez (1999: 201) comments upon the Lopo Homem-Miller Atlas; the erroneous diagonal alignment of islands ranged east of Sumatra may be rooted in Arabic navigational texts. To be sure, this is a highly stylized rendition, including many arbitrary island placements. With its rhumb lines bursting out of nodal wind rose points, it is more reminiscent of the Catalan Atlas of a century and a half earlier than the map produced two decades later by Diogo Ribeiro. More generally, this baroque Atlas falls into a hybrid portolan style adapted to *mappaemundi* scale. Region is framed as with the *Magnus Sinus.* The China Sea is revealed as with the Moluccas but at a very high level of generalization, as with many of the mixed portolan genre. Additionally, Lopo Homem's *Planisfero* (world map) of 1554, with the original held in the Instituto e Museo di Storia della Scienze in Florence, offers a number of trading places frequented by the Portuguese on the coast of China in the 1530s to the 1540s, as with Liampo a site possibly related to the Ningbo maritime region in present-day Zhejiang province. Additionally, Chinchew (Quanzhou) in Fujian, another trading port frequented by the Portuguese in the 1540s, is identified (Barreto and Martins 1999: 20).

As with a number of other maps, including that of Ottoman admiral and mapmaker Piri Reis (ca. 1515), a Ptolemaic geographical conception of a land connection between "Southeast Asia" and the "southern continent" appears also to be present or at least implied on a map of the Moluccas in the Lopo Homem-Reinel Miller Atlas (Suárez 1999: 201; 221; McIntosh 2000: 50). Notably, the Atlas adds the coastline of Brazil to an encircling southern continent ("*Mundus Novus Brazil*" linked to the Asian continent by an imaginary continent identified generically as "*Mundus Nouvu*"), enclosing the Indian

Ocean in the classic Ptolemaic version, although Africa is now a continent and with the sea route to India not only feasible, but already accomplished by da Gama. Sumatra and adjacent islands display more form, as does Ceylon. As discussed in chapter 3, the Behaim Globe, believed to have been consulted by Magellan, updated information on the Portuguese voyages in the Atlantic and the sea route around southern Africa to India.

Known for the abandonment of Ptolemaic influence in cartographic representation, especially in the light of the Portuguese discoveries, the Magellan voyage included, Fernão Vaz Dourado (1520–ca. 1580) set a new standard. Among other charts and atlases, he was author of a codex made in Goa in 1571 commissioned by Francisco da Costa, a captain of Malacca in 1564. Dourado included the following geographic maps, folio 3: the coast of Africa, from Guinea to São Tomé; folio 4: from the Cape of Good Hope to the equinoctial east to west; folio 5: islands south of the equinox; folio 6: the coast of India up to Bengal; folio 7: from Cape Comorim, Japan, the Moluccas and north; folio 8: Pegu and all the coast discovered by the Magellan expedition up to the coast of Java; folio 9: the coast discovered by Magellan; Technically, in an age before Mercator's famous projection, he plotted latitudes and magnetic directions directly onto a plane with a constant scale, as if the Earth were flat. As with folio 16: rules of solar declination, and tide tables; folios 17 and 18: board of solar declination for four years, he did not neglect celestial observation and the need to collect data. Using color, his maps also encoded a fair amount of visual information and nomenclature. Besides specific maps of Japan and Korea, one of the six atlases he produced is the first to show Macau (Barreto and Martins 1999: 25).[12]

CONCLUSION

Essentially, this chapter sought to elucidate impacts with respect to the construction of regions following the Iberian discoveries and how knowledge of this meeting across civilizations was received back in Europe. As explained, Portuguese and Spanish mastery of the nautical astrolabe, cross-staff, quadrant, and other technical innovations, many of Arabic origin or derivation, was crucial to the shift accompanying Atlantic navigation alongside the appearance in fifteenth-century Europe of stellar navigation, necessary for sailing beyond sight of coastlines. Such an evolution permitted Portuguese cartographers to advance over the use of portolan charts which, as explained, typically represented coastal features in accordance with the dead reckoning employed in coastal navigation. The new Portuguese sailing charts that evolved in tandem with Atlantic explorations combined the traditional grid of magnetic courses with the introduction of scales of latitude, such as lauded by Barreto (n.d.: 5) as a major technical innovation made by the Portuguese in sixteenth-century cartography.

All told, by the 1560s a fairly complete picture of Asian coastal geography had been assembled in Lisbon. This we saw through the lens of Portuguese mapmakers. Marking this general progression, we revealed how prototype Mediterranean *roteiro* morphed into portolan maps of the Indian Ocean and even the China Sea/western Pacific servicing future voyages of conquest and commerce. Cartographic representations followed suit. It is now possible to see how the European territorialization of Asian space proceeded in tandem with colonial "expansion," the preferred term adopted by many modern European historians. Nowhere was this better expressed than in the production of atlases, *mappaemundi*, and coastal charts of which the Iberians were undoubtedly pioneers. As the navigational science of map coordinates and grids developed apace, so it would appear the increasing irrelevance and even redundancy of the age-old Ptolemaic template (although, as demonstrated in following chapters, not entirely in other realms).

NOTES

1. This is not well documented but see Agência Nacional para a Cultura Científica, Ciência Viva, Descobertas e Invenções Portuguesas (in Portuguese) http://www.cienciaviva.pt/projectos/concluidos/inventions/instrumentos.asp

2. The *Tratados da Esphera* could refer to the influential work of João de Sacrobosco dating from the thirteenth century or, more likely, the version by Pedro Nunes of 1537.

3. It is also believed that the phenomenon was theoretically determined by the Arab geographer Abu'l-Fida, as introduced in chapter 4.

4. As with Columbus, the Magellan historiography is long, especially if we go back to the first sixteenth-century accounts of the voyage as with Maximilian of Transylvania, *De Moluccis insulis* (1523). Modern scholarship also balloons. The following is but a sample, namely, Guillemard (1971); Meltzer (2001); Nowell (1962); and Stefoff (1990).

5. Skelton (1969) offers an English translation (with facsimile volume) of a French version of Pigafetta's lost original journal *Relazione del primo viaggio intorno al mondo* (Relation on the First Voyage around the World), dated about 1525 and now owned by Yale University. Skelton also discusses three other surviving "copies"—two in French, one in Italian.

6. Transcribed and translated in Marie Armand Pascal d'Avezac, *Martin Hylacomylus Waltzemüller, ses ouvrages et ses collaborateurs* (Paris, Challamel, 1867, 39).

7. Needless to say, the Waldseemüller map has generated wide scholarly discussion since its modern discovery in 1901 and its removal to the Library of Congress in 2001. See <http://www.loc.gov/rr/geogmap/waldexh.html>. See Nunn (1932: 57,

12–13, 49–51), pointing to the theoretical thrust of the Waldseemüller project, as opposed to the empirical. The scholarship continues as with Lester (2010).

8. A 1604 Japanese copy of the 1602 Chinese map *Kunyu Wanguo Quantu* depicts 墨瓦喇尼加 *mòwǎlǎníjiā* ("Magellanica") as a large continent in the South. This, however, is derivative of the Ricci-Ortelius template. For another example, see Figure 6.1.

9. Fondazione Istituto Internazionale di Storia Economica e Sociale "F. Datini" http://www.istitutodatini.it/biblio/images/it/lazzer/munster/htm/carte.htm

10. The Iberian literature on the conquest of the Moluccas is summarized in Lach and Van Kley (Vol. 3, Book I, 1993: 311–12) beginning with Bartolomé Leonardo de Argensola's, *Conquista de las Islas Malucas* (1609).

11. This album is presently located in the Biblioteca Nacional do Brasil.

12. Yet another Atlas in this vogue was the "Livro universal das navegações do mundo do capitão Francisco Ferroni," by João Teixeira Albernaz (d. 1652). Executed in 1643, it draws upon his various other maps and atlases to offer a hydrographic or navigational spin. Lisbon-born, he learned map-making from his father, Luís or Ludovico Teixeira (Arquivo Nacional Torre do Tombo 2014).

Chapter 5

The European Discoveries and the Print Revolution Reprised (1500–1700)

Europe's "Gutenberg revolution," publication in the vernacular, and the more general dissemination of knowledge of the great discoveries not only fed an appetite for even more discovery literature, in part linked with early colonial expansion, but also challenged many late medieval and neoclassical views of the universe. The travel and travel collection literature emerging in Europe from circa the 1560s through to the 1640s is large but still finite compared to texts wielded by Enlightenment-era authors of the following century. Nevertheless, as this chapter explores, several seminal texts and accompanying map images from the pioneering era of print lend themselves to deeper exploration, especially as to their precocious Renaissance framings of world regions.

This chapter examines four-plus-one such framings. The first is a collection of travel voyages with commentaries, practically a pioneer of the genre, that of the Venetian, Giovanni Battista Ramusio, drawing upon late medieval and early Iberian journeys. A second section introduces Portuguese poet laureate Luís Vaz de Camões's *Os Lusíadas* (1572), a work whose literary "itinerary" East bears referencing against Fernão Mendes Pinto's hyped "real-world" *Peregrinação* (1614). The third framing, a single-authored synthetic text, is that of Englishman Peter Heylyn (Heylin) entering print a half century on but also drawing upon the English and Dutch voyages, just as knowledge of the New World massified in the intervening decades. Emerging from the first wave of the Portuguese "discoveries" of the Far East in the mid-to late sixteenth century, if taken together these works are singular in the way of imaginatively mapping new "Oriental" worlds. A discussion around fantastic or imagined worlds and continents is added as a foil to illustrate the point that the late Renaissance mind had not entirely abandoned premodern notions and superstitions or at least sought escape from the strait-jacket of convention.

In sum, this chapter seeks to draw out certain conclusions as to the longevity of the Ptolemaic template in the face of startling new "choreographic" evidence. This it does by examining the vanguard role of Venice in servicing the print revelation of new discoveries, the Spanish conquest of New Worlds alongside that of Portugal also reconnecting with the ancient maritime silk roads, as well as the sequential role of England and the Low Countries in positioning themselves as platforms for the winning of global supremacy over Iberian rivals. With particular reference to Asia, it also seeks insights into how a regional and subregional template came to be applied.

RAMUSIO AND THE VENETIAN PRIORITY

In discussing the achievements of Giovanni Battista Ramusio we cannot ignore his native Venice or the general context of the Mediterranean hub at a particular conjuncture in European world history. It was in Venice where his father, Paolo, was a magistrate and where he started his career in 1505 as the personal secretary of Alvise Mocenigo, Venice's ambassador to France, going on to hold office in the employ of the Venetian state. As Jerome R. Barnes (2007: 9; 45, 50) points out in a dedicated study of Ramusio, Venice from 1503 had been in full retreat from the Turks. At the same time, the Portuguese, Spanish, and French were surpassing Venice as a maritime power. Venice had a prior history of exploration through the Straits of Gibraltar and down the coast of Africa, arriving in Cape Verde in 1455, and Ramusio was well aware of the fact.

Although the Gutenberg revolution had its origins north of the Alps, by 1482 Venice was the printing capital of the world (and with the art of paper making introduced to Europe from its point of origin in China by Muslims in Iberia and Sicily and firmly established in Spain by the twelfth century). Italy was also the first to revive an interest in classical geography during the Renaissance and, as mentioned in chapter 3, the first revival editions of Ptolemy were printed in Rome, Bologna, and Florence. From around 1544 Rome and Venice were both important centers of cartographic activity. An important cartographic school was also developed in Ancona on the Adriatic by navigator cartographer Grazioso Benincas, the first (in 1468) to show the West African coast down to Liberia, along with the Cape Verde Islands. As mentioned below, Venice-based Giocomo Gastaldi was another exponent, specializing in New World mapping through the medium of copper engraving. Yet another was Antonio Lafreri publisher of composite atlases around 1570, contemporaneous with the works of the leading Antwerp mapmakers. One product of the age was the world map by Giovanni Paolo Cimerlinus (Venice, 1566), drawn on a cordiform or heart-shaped projection, which allows the world to be illustrated in one hemisphere.

The vanguard role of Venice in Mediterranean seafaring has been noted and with Venetian merchants also connecting up with markets in the Low Countries directly. Notable in this sense is the role of Alvise Cadamosto (ca. 1432–83), a Venetian trader-adventurer en route to Flanders (second voyage on a galley in the employ of Marc Zen), who was engaged by Henry the Navigator to explore the West African Coast. Supplied with a caravel, this he did via Madeira and the Canaries in two known journeys in 1455 and 1456, accompanied by the Genoese captain Antoniotto. Together, they are credited with the discovery of the Cape Verde Islands and the Gambia River area along the Guinea coast, among other coastal points. As Lach (1965: 158–159) points out, Cadamosto's account is particularly valuable insofar as practically nothing remains of the contemporary written record of the Henrican discoveries.

Ramusio and the Discoveries

As Barnes (2007: 9; 45; 50) reveals, in preparing his three volume *Navigationi et Viaggi* (Navigations and Travels) published in Venice between 1550 and 1559, Ramusio already had some established models at hand. One was Francazano Montalboddo's *Paesi nouamente retrouati et Nouo Mondo da Alberico Vesputio Florentino intitulato*, 1507, a travel collection written in the vernacular Italian as opposed to Latin.[1] Even prior to preparing *Navigationi,* Ramusio collaborated with Peter Martyr and Gonzalo Oviedo on their *Summario de la General Historia de l'Indie Occidentali*, published in Venice in 1534 (and with both authors pioneer specialists on the Spanish discoveries). Ramusio was also undoubtedly au courant with travel itineraries collected by Alessandro Zorzi at Venice between 1519 and 1524, which included not only that of Fernando Colon but also accounts of Ethiopia.

Content Analysis of Navigationi et Viaggi

With the first volume of *Navigationi* issued in 15505, followed by the third volume in 1556, the second volume only appeared in 1559, owing to a fire which destroyed the manuscript prior to being sent to the printer. Translated into several languages and reprinted in Venice a number of times, *Navigationi* undoubtedly gained major attention across Europe. Importantly, the Ramusio collection offers the first complete account of such late fifteenth-century voyages as those of Cadamosto, Columbus, da Gama, along with Jacques Cartier to the eastern seaboard of what is now Canada in the 1530s and 1540s. The first volume relates almost entirely to Africa, including the description by the Muslim merchant traveler, Leo Africanus, of the "Barbary" coast. The second relates to Asia and includes the accounts of Polo, de' Conti, Giosafat Barbaro (the Venetian ambassador to Persia), and Magellan.

The third volume relates to America matched by the earliest obtainable map to accurately depict the Americas, including such accounts as those of Martyr, Oviedo, and Cortes.

In publishing the account of the Magellan circumnavigation by Antonio Pigafetta, Ramusio achieved a priority. Though deemed highly corrupt as to detail, readers were nevertheless introduced to the Antarctic Pole, the *mar oceano*, the *mar Pacifico,* the Archipelago de San Lazaro (as Magellan dubbed the Philippines), the Sultanate of Brunei, the Moluccas, along with the Tidor-Ternate polities engaging complex Portuguese relations at the source of the fabled spice trade, Timor Island, source of prized sandalwood, and a cornucopia of anthropological and navigational lore, not excepting the revelation by Pigafetta that the westward-sailing *Victoria* lost a day in the circumnavigation.

Ramusio also published an excerpt of Tomé Pires's *Suma Oriental* (1512–15), a highly factual account of the Portuguese maritime trade on the Indian coast, the Indies (including the spice islands), and the route to China, and with Malacca-based Pires accompanying the first Portuguese mission to Beijing in 1516. Obtained in Lisbon between 1525 and 1528 by Andrea Navagro, Ramusio's emissary, the manuscript lacked both Pires's name and, crucially, the section dealing with the Indonesian archipelago and the spice trade (facts of secrecy).[2] As Rui Loueriro (1996: 32–33) explains, undoubtedly at a time of extreme diplomatic tension between Portugal and Spain over the Treaty of Tordesillas, Lisbon was unlikely to have released such a critical document.

Ramusio's Commentaries According to Barnes

As Barnes (2007: 47–48) points out, little attention in the literature on Ramusio has focused upon what he actually wrote. In fact he made some twenty commentaries, from one to twenty pages in length and with several dealing with trade issues. The most important of these is a section in Volume I, mostly dedicated to Asia and Africa, titled *Viaggio Verso La India Orientale Descittoda Plinio* (Travels in the Vicinity of the East Indies as Described by Pliny) contained inside a longer commentary "*Discorso Spra li Viaggi dell Spetier.*" Barnes terms this section as probably Ramusio's "finest writing" both dealing with Pliny and the then present Portuguese role. Relying upon Pliny as a source and entry point into the discussion on the spice trade, he also shows how the Portuguese entered this traffic.

On Venice

As Barnes (2007: 81–83) points out, while the ruins of the Roman empire at the hands of the barbarians brought an end to the ancient sea route trade between India and the Mediterranean, Ramusio reminds readers that the trade was in fact revived by Genoese through the Caspian Sea route trade, based mainly at Tana. He writes in a way that emphasizes the "limitations" of this trade, as with the difficulty of getting products to market via river traffic alongside political difficulties. He structures this discussion in the way of highlighting the achievements of the Portuguese alongside the difficulties encountered by the Venetian merchant trade networks, holding out perhaps a model for Venice to emulate. "Part of the "discovery" was the "rediscovery" of the ancient knowledge of the Indian Ocean . . . The Portuguese are upheld as an example of what could be achieved." The Portuguese routes were there to follow, just as the Italians and especially Venetians had achieved their own discoveries over lands and seas. Only the will was needed.

Ramusio's Use of Maps

In discussing Ramusio's use of maps, it is important to understand that he simply did not entertain the reproduction of Ptolemaic maps (even if his map-maker came from such a background). As Keith Lilley (2013: 157) points out (and as taken up below), Ramusio was part of a "mapping through words" tradition reaching back to Pliny. At the same time, his interest in cartography cannot be denied. In this regard, as Barnes (2007: 3) points out, [Ramusio] "seems to have an appreciation of Ptolemy's distinction between the cartography and chorography included on a map."

As partner in the production of *Navigationi*, Ramusio chose mapmaker Giacomo Gastaldi, a Piedmontese in the employ of the Venetian state. A pioneer in the use of copperplate maps as opposed to woodblock, Gastaldi also prepared the maps for the 1548 Venice edition of Ptolemy's *Geographia* (see MapForum.com 2014). Notwithstanding a background in the classics, he was one of the earliest to introduce the Portuguese discoveries in Asia, as with his map reproduced in Girolamo Ruscelli's *India Tercera Nuova Tavola* (Venice, 1598). Though still a transitional work, Gastaldi's map is actually foundational in setting a template for what would later be glossed as Southeast Asia, albeit allowing for the Bay of Bengal and the coast of China.

As Barnes (2007: 216) declares, "The ability of Gastaldi and Ramusio to use maps and texts together was innovative." The production of a single set of maps from multiple sources was probably the first in cartography. In fact it was the nature of the multiple sources which distinguished Ramusio (and Gastaldi) from Ptolemy, allowing both to "displace" Ptolemy, as discussed below. As Barnes (2007: 218) asserts, "it was inconceivable that Ramusio could have greater knowledge than Ptolemy, Strabo, Plato and Aristotle, yet it was a reality." As his publisher, Lucantonio Giunti, pointed out, the ancient scholars were becoming irrelevant in the field of geography. Nevertheless, Ramusio frequently refers to the classics as with Ptolemy and Strabo alongside Renaissance scholars. He was also acutely aware of the role of Rome in the ancient overseas trade.

Even without the complete *Suma Oriental* manuscript of Tomé Pires, the knowledge added on the maritime route spanning the western seaboard of China, the Bay of Bengal region, and the importance of the Malacca trading hub reaching to Vietnam and China, is extraordinary. Accordingly, Ramusio's second edition of *Navigationi* (1554) also incorporated the Duarte Barbossa and João de Barros narratives touching the sensitive Sumatra, Java, and Moluccas region. Importantly, the 1555 edition of *Navigationi* (repeated 1563–59–65) (Vol. 1) contains a map named *Terza Ostro Tavola*, essentially a rendition of an enlarged Southeast Asia centered upon the Malay Peninsula but in inverted or southern orientation.

Ramusio and Ptolemy

The question arises as to whether Ramusio radically overcame or just advanced upon Ptolemy certainly merits attention. As Barnes (2007: 2; 104) points out, Ptolemy's *Geographia* had only been "rediscovered" barely a century before Ramusio's work and that its first printing had only appeared fifty years prior to Ramusio's map. Therefore the influence of Ptolemy and the classical tradition upon Ramusio and his generation was obvious. The influence of Ptolemy can be seen in the embrace by Ramusio/Gastaldi of lines of longitude and latitude "one of the great distinguishers" between medieval and modern mapping. His charts in Volume 1 were much in the Ptolemaic style (Barnes 2007: 99). This much is true but, as alluded below, it is important as well to uncouple Gastaldi the graphic mapmaker and Ramusio the textual mapmaker.

To further make his point as to the weight of the classical tradition in *Navigationi,* Barnes (2007: 74) contends that Ramusio gets "very specific in listing the exact locations of the Roman voyages," making a direct comparison with the Portuguese voyages in the second part of his discussion on the spice trade. Pliny is used to reference and describe the ancient trade to India by way of Egypt and the Red Sea. Ramusio measures the distance of the ancient trade right through to the Southeast Asia area. The ancient Suez canal favored this and he writes of the efforts of the ancient Egyptians and Romans to facilitate trade with the Indian Ocean and, citing Pliny and Strabo, observed how they built canals to the Red Sea from the Nile to facilitate the trade. Ramusio writes that the locations of Africa and India "named by Pliny are the same ones where today practice the Portuguese" (Barnes 2007: 84). In a sense, the practice of referencing Pliny and Strabo would die hard, indeed was long lived and even survives in some archaeological contexts.

As Barnes (2007: 50) interprets, Ramusio well understood that Ptolemy's tables as they related to Africa and India were "quite imperfect, to the knowledge we now have of the region." Further, as Barnes (2007: 66) embellishes, Ramusio must have relished the opportunity to make use of the tools provided by Ptolemy to actually locate a land "unknown to Ptolemy himself." For Ramusio and his emulators, the revived Ptolemy presented a secular method or way forward to advance over long-standing medieval stereotyping of knowledge based upon religious myth. In this interpretation, Ramusio undoubtedly followed Ptolemy in making a distinction between the "geography" of the map as scientific and the qualitative "chorography." This is an important point and allows us to see the role of Ramusio as a transitional figure in the progression of knowledge.

In somewhat contradictory fashion, as Small argues (2013: 153), Ramusio "ultimately resorted to a textual mapping of the world." It was not that he was disinterested in maps but just that Ptolemaic type maps did not reflect the changing world such as he knew from his Venetian associations. While Ptolemy was highly esteemed in Ramusio's age (and Ramusio was no exception), and while his theories on how to project a sphere onto a plane surface and to divide the world into a longitudinal and latitudinal grid was still relevant, his data was flawed and this was becoming increasingly obvious (Small 2013: 158). Both radical and traditional at the same time, Ramusio's textual mapping displaced Ptolemy's mathematical and cartographic model (in the way of practical mariner's charts it might be added). While he published maps, these were emphatically *not* Ptolemaic maps. In a world of flux, he updated his new editions with new knowledge. As a transitional figure preceding the Hakylut, Purchas, and van Linschoten travel collections by decades, Ramusio's *Navigationi* would, in turn, be superseded by even newer empirical knowledge entering Europe, as with travel accounts from India, China, Japan, and the islands entering Jesuit letters sent by the missionaries. Echoing Small (2013: 153), as this small example illustrates, we should be alert to broad distinctions between cartographic and narrative forms of geographic descriptions across the ages, even if the classical and revived classical tradition embraced both.

TWO PORTUGUESE DISCOVERY NARRATIVES: *OS LUSÍADAS* (1572) AND *PEREGRINAÇÃO* (1614)

In the wake of the epochal Vasco da Gama voyage to India (1497–98)—at least for the victors in this age of conquest and plunder—direct confrontation with Muslim power including the Portuguese capture of Malacca in 1511, and entry into the China seas, Portugal in Europe emerged as a broker on information on the Orient in much the way that Venice served this purpose in prior centuries. In part owing to secrecy not too much hard data entered print, although some of it was leaked by rivals as evidenced by Van Linschoten's publications (and maps), and with part of the *Suma Oriental* of Tomé Pires printed by Hakluyt. The turgid chronicle style writing of João de Barros's *Decadas de Asia* (1552; 1553; 1563; 1615) held little appeal but what did strike resonance in late Renaissance Portugal was the writing of two of its sons who had returned from the Orient and both of whom produced enduring epics, respectively Luis Vaz de Camões and Fernão Mendes Pinto.

Luis Vaz de Camões's *Os Lusíadas* (1572)

No other work quite symbolizes the geographical imagining by Europe of Asia in the wake of the "discoveries" than *Os Lusíadas,* the epic poem of the Portuguese Luis Vaz de Camões (1524–80), first entering print in Lisbon in 1572. Composed in ten *cantos* or chapters, *Lusiadas* is modelled upon Virgil's epic but with the heroes Portuguese and the subject matter da Gama's voyage to India. As one translator observes, the poet's experience of da Gama's trailblazing voyage over a half century later "supplies a thousand intimate touches" (White 1997: 10). As the first acclaimed European artist to travel East, Camões's exact itinerary remains obscure, although it is known that he returned to Lisbon in 1570 after seventeen years of picaresque travels. Whether or not the poet sojourned in Macau is a question that has generated more heat than light over the centuries.

Publication and Description

Following the first edition of 1572, *Lusiadas* then appeared in several subsequent Portuguese editions alongside Spanish, French, Italian, and Latin, suggesting a quick and ready reception. Long time would pass before the appearance of the first English edition by Richard Fanshaw (1655), the Restoration ambassador to Portugal. Held in some esteem, his effort was followed by a long line of translations, with an upsurge in the nineteenth century and, in the words of Michael Freeman (2000: 222), each version palpably more difficult and removed from the cultural-literary context of its writing with the passage of time.

Preceded and informed by de Barros's *Decadas, Lusiadas* was not only a peon to Portuguese exploits in the East, but also offered a description of the flora and fauna of Asia alongside ethnographic details of Asian peoples. However, it is the geography of the region informed by Camões's own experiences as well as familiarity with Ptolemy and de Barros that we seek to highlight. In other words, while *Lusiadas* can and has also been read as a national Portuguese poem par excellence in defense of home and empire or even as singing the praise of Christian vigilance against an Islamic other, we are concerned here with the spatiality of the sprawling work and especially as to the poet's configuration of Asia.

One of the central themes of *Lusiadas*, a virtual "empire of print" in Freeman's (2000: 222) phrase, is da Gama's discovery of the maritime route to India. But, significantly, the itinerary is extended east of India to embrace what today is known as maritime and mainland Southeast Asia and the South China Sea area. Writers such as the Brazilian Isócrates de Oliveira (1975) have offered close detail on the iteration of Asian toponyms in the *Lusiadas* matched by historical context. Among the many names east of India "mapped" in *Lusiadas* we can include, Cape Comorin (Cori), also Taprobane revealed as Ceylon; the Malacca emporium; Sumatra Island; the Singapore Straits, Siam, and the Menam (Chao Phraya) River; Catigão (Chittagong); Tavai (Tavoy); Arakan; Pegu; Tenasserim, Kedah, Pahang, Patani; "names of a thousand nations" (a reference to tribal federations of the interior mainland); Laos, Ava, Brahmas; Cambodia on the Mekong; Champa on the coast, Cochinchina, Hainan, the route to the Chinese empire; China-Japan (directly associated with silver and trade); Tidore-Ternate; Banda, Borneo (camphor); and Timor (sandalwood). The Antarctic Pole is also referenced as is the Magellan circumnavigation.

Os Lusiadas as a Transitional Ptolemaic Text

A number of writers, such as Landeg White (1997), have also pointed out that Camões was ignorant of the theories of Copernicus, and it is the Ptolemaic system of concentric spheres which is expounded in Canto 10. As Shankar Raman (2001: 39) interprets, Camões's cosmos evidences a renewed sixteenth-century European concern with Ptolemy via Aristotelian astronomy. The work opens with cosmological references. On the voyage, the description of the cosmos by Tethys, sister and wife of Oceanus in Greek mythology, runs to fourteen verses. Vasco da Gama and his men were led to the top of a mountain where they behold a suspended globe (an astrolabe), a reduction of the world or the image of God's creation. Dubbed the *Máquina do Mundo* or "great machine of the world," as Raman (2001: 43) interprets, Camões' miniaturized cosmos "corresponds meticulously to the structure of the Aristotelian universe as it had been revived by certain Scholastics."

To elaborate, Camões's *Máquina do Mundo* demonstrates a fairly accurate knowledge of cosmogony of the time, even without the heliocentric conceptions of Copernicus. According to a modern Portuguese source, he may have obtained this knowledge from Pedro Nunes or from such works such as Sacrobosco's *De Sphaera.* The *World Machine* relates to the cosmos and to the world system. Derived from Aristotle-Ptolemy, the *World Machine* had the Earth at the center (the geocentric view) and was surrounded in concentric circles by the Moon, Mercury, Venus, the Sun, Mars, Jupiter, and Saturn, all rotating around the Earth. In the Ptolemaic view, the heavenly bodies remained suspended in crystalline translucent spheres with each larger one encompassing the others and with the furthest sphere the prime mover (*Infopédia* 2003–2014). In simple language, Camões *Máquina do Mundo* is a veritable Ptolemaic model of the universe with Portugal at the center. Yet, he may also have been informed by Abraham Zacuto, as introduced. In Canto IV, v. 94–104, Camões possibly offers a poetic interpretation of a meeting between Vasco da Gama and the Jewish astronomer (Buchmann 2015: 9). In Canto V stanza 25, Camões describes the "new astrolabe" as a "sage and wide invention."

Illustratively, in Canto V, Camões has the fearsome Adamastor character from Greek mythology proclaim, "I am that mighty hidden cape, called by you Portuguese the Cape of Storms, that neither Ptolemy, Pomponius, Strabo, Pliny nor any other of past times ever had knowledge of. This promontory of mine, jutting out toward the Antarctic Pole, marks the southern extremity of Africa." No other statement could be so triumphant, just as Adamastor was symbol of the forces of nature that Portuguese navigators had to overcome during their discoveries. While affirming the Alexandrian conception of the world, Camões is shy of transcending that vision no matter the new geographic knowledge falling into place in his age.

Ptolemy, Pliny, and other classical texts are also referenced with respect to the voyage down the West African Coast, as well as the Red Sea, Mecca, and Suez. In Canto I, stanza 45, as the navigators approach Mozambique Island, they pass a cape which Camões identifies as Prasso or Ptolemy's *Promontorium Prasum*, the furthest point south known to the Greeks. According to White (1997), here Camões is making an ideological point, namely that these sea routes and lands occupied by the Moors, otherwise conquered in European imagination long before Islam, were about to become a battle zone.

As Camões explains of the oversized Golden Chersonese or Peninsula of Ptolemy (54), "the noble island of Sumatra was said in ancient times to be joined to the mainland, until ocean breakers eating into the coast drove wedge between" (see Suarez 1999: 85). In describing the fragmentation of Sumatra from the Malay Peninsula, Camões evokes a biblical image of Ophir, inter alia associated in the classical world with a site in eastern Africa. However, with the Portuguese discoveries, Ophir came to be localized eastward linked with the Aurea Chersonese (and subsequently reinterpreted by the Spanish as having a Pacific Ocean location). As Oliveira (1975: 80) explains, in describing the tip of Singapore around which the caravels swung north, he evoked the "costa de Cinosura" or the constellation of Ursa Major (Polaris) known to the Greeks as indicating a passage north. By also indicating a curving route east, he offered a cartographically correct imaging of the Gulf of Siam or, the Ptolemaic *Sinus Magnus*?

One wonders as well, why Camões referenced Chiang Mai in the upper Menam basin, long since eclipsed by the maritime-oriented Ayutthaya kingdom that existed from 1351 to 1767. But by associating Chiang Mai with a lake of the same name, he was actually recycling a fiction set down in Ramusio/Gastaldi woodblock maps. As Suarez (1999: 153–56) muses, the conception of the fictitious Chiang Mai Lake (as with other inland lakes studding earlier European cartographic images of Eurasia), palpably emerged from Christo-Judaic beliefs in interior lakes as sources of major rivers linked with conceptions of an earthly paradise. Possibly, but Ming-era Chinese maps coming to Portuguese and Spanish attention such as mentioned below also included large inland Chinese lakes.

As Raman (2001: 60) points out, *Lusiadas* is a transitional work between the pagan/classical world and the future. However, he also sees the epic as caught between conflicting Catholic Christian visions, noting that later versions of *Lusiadas* had certain pagan allusions edited out by the Holy Office of the Inquisition, set up in Portugal in 1536. As he further points out, Camões hails the Portuguese as a people who had sailed the oceans "none had sailed before" and who sailed "past Taprobane's far limit." And so, the pagan gods connect Camões to the "antique" cosmos. Cosmos thus becomes a metaphor to "reinscribe" da Gama's inaugural voyage to India "as the sign of a new, self-assertive relationship between the Portuguese subject and the world," "in effect attempting to create a new world." But the stability of the ancient cosmos to which early Christianity adopted was, by Camões time, being eroded (none the less by the Reformation). But for Camões, voyage had come to replace ordered cosmos, just as the Earth would be remade and possessed by the colonizer. In other words, Camões is embedded in a theological debate engaging Reformation and Catholic Counter-Reformation, and with Black Death, peasant rebellion, widespread fear and anxiety, and loss of stability and order in the background (Raman 2001: 35–36). Lacking here is the context of defense of Latin Europe from the Moorish threat and, as White (1997: 19) points out, *Lusiadas* is full of otherings against the Muslim enemy, notwithstanding the Islamic heritage upon which the Iberian seafarers drew.

Fernão Mendes Pinto's *Peregrinação* as a Transcending Ptolemaic Text

In this discussion we cannot ignore another Asian itinerary, that of Fernão Mendes Pinto (1509/11–1583). With his *Peregrinação* entering print a generation later than Camões, Pinto offers an even more elaborate—and bewildering in its vagueness—toponymy of China, the China coast, Japan, and Southeast Asia.[3] Outside of the classical erudition of Camões, more than any Portuguese author of the age, Pinto was also part of a real world, no matter the confabulations entering his text. Possibly more so than Camões, his text was widely read in Europe in its age. Setting aside questions of editorial manipulation of his manuscript text prior to publication and license taken by Spanish and other translators, it is significant that a partial translation entered the Purchas collection of 1625 (as history) (see Santos Lopes 2010: 262). Unlike Camões, Pinto knew Macau and China firsthand. First, a letter written from Macau by then Jesuit Mendes Pinto dated November 20, 1555, survives and, second, the *Peregrinação* describes the protagonist setting out from Macau on his last voyage to Japan on this date.

Also unlike Camões, Pinto extends his spatiality to include the Ryukyus, the smaller Tanegashima, the larger Kyushu, Ningbo on China, and the broader China-Tartary area. Still, his account of China is both "utopian" and "conflicted" in the words of one translator (see Catz 1989: xi–xiii). As Suárez (1999: 86) points out, Pinto was one who went in search (in Southeast Asia) leading or misleading with his own partly-invented text descriptions. Yet mapmaker Ortelius would also reinvent silver islands as with his map, *Maris Pacifici* of 1589 placing them squarely—and accurately—in Japan. (As late as the mid-eighteenth century the Dutch sent ships in search of legendary silver and gold islands off the east coast of Honshu.)

But Pinto is also self-conscious that he and his generation of adventurers were pioneering in transcending the stereotypic Ptolemaic known world. This is explicitly stated. Notably, in his account of the Léquios Islands (Ryukyu), he declares, "according to what we have seen and read in the works of Ptolemy as well as the others who have written on geography, not one of them ever went beyond the kingdom of Siam and the island of Sumatra" (chap. 143). But the Portuguese, he alludes, "have gone a little further and have already described the Celebes, Papua, Mindanao, Champa, China and Japan, though they have not yet mentioned the Ryukyus" (chap. 143), of which he purported to have been the discoverer (cited in Loureiro 2010: 245).

According to Jin Guoping (2010: 319–320) who researched Pinto's sources on China, de Barros's *Decada II*, letters from Portuguese captives in Canton, and *Tratado das Cousas da China* by Gaspar da Cruz reveal elements of intertextual contact with *Peregrinação* (and this is confirmed by Rui Loureiro). In Jin's estimation, it is impossible to clarify the titles of Chinese books or chronicled that Pinto claimed to have consulted. But a likely title is *Zichi Tongjian* (Comprehensive Mirror to aid in Government) first printed in Hangzhou in 1086 and acquired by Portuguese in Fujian subsequently becoming part of a collection of Chinese classical works brought to Portugal lodged in the royal library of Escorial in Spain. But, conceivably, Chinese cartography could also have been consulted by Pinto in order to orientate his *roteiros* (routes) leading to inland places in China, namely his evocation of a purported inlet reaching from Nanjing to Beijing and the Great Wall.

The number of possible Chinese maps that Pinto (and, separately, de Barros) may have consulted to enhance their knowledge of China are various. One cited by Jin (2010: 327–328) is the *Guangyu Tu* (Comprehensive Atlas of China) by Luo Hongxian, 1555, drawn with traditional Chinese measuring methods to reveal a grid scale. As he notes, this comprehensive map presenting topographic features, administrative units, and frontier garrisons of the Ming dynasty, was the base for Jesuit Michele Ruggieri's *Atlante della Cina.* Another candidate map is the *Gujin Xingsheng Zhitu* (map of the ancient and modern strategic places) published in 1555, close to Yuegeng, the main East China Sea port of the "Chincheo" or Fujianese people frequented by the Portuguese arriving from Malacca.

By contrast, as Loureiro (2010: 252–253) suggests, Pinto's sense of China's spatiality was actually drawn from trailblazing Portuguese mapmakers as with the Diogo Homem (charts of 1558–1561), the Bartolomeu Velho (chart of 1561), the Lazáro Luís (chart of 1563), and the Fernão Vaz Dourado (chart of 1570–1580) (actually notable for its abandonment of Ptolemaic conventions). More certainly, Pinto used the Luís Jorge de Barbuda map of China, which represented the interior "with considerable accuracy" (a reference to his *Chinae . . . nova descriptio,* such as profiled in chapter 4. Even so, he concedes, both Barbuda (known under the Latinized name Ludovicus Georgius) and Pinto could have consulted a Chinese map. But, either way, his conceptualization of China should be seen more as a "reflection on sixteenth-century Portuguese representations of Asia than as an accurate depiction of the Asian continent."

More generally, maps of the Great Ming became known to Europe by diverse sources. The missionary route was one. But others were brought back to Europe by English, Dutch, Portuguese, and Spanish officials and seafarers. This was the case of the *Huang Ming yitong fang yu bei lan* (Comprehensive View of the Imperial Ming), brought to Europe by John Saris former head of the East India Company in Banten on Java, who had acquired it from a Fujianese merchant. Delivered to Richard Haklyut, the map was published in Purchas's *Hakluytus Pothumus or, Purchas his Pilgrimes* (1625). The Chinese characters representing local place names were removed and province-level names in English substituted as with Canton and with Macau added. To be sure, as Robert K. Batchelor (2013: 102) has commented, the Chinese merchant who possessed this map undoubtedly held to a sense of Chinese grandeur or empire just as the Ming themselves assembled data with broader East Asian strategies in mind. Another example of an imported Chinese map was the "Selden map" also brought to England by a merchant mariner. Even so, as revealed in chapter 8, the Spanish were likely the first to view a classic Ming-era map of China especially revealing for its detail. At a minimum, we can conclude of *Lusiadas* and *Peregrinação* that, separately, they evoked a veritable hybrid European-Asian toponymy from Africa to India and Asia, actually giving print fixity to many familiar names from Siam to Japan.

PETER HEYLYN'S *COSMOGRAPHY* AND THE ENGLISH PRIORITY

Having brought to the fore a pioneering collection of travel literature alongside a powerful epic work of imagination, albeit grounded in a classical framing of the cosmos, we turn in this section to a single-authored world history text, namely Peter Heylyn's (1600–62) *Cosmography in Four Books: Containing the Chorography and History of the Whole World, and all the Principal Kingdoms, Provinces, Seas, and Isles Thereof.* Departing from the collection genre carried through in England by Hakluyt, Purchas, and others, Heylyn's was a synthetic work of global history—a veritable "cosmography." Comprising 1,000 pages ranging over the entire world according to continent and region, we probably have to await the labors of French Enlightenment Encylopedists before this solo act was repeated. Heylyn's work was one of the earliest attempts to describe the entire world in English, offering the first known description of Australia and among the first of California. Published some half century after Ramusio, in an age when the great "discoveries" were well revealed, and with a new school of cartography well developed in Flanders, Heylyn's work also offers a convenient benchmark to test our thesis of Cartesian innovation in the face of neoclassical and medieval influences.

There is no doubting Heylyn's standpoint as clergyman and as patriot. As he boldly states in an introduction to the work, "I am an Englishman; and, what is some-what more, I am a church-man" (in other words, Church of England). At the age of eighteen, Heylyn was elected a fellow of Oxford University's Magdalen College where he taught historical geography. Largely drawn from his lectures at Oxford, in 1621 Heylyn published *Microcosmos, or a Little Description of the Great World*, which went through several editions, notwithstanding its "errors," but serving as a prototype to an expanded version. In 1624 he proceeded with theological studies, also becoming embroiled in theological disputes, and with his enemies removing him from ecclesiastical posts. As recalled in the introduction, in 1640 he was unjustly and ignominiously hauled before the "Committee for the Courts of Justice." Among other humiliations, he was enigmatically warned that "geography is better than divinity" and, in his own words and in a moment of self-doubt or renewed confidence, began to work on an enlarged and "corrected" edition of his *Microcosmos* called *Cosmographie* (1652).

Entering a first edition in 1652, the printer-publisher was Henry Seile of Fleet Street. Seile also collaborated with mapmakers. Upon Seile's death in 1662, his widow Anne issued another edition of *Cosmographie*, with frequent republication with text amendments (1657; 1660, 1666, 1669, 1670, 1673, 1674, 1677, and 1682), suggesting both the popularity of this work and its cross-generational influence. Unlike Ramusio and many other Renaissance and Enlightenment figures, Heylyn simply lacked sponsorship. As he wrote in the introduction, "Lastly, I am to tell the reader, that I have now no other Patron than himself to whom to dedicate this work and from who to receive protection and encouragement for it." In other words, in the absence of royal or even church patronage, he dedicated his work to the "public" (insofar as they could read). Striking a resonance with a burgeoning bourgeois print-literate market, Heylyn wrote for a living until blindness foreclosed that activity.

Engraved by Robert Vaughan, the allegorical engraved title page shows the creation of the world, the four continents symbolized three times by four women, four men, and four animals. In the center left and right are female figures representing Europe, Africa, Asia, and America, below which are corresponding male warriors, with horse, lion, elephant, and dog. Comprising four sections (each veritable separate books and separately numbered and indexed in some editions), *Cosmography* examines the geography, politics, and cultures of Europe, Asia, Africa, and America. I have consulted both the second edition of 1657 and the final edition of 1682 editions and with page numbers referring to the latter.

The work opens with a general introduction to geography. This is followed by two books on the history and geography of Europe. A third book deals with Asia, namely, "Chorography and History of the Lesser and Greater Asia and all the Principal Kingdoms, Provinces, Seas and Isles, thereof." Africa comprises book four, part one, and America (book four, part two) concludes the description of the known world. Part I of the fourth book contains the "Chorographie and History of Africa and all the principal Kingdoms, Provinces, Seas and Isles thereof," along with "An Appendix to the former Work endeavouring a Discovery of the unknown parts of the World. Especially of the Terra Australis Incognita, or the Southern Continent."

Sources

Heylyn only hints at source materials, but mentions that "the greatest help I had, was from Oxford library," nine miles off but increasingly inaccessible as age proceeded. Founded in 1602, the Bodleian library rapidly expanded in the following decades, coming to include large collections of works on the Middle East, China, Java, and so on, many sourced to London's Levant Company and the East India Company (see Batchelor 2013: 4–5). As Heylyn explains, he set 1648 as his cut-off date prior to publication. Simply, his sources are both classical and theological, as with his opening "creationist" chapter replete with Old Testament/Biblical (Genesis, St. John's Gospel); classical (Aristotle, Democritus, Pomponius Mela, Tully, *De natura deorum*, Valesius, Strabo, Ptolemy, Marcellinus, Stephnus de Urbibus); Late Medieval (de Barros on colonies, race, and trade goods); and "modern" sources (Mercator, Sir Walter Raleigh). However, in summing up, the key sources for his or any study, he pronounces, are "Ptolemy, Pliny and Strabo for the elder times, Mercator, Maginus and Ortelius for the later ages" (Heylyn 1682: 24). However, Hakluyt's successor or beneficiary Samuel Purchas (d. 1626) had already published his magnum opus *Hakluytus Posthumus, or Purchas his Pilgrimes* in London in 1625, so it would be surprising if Heylyn had not consulted this source.

Though buried in reductionist Christian fundamentalist language, Heylyn (1682: 17) still had method. "History and Geography," he adverts, are like "the two fires or meteors which philosophers call Castor and Pollox [twin brothers in Greek and Roman mythology], if joined together crown our meaning with delight and profit; if parted threaten both with shipwreck; and are like two sisters dearly loving, not without pity (I almost said impiety) to be kept asunder." He even has a definition of history, to wit: "A perfect relation of all occurrences observable, happening in the state whereof it is written, described by the motives, pretexts, consultations, speeches, and events; a special care being had both of time and place" (Heylyn 1682: 17). But for a definition of geography, he simply cites Ptolemy, "a description of the whole Earth imitated by writing and delineation with all other things generally annexed unto it.... in which we look to the Earth not simply as it is an element, for so it belongs to *philosophy*; but as a *spherical* body proportionally composed of earth and water, and so it is the subject of *geography*."

Maps/Geography

Perhaps learning from Mercator, Heylyn accepts that the Earth's circumference is 21,600 miles (against modern measurement of 24,901 at the Equator). He accepts greater and lesser circles of 360°, divided into 60° (therefore 21,600). Latitude is north or south of the Equator, and must be measured on the globe by degrees in the meridian. "The meridian is a great circle rounding the earth from pole to pole." There are many meridians but the "chief" or prime meridian runs through St. Miguel island in the Azores. As explained, the use is to show longitude of any place, east or west of the first meridian.

He acknowledges the Equator as a great circle going around the terrestrial world from east to west (passing Havana, Sumatra, Guinea, etc.), along with lesser circles as with the Tropic of Capricorn and Cancer. He acknowledges an Arctic Circle and an Antarctic Circle ("so-called because opposite to the other") passing through *Terra Australia Incognita* (about which more below). He acknowledges frigid, temperate, and torrid zones, according to latitude. Parallels or equidistances also circle the Earth from east to west, commonly spaced at 10° distance, such as set down on maps or globes. However, he also reintroduces the Ptolemaic notion of climes (24 north or south) relative to seven in the classical version borrowing from a book on Sacrobosco's *Sphaeram* (1570) authored by Jesuit mathematician Chistoph Clavius, also serving as adviser to Pope Gregory XIII (1572–85) on the switch to the Gregorian calendar. Accordingly, he offers a table of "climates" occupied by three categories of inhabitants "as they are treated of in Geography" according to the way their shadows are cast.

Finally, Heylyn draws attention to the distinction made by Ptolemy between geography and chorography: "Chorography is but the description of some place or country, and not of the whole earth universally, which is the proper work of a geographer." Notably, chorography is distinguished from geography "as a part from the whole," "Geography in its full latitude comprehends not chorography only but topography and hydrography also." Both topography and hydrography have classical precedents (see Figure 5.1).

Heylyn Content Analysis (Book Three, Asia)

The "Chorography and History of the Lesser and Greater Asia and all the Principal Kingdoms, Provinces, Seas and Isles, thereof" runs to 230 pages. As such, this is a veritable self-contained and indexed book-within-a-book spanning the Eurasian landmass, commencing with Anatolia (Asia Minor), and proceeding with Cyprus, Syria, Palestine, Arabia, Persia, Tartary, including Cathay, and with China in the east (and with Korea across the Yalu River), not neglecting India including *India Extra Gangem* (Southeast Asia [mainland and maritime]), and other zones, kingdoms, and countries and subregions. In an attempt to add verisimilitude to his account, perhaps again taking cues from Ptolemy, Heylyn adds a table of longitude and latitude of all chief towns mentioned in the book, literally from Aden and Amboina to X, Y, and Z. All places mentioned in the book are indexed in a very modern style, Australia included (Heylyn 1682: 149).

To summarize this book-length treatise would be pointless but, having identified a regionalizing geographic framing in this text, we also seek to know how Heylyn departed from or improved upon Ptolemaic conventions. For economy, we draw examples from within regions, so identified, namely

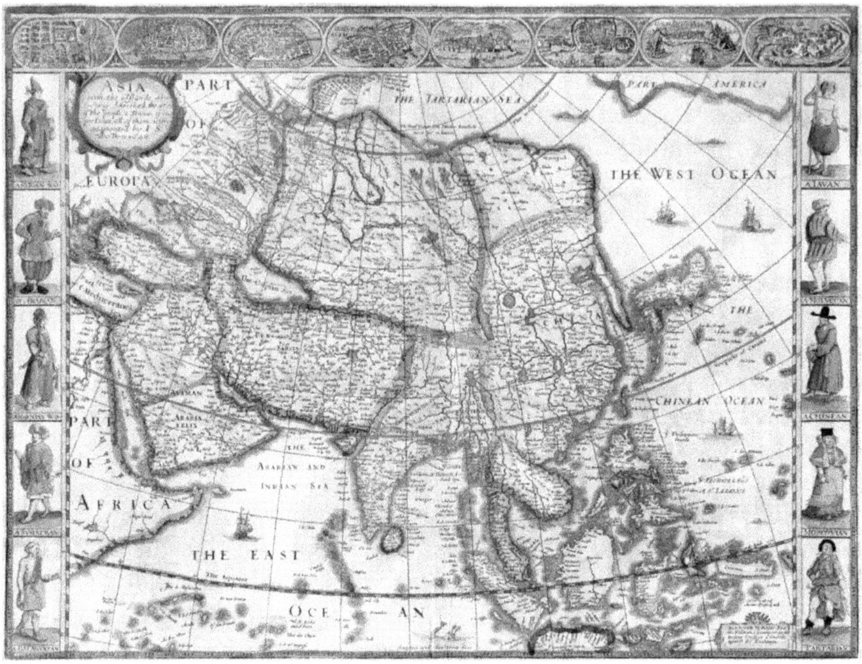

Figure 5.1 Asia: With the Islands Adioyning Described, the Atire of the People, & Townes of Importance, all of them Newly Augmented. John Speed (1626). Published in *A Prospect of the Most Famous Parts of the World* (London, 1662). City maps in upper border: Candy—Goa—Damascus—Jerusalem—Ormus—Bantam—Aden—Mac ao. Scale [ca. 1:27,000,000] at Equator. Impressive with its longitude and latitude grid, here Speed creates his own world regional delineation of Asia with boundaries imposed. China is named, as is Tartaria, West Ocean, East Ocean, and Chinean Ocean. The American continent is acknowledged. *Source*: Map reproduction courtesy of the Norman B. Leventhal Map & Education Center at the Boston Public Library.

the zones now understood as mainland Southeast Asia, maritime Southeast Asia, and China. In fact, Heylyn is framing macro regions on the basis of shared history, civilization, classical precedents, namely Ptolemy, and his understanding of geography. Mostly he has picked winners because, as mentioned below, state consolidation, especially in mainland Southeast Asia, but also in central Asia has swept away many smaller polities, just as many larger ones (Tibet) have lost their autonomy.

Tartaria

Undoubtedly one of the largest regional entities that Heylyn framed was "Tartary," essentially spanning the Mongol empire of the fourteenth century at a time of disunity and fragmentation (although the term "Mongol" is not part of his vocabulary). He thus speaks of region, and a politically bounded realm, that of the Tartars, as defined below. This vast region, essentially the Eurasian landmass east of Podoloia in Ukraine (only liberated from the "Golden Horde" in 1362), to the Bering Straits (then unknown), was 5,400 miles from east to west and 3,600 from north to south. From longitude 50° to longitude 195° (which exaggerates the easterly extent into North America), and from 40° north to 8° north. As described, this region straddled the sixth clime right to a zone "where they cease measuring climes." Acknowledging no single geographical feature across this vast zone, he did recognize a race; "Tartars," utterly barbarous, given to horseback riding, some pagan, some Muslim, and with some evidence of Christians, but below "Tartar" alongside a great ethnic heterogeneity of "tribes" and "hordes." The Great Cham and his court were acknowledged.

Tartaria, in turn, was divided into five main subregions (with additions), Tartaria Preompensis (Tartaria Minor, wholly in Europe); Tartaria Deserta (Asiatica); Antiqua; Zagathay (*Sythia intra montem Imaum* of the ancients); Turkestan, and Cathay (seat of the Great Cham with his court in Cambalu [Khanbalik]). Additionally, he expounds upon *Serica* and *Synthia extra montem*, "in which I follow not the bounds laid down by Ptolemy, who for want of that knowledge in these remote countries which these later ages have afforded, if fain to shut it up on the north and east with Terra Incognita." In other words, Heylyn presumes to know more about these eastern and northerly zones than the ancients. "At the present it is called Cathay." As for *Serica/Sera/* Silk, he offers no advance on Pliny, fictionalizing a race of people said to live to up to 200 years of age. But Ptolemy locates this place too far north, he asserts. This excursion on Tartary is important as the issue came to beset Czarist Russia and the Qing dynasty at a point in history when both expanding empires came into collision over ill-defined boundaries or, actually tracts of steppe virtually unowned apart from nomadic inhabitants.

Cathay or China

On Cathay and China, as Heylyn continues, by contrast with Tartaria:

> China is bounded on the North, with Cathay, and the eastern Tartars, from which separated by a continuous chain of hills . . . and where that chain is broken off, or interrupted, by a great wall extended 400 leagues in length, on the South partly with Cochinchina, a province of India, partly with the Ocean, on the East with the Oriental Ocean; and on the West with part of India, and of Cathay. (Heylyn 1682: 181)

As he continue, with allusion to the name of China:

> The Ancient inhabitants of *Sine,* in the time of Ptolemy, were towards the North. . . . United into one body by the name of *Sina*, and known by that name to the Romans, in the time of Ptolemy who bounded them on the north with *Serica*, on the west with *India extra Gangem*, and on the East and South with *Terra incognita*; which though it makes up the least part of modern China, Serica touching only a corner of the Northwest of it, and the main Ocean washing it on the East and South. (Heylyn 1682: 187)

He does allow, however, that the Arabs and (modern) Latin authors do reveal the dimensions of this "mighty kingdom." But, "How far they stretched beyond the bounds laid down by Ptolemy, and whether they subdued the eastern or northern nations or were conquered by them, it were hard to say, nor is it easy to resolve." For geographer and historian Heylyn, ancient Chinese history is indeed difficult to recover, although that does not deter him. However, without an accurate sense of that history, he really cannot overcome Ptolemy and fill in the gaps relating to *terra incognita.* After all, in his own phraseology, the "two lovers" geography and history are inseparable.

Allowing that the northern parts were utterly unknown to the ancients, the southern part was known to Ptolemy as *Sina,* hence *Sinarum Regio*, the Latin name applied to the whole country "and from whence China seems to be derived." Rather than imposing his own frame upon China, Heylyn appears to have borrowed from (Jesuit?) sources a fifteen-province divisioning of China, each one of which was larger than the greatest realm (nation) of Europe. In other words, and in contrast to many other polities he describes, he allows an indigenous make-up of internal boundaries. As he mentions, Ptolemy only named but one mountain in his description of *Sinarum Regio*, along with four rivers, but vague. Heylyn adds several others, including the Yalu in the northeast of the country entering the ocean at the "island" of Korea. He also acknowledged the major south and southeast flowing rivers (although the nomenclature is obscure). Notably, he mentions other rivers flowing into "lake of Quinsay" (as also mentioned by Polo). Quinsay (Quinsai) was Hangchow (Hangzhou) on the Grand Canal and capital of Southern Song (1127–1279) when Polo entered China. (Samuel Purchas also described Quinsay in *His Pilgrims*, 1626, III, 29) (see Moule and Pelliot, 1938: 28, 519).

Heylyn's is a largely benign even laudatory view of China, fairly typical of that age, albeit a highly fluid image in Europe. In any case, the moralizing and horrors he bestows upon the Tartars, among other "pagans" is largely absent in his account of China. On governance, however, he views China as ruled by a tyrannical god-like king. "Common people are kept in such awe and fear, that they are rather slaves than subjects, calling their king, the *Lamp of the world*; and *Son of the everlasting sun*, without whose light they were able to see nothing." China's naval prowess was also observed as with the comment recycled from a Jesuit source, that it was capable of supporting a "bridge of ships from China to Malacca."

Specifically, as Heylyn (1682: 189) reveals, "according to the general current of received opinions, have I followed the descriptions and affairs of China, as a different country from Cathay; not ignorant that there are many probable reasons which might incline us to believe them to be one and the same." The voyage of Jesuit Bedetictus de Goës (Bento de Góis) who entered China in 1605 makes no reference to Cathay. Neither did he mention the Great Cham. The Jesuit Martini Martinus author of *Bellum Tartaricum: Or, The Conquest of the Great and most Renowned Empire of China, by the Invasion of the Tartars* (1654a), also ignored Cathay. Jesuit letters from Beijing also confirmed the fact, as did more credible Muslim sources. Cathay has dropped off the map. It was actually Ramusio's commentary on Marco Polo meeting a Persian traveler that Cathay was noted "in the general notion." In the light of recent voyages made known in Europe, he continues, "The name and empire of Cathay has grown out of knowledge." "Peking, the chief city of China and Cambalu [Khanbalik] the supposed chief city of Cathay, being two words of the same sense and signification, though of diverse sounds, for Peking signifies the court or the city of Peking in the Chinese language, and Cambalu signifies the court or city of the Cham, in the language of Tartary." Muslim travelers and Jesuits alike made the statement that Cathay and China were one (and that maps placing the two as adjacent were false). For Heylyn (1682: 191), this was a case of Penelope "unraveling all the web which before was woven."[4]

As a way out of this tangle of geography and history, Heylyn agrees that the "Chams of Tartary" (Mongols) should be brought "a peg lower." "It must be taken for granted that the Tartars, having conquered the realm of Cathay, and afterwards, with the help of that people, subdued China also, caused all the northern parts of China to be called Cathay." Writing of the Mongol conquest, he observes, "And having thus extended their name and territory of *Cathay,* the first foundation of their greatness, they fixed their regal seat in the city of Peking, which by a new Tartar name (but of the same sense and signification), was from thence called Cambalu." But with the Manchu conquest, and expulsion of the Mongols first at Nanking (Nanjing) and victory over Tamerlane, the new conquerors installed themselves in Peking making it their capital. Over time, with loss of rank and parceling out of regional power, the "true Cathay might lose that name by little and little which formerly it had communicated to those parts of China."

Khanbalik (City of the Khan), from the Mongolian or Uyghur (also known as Dadu), was the capital of the Mongol-Yuan dynasty (1271–1368) founded by Kublai Khan, located in the center of what is now modern Beijing. Traditionally written as Cambaluc (or Cambalu) in English, after its spelling in the Italian version of Marco Polo, in the era when Heylyn wrote, the exact location or identity of the city was frequently confused by European cartographers and its identity with Peking had yet to be fully clarified. The common pattern on maps of the early 1600s was to offer two Khanbaliks, as with "Cambalu" north of the Great Wall and one labeled "Paquin" (Peking) at its correct location.

Arabia/India

The bounds of Arabia are also described with reference to the Red Sea, some part of Egypt, the Euphrates River to the north, and the southern sea ("called by some "*Oceanus Indicus*, by others *Mare Australe*") to the south. But, as Heylyn (1682: 94) also observes, Pliny, Xenophon, and other authors enlarged it further northwards, and so on. It is a "circuit" of 4,000 miles, internally divided into Arabia Deserta, Arabia Petraea, Arabia Felix, and the Arabic Islands.

Allowing India to be the largest country in the world next to China and Tartary, and referring back to Pomponius Mela, Heylyn reckoned it would take a ship forty days to traverse the coastline. He locates it by latitude, longitude and "clime" (from the beginning of the first to the end of the sixth). He also teases the reader by observing that facts mix with fiction in describing this country (as of classical times). Alexander the Great and Pliny are proffered. Describing John Mandeville as "the greatest fabler in the world," Heylyn nevertheless concedes that his account is supported by evidence supplied in the wake of the Vasco da Gama voyage.

India Extra Gangem

In Book Three on India, Heylyn (1682: 211) describes *India Extra Gangem* as:

> That part of the great continent of India, which lies on the furthest side of the River Ganges, from the spring or fountain of it, wheresoever it be, to the fall thereof into the sea by the sixth and last mouth thereof which is called Antiboli, the other four being reckoned into the other part of India, on this side of that river. From hence extended eastward as far as China and the Oriental Ocean, on other parts as bounded before.
>
> The country in those olden times so renown for wealth hath the name of the Silver Region, and of another the Golden Chersonese, this last supposed to be the Ophir of Solomon of which more hereafter.

As Heylyn (1682: 211) explains, the people on the far side of the Ganges were not well known to the ancients owing to the remoteness of the place, and moreover not yet so well known. However, he would illuminate with reference to the classic knowledge and with reference to the known. He goes on to enumerate a list of names of important mountains and rivers (all obscure), and three of the rivers touching the Golden Chersonese. Names of chief towns defy interpretation. Of the kingdoms, he cites Ava but references Alexander the Great's trepidation at the most powerful of the kings beyond the Ganges, able to muster 20,000 foot soldiers, 20,000 horses, and 3,000 elephants. These were pagan people, unknowing of Christianity prior to the arrival of the Jesuits. Islam did touch the seacoast of Bengal, tapping into the Arabian trade, but Islam was as unknown as Christianity on the coasts of China. In that he was wrong. As mentioned, Islamic communities had followed Arab trade routes to China from the times of Southern Song.

Notwithstanding this obscuranta, under the rubric "*India extra Ganges*," Heylyn identifies and describes five kingdoms and elaborates on each with surprising detail. These are Burma, Cochinchina, Cambodia, Laos, Siam (and its tributaries on the Malay Peninsula, including Malacca), and Pegu, a precocious enumeration of the charter kingdoms of modern mainland Southeast Asia as it were. Only Pegu has dropped off this list becoming part of modern Burma, and with a unified Vietnam superseding a more restricted sense of Cochinchina, including Champa. Running for many pages it would be tedious to even summarize this mass of data.

Although Heylyn acknowledges a Portuguese presence in mainland Southeast Asia and occupation of Malacca, he nevertheless treats Malay/Malacca "the southern part of the Golden Chersonese," the cape of promontory of which was dubbed "Malanconia" by Ptolemy, as a virtual satrap of Siam. Ludovico di Varthema, the Bologna-born medieval traveler who succeeded Marco Polo in visiting Malacca between 1502 and 1508, described the trading port as receiving "more ships than anywhere in the world," and with Malacca as "the most noted empory of the East." Malacca, Heylyn reveals, was identified by Strabo as "Musicana," or Musican Terra, sited on the banks of the Gaza River running 15 kilometers breadth and flood prone. Obviously this evocation is highly muddled and could be conflating with historic Singapore which he describes as "the mother of Malacca." Deceptively, he describes Sincapura as located "east of Malacca, near the promontory of old called Magnum, supported by some to be the Zaba of Ptolemy." Heylyn continues that this designation would be more probable than Palura "as Maginus" would have it, Palura being a city of the Hither India, and different at the least 20° longitude from any part of this Chersonese." Maginus (Giovanni Antonio Magini, 1555–1617) was an Italian astronomer, astrologer, cartographer, and mathematician who also produced

Ptolemaic map renditions. "But whatever it was called in the former times, it was in these late ages the mother of Malacca, the greatest part of the trade and people being removed from thence to this newer foundation; before which time it was the best frequented emporium in these parts of the East." He also embroiders a narrative of the foundation of Malacca by a Javanese prince, leading to the removal of the Singapore trade to the new city, subsequently coming under Muslim domination and with Siam eclipsed as overlord. Historians now agree that this was Parameswara (1344–1414), also called Iskandar Shah. Johor and Patani "governed of late by queens" are also profiled.

Heylyn's Imaginary Geography

Imaginary geographies and literatures also with roots back to classical mythology played an important conditioning role across many societies, sometimes with utopian overtones. Plato's *Atlantis* was one. But in late Renaissance–early Enlightenment times, such ideas were sometimes transmuted into a literature of the fantastic. David Fausett (1993: 37–39) has tracked the renaissance of utopianism into the sixteenth century, certain of it fueled by the new discoveries, as with Thomas More's *Utopia* of 1516. This was a small step toward the production of a series of "baroque allegories" or satirical works playing with imagined space, even *faux* lands, peoples, and continents of which the Great South Land was a major trope. To return to Heylyn, as he wrote, "for men, the Earth is divided into the left side and the right side," an allusion to Ptolemaic geography but, "in respect of itself," it divides into "parts real and imaginary." "The imaginary parts of Earth are such which not being at all in the Earth, must yet be supposed to be so, for the better teaching and learning of this science; and are certain circles going about the Earth, answerable to them in Heaven, in name" (Heylyn 1682: 21).

Heylyn served early notice of this literature, not entirely dismissive, especially as he warped potted summaries into an appendix to his *Cosmography* under the rubric, "Endeavouring a Discovery of the Unknown Parts of the World, Especially of *Terra Australis Incognita* or the Southern Continent." This was actually indexed as "Australia." Heylyn possibly went further than any other author in print in his age to pin down a southern continent. However, he approached the problem in two ways, not entirely separately, one "geographic" and one "cosmographic." With respect to the "geographic," he scanned a small mass of travel literature or voyages seeking evidence of a southern continent from Tierra del Fiego to the Cape of Good Hope. But in a major section of the work, labeled

Terra Incognita, he explores *Terra Incognita Austalis* in the framework of fantastic literature. Among other examples, he cites Joseph Hall's *Mundus alter et idem* (1605).

As Heylyn admits, he worked from conjecture, not received knowledge of the Dutch discoveries. While, a century earlier, as discussed in chapter 7, the Malacca-based "cosmographer," Manuel Godinho de Erédia, had also conjectured a *Nusa Antara*—a synecdoche for Australia—informed by local sailors and sources, this was unlikely to have been known in London (see Gunn 2016: 142). As mentioned, the existence of a southern continent recalls Ptolemaic images of an encircling southern landmass. The Polo journey was of no use in this respect and made no reference to a land south of Java. In any case, as revealed in the concluding pages of this book, the disambiguation of the continent of Australia from the imagined *Terra Australis Incognita* was only accomplished through voyages of discovery in the late eighteenth century. (see Figure 5.2).

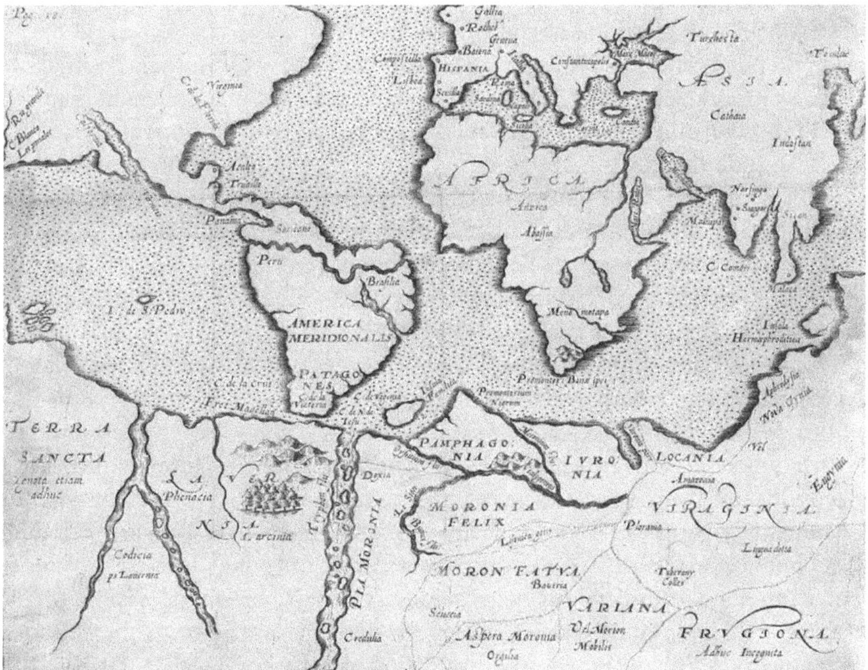

Figure 5.2 Map of Imaginary Southern Lands. From Joseph Hall, *Mundus alter et idem* (1607). *Source*: Cornell University, Division of Rare & Manuscript Collections: Persuasive Cartography: The PJ Mode Collection, PJM_1009_01.jpg.

CONCLUSION

What then do the texts under discussion in this chapter tell us about the longevity of the Ptolemaic template in the face of empirical truths derived from the great discoveries—the Ptolemaic trap—and, how do they inform as to the construction of Asian regions? First, the half century separating, respectively, the Venetian, the Castilian, the Portuguese, and the Englishman hardly makes them symmetric, just as their lives and social-political milieu were vastly different. First, Ramusio was writer and statesman, part of the Venetian officiate and, literally embedded in the city-state's diplomatic and commercial world. Camões, as revealed, was a classicist, a poet, and a romantic. He also lived the Portuguese discoveries as had his near contemporary the swashbuckling Fernão Mendes Pinto. Whereas Camões validated a Ptolemaic world, Pinto celebrated its actual transcendence such as achieved by his band of Portuguese adventurers literally traveling off the end of the map. Heylyn, on the other hand, was a retiring religious controversialist and Oxford don. At least, Heylyn was not drumming up colonial conquest, not overtly, although avid readers of Ramusio might have been so fired.

Of Ramusio and Heylyn, there are differences as well in their approaches to history and geography. One, the product of a Renaissance humanistic education, the other, a pompous classicist and an Old Testament reductionist. Both rode the early wave of print and both worked with mapmakers and engravers, indeed becoming part of a tradition. True to his Mediterranean traditions, Ramusio was still part of the world of *portolan* charts and associated merchant venturing, yet he was also connected up with commercial and political intelligence derived from his contacts, agents, and friendships in Spain and Portugal. North of the Alps, Heylyn and his publisher was part of the Cartesian "revolution" in maps and print, yet was detached in the way of a modern Oxford scholar. Stated another way, a world of difference stood before Venetian mapmaker, Giacomo Gastaldi, and Fleet Street printers. With confidence and pretension Heylyn matches place names with coordinates gleaned from myriad sources, just as map engraver Vaughan adopted a Mercator projection. There is also more to wonder in Ramusio for a mid-sixteenth-century audience than in Heylyn. Not only was he first, a priority that cannot be trumped, his was a humanistic Renaissance vision of the discovery of new lands, souls, mainly through the agencies of the Catholic kings of Portugal and Spain, and the optics of their servants. Heylyn's early seventeenth-century English audience had read it all before in Hakluyt and Purchas, hence perhaps the clergyman's need to embellish, titillate, and "conjecture," as with the discussion of the "southern continent."

In the volume dedicated to Asia, we have seen how Ramusio used Pliny as a reference point. Pliny is a discursive text, Ptolemy is a wooden framework.

As mentioned, Ramusio was alert to the old Caspian Sea trade route via Tana. As a Venetian, with a knowledge of the Venetian connection with the Eurasian silk roads, Ramusio was uniquely informed to comment upon the geopolitical shift occasioned by Portugal's advance across and down the Atlantic coast, as much the Turkish threat to the Adriatic. Informed by his Venetian diplomatic sources, he heralds to Europe the new emerging geopolitical alignments surrounding Iberia as the global core (he knows nothing of India or China in this regard), and with Africa, the Americas and, who knows, with Asia as a potential periphery on the Moluccas, Timor, and Philippines model. By contrast, Heylyn is simply out of the loop as diplomat or commercial attaché. His knowledge is second hand and dated, although he attempts to catch up. Frequently, he abandons Pliny in favor of Ptolemaic renditions, proudly offering corrections but always referring back to the classical *India Intra Gangem-India Extra Gangem* template, even though the empirical knowledge he was dealing with had massified in what was a transitional age looking ahead to the new Enlightenment geographies such as would soon appear in France.

We acknowledged that *Os Lusiadas* and *Peregrinação* actually accorded print fixity to many familiar names from Siam to Japan. With respect to *Lusiadas*, there is no doubt that Camões anchored his epic to antecedents familiar to classicists of the time, not excluding a veritable Ptolemaic world-in-motion view, as it were. Neither would we deny conflicted theological understandings and ideological inferences whether Christo or Portugal-centric in orientation. Not part of the scientific revolution that would await Galileo and Copernicus, the Camões narrative is also transitional from and between the Aristotelean-Ptolemaic referencing of space *and* the spatial revolution that the Indian Ocean and Magellan voyages revealed and which the poet apparently experienced firsthand.

NOTES

1. According to Lach (1965: 164), Montalboddo asserts a double purpose in producing his collection. First, he sought to demonstrate that Pliny's history of the Afro-Eurasian world could be vindicated with reference to the new data emerging from the maritime voyages of discovery. Second, he sought to give pleasure to the astute reader. It is important that Montalboddo referenced his work against classical authors, albeit premature or wanting in testing the Pliny thesis in the light of the discoveries of the Americas and with more information to come. Ramusio's use of sources is also discussed in Parke (1955).

2. In fact, the complete *Suma Oriental* composed between 1512 and 1515, entering the library of the French National Assembly (Ms. 1248 Ed. 19), was only "rediscovered" in 1937 and published by Armando Cortesão in a Hakylut edition of 1944.

3. The full title of the book (from the English edition by Henry Cogan of 1653) is indicative of both Pinto's spiritual journey and his sense of spatiality, namely, *the voyages and adventures of Ferdinand Mendez Pinto, a Portugal: during his travels for the space of one and twenty years in the Kingdoms of Ethiopia, China, Tartaria, Cauchinchina, Calaminham, Siam, Pegu, Japan, and a great part of the East-Indies, with a relation and description of most of the places thereof; their religion, laws, riches, customs, and government in the time of peace and war, where he five times suffered shipwreck, was sixteen times sold, and thirteen times made a slave.*

4. According to Henry Yule (1866: 530), the modern translator of Polo, Góis (Goes) made "*Cathay* . . . finally disappear from view, leaving *China* only in the mouths and minds of men." In fact, Portuguese Jesuit Bento de Góis (1562–1607) was the first known European to travel overland from Goa in India to China, via the Pamirs, Kashgar, and Xinjiang, departing in 1603. In this quest, the Persian-speaking Goes was supported by the Mughal court at Agra. Inspired by controversies among the Jesuits as to whether the Cathay of Marco Polo was the same country as China, his expedition conclusively proved that the two countries were one and the same. Matteo Ricci who reached Peking in 1598, already knew that China *was* Cathay.

Chapter 6

The Ptolemaic World-in-Motion

Enlightenment Science and the Jesuit Advance on China

With the albeit inconclusive Columbian, Vasco da Gama, and Magellan "revolutions" explained in the previous chapter, the age of the great European discoveries was fading (even though European colonialism was fast tracking empty spaces on the map, not excluding the Pacific Ocean, alongside the consolidation of imperial outposts and proto-colonies). As hinted in the previous chapter, the discoveries were not neutral but pushed to the fore the debate between defenders of a geocentric earth such as bequeathed by Ptolemy and as adhered to by the Catholic Church versus the revelation of a heliocentric universe coming out of the new science and this was just one of the philosophical issues besetting late Renaissance thinkers and theologians. It was also an age where direct observation, surveying, and the incorporation of local knowledge led to major revisions to the Ptolemaic template, such as stubbornly dominated the mind-set of the sixteenth- to seventeenth-centuries generation of Iberian and Dutch mapmakers from Mercator on. Hence, the attention given in this chapter to technical innovations in measurement at least as facilitated by the adaptation of the telescope to astronomical observation as well as advances in timekeeping. The upshot, as described in this chapter, was a Ptolemaic world-in-motion, increasingly fed by Enlightenment science and the application of reason.

It was also an age when disparate civilizations and cultures came into contact or even conflict in unprecedented ways. This was especially the case in the Western encounter with Ming China (1368–1644). While the introduction of European cartography into China by the first-arriving Jesuits in the late sixteenth century answered the question for a curious elite as to where they came from, Jesuit observations and letters back to Europe offered the first convincing post-Ptolemaic imaging of China's actual space since the highly generalized reports of the Franciscan missionaries of the thirteenth century along with the Polean account. Arriving in China along ancient caravan routes, Czarist envoys from the court in St. Petersburg would gain special privileges in Beijing at a time when Russia was aggressively pushing east into "Tartary" redefining boundaries with the Qing across the sprawling steppe. Mapmaking would follow. European national rivalries cannot be discounted and the French Crown would give especial impetus to the application of astronomy to the measurement of the Eurasian landmass. The new version of the age-old Eurasian encounter also came as a wake-up call to Qing China.

This chapter leads with a discussion of the (European) scientific revolution as it came to revise classical and medieval conceptions. In a second section it elaborates upon the initial Jesuit advance on Ming China highlighting the European disambiguation of Cathay/China, thus settling an issue outstanding since the days of Marco Polo. A third section examines the ancient Chinese astronomical tradition as confronted by the first-arriving Jesuits in China. A fourth section deals with attempts by the Cassini dynasty of astronomers backed by the French Sun King to harness Jesuit astronomical prowess to their ambition to measure the globe (and to definitively lay to rest confusions over the easternmost coordinates of the Eurasia landmass such as sowed by Ptolemy). A fifth and final section highlights pioneer Russian mapping of "Tartary" and new Qing/Czarist validations of territory/boundaries/sovereignty.

PTOLEMY IN THE FACE OF THE SCIENTIFIC REVOLUTION

We have inferred a long transition period in early Renaissance Europe that witnessed an embrace of hitherto lost classical knowledge, Ptolemy included, the early reception of this knowledge in the centers of medieval learning, and the replication of this knowledge through the print revolution. However, the humanistic revolution of this age matched by a turn toward secularism, experimentation, and early science eventually gave way to what became known as the Copernican revolution that would, in turn, not only challenge Ptolemaic conceptions of an Earth-centered universe but also key tenets held by the Catholic Church.

The seeds of the humanistic-scientific revolution should not detain us here but the sprouts obviously involved numbers of individuals working in disparate European locations. As Woodward (1987: 316) observes, it was actually the Vienna school of science around the University of Vienna mathematician and astronomer Johannes de Gmunden and the Augustinian monastery of Klosterneuburg, especially between the 1420s and 1442, that led to a veritable "seedbed of scientific innovation" relating to astronomy and mathematics including coordinate tables, and cartography. Be that as it may given the long conversation between diverse schools and civilizations such as described, scholars have long debated the scientific revolution stemming from the revelations of Galileo, Copernicus, Kepler, and the Danish astronomer Tycho Brahe, among others (see Sharrett 1996: 9–12). For example, in his study of Brahe, Adam Mosley (2007: 291) has drawn attention to the "astronomical community of the late seventeenth century," the movement of individuals involved in astronomical studies and the significance of the sites to which they migrated. In other words, collaborations and communications were also important to the new science alongside instruments, methods, and theories.

Rightly, Galileo Galilei's (1564–1642) improvement upon and pioneering use of the recently invented telescope was a game changer in celestial observation. Simply, his observations of the eclipses of Jupiter's satellites, interpretations and published works were seminal in challenging the Aristotelean-Ptolemaic version of a geocentric universe even if scientists and theologians would remain divided. While, as this book argues, Ptolemaic cartography would prove far more resilient at least until the continents and oceans were explored and mapped, the notion of a stationary Earth around which the planets circled was no longer sustainable outside of strict Catholic orthodoxy. In the words of Maurice A. Finocchiaro (1997: 2), the transition from a geocentric to geokinetic worldview kicked off by Galileo involved "not only many branches of science but also other disciplines," including philosophy, theology, religion, art, literature, technology, industry, and commerce. Indeed, he continues, "it changed mankind's self-image in general."

As R. H. Van Gent and A. Van Helden (2007: 134) have written, with each increase in the light gathering power of the telescope, new celestial discoveries were made. "Most important for terrestrial cartography" they contend was an improvement in astronomical measuring instruments allowing the first accurate determination of a degree and the shape of the Earth. This in turn was the subject of considerable controversy until the middle of the eighteenth century. The eclipses of Jupiter's satellites provided for the first time a convenient method of determining longitude on land "and thus lay at the heart of the revolution in geodesy and cartography" (even though the problem of calculating longitude at sea was never solved in this age).

Still, as popular historian Dava Sobel (1996: 27) expands, the success of Galileo's method quickly translated into the arena of mapmaking. While earlier maps tended to underestimate the distances to other continents and the outlines of individual nations, "now global dimensions could be set, with authority, by the celestial spheres." Still, arguments over the prime meridian remained. Also confounding modern historians, national differences remained over the measurement of nautical leagues, or length corresponding to an angle degree of a meridian arc, though Portugal and Spain agreed on a standard at least. With Galileo's method still wanting for sea journeying, European monarchs vied to harness the new science to practical ends.

Changing Conventions

It might be trite to suggest that navigation was the handmaiden of cartography but there was obviously a correspondence. While conventions were well established in both fields, there was also innovation and experimentation as well. In his *Almagest*, Ptolemy took the "Fortunate Islands" associated with the Canary Islands (13°–18° west of Greenwich), as his first or prime meridian (meaning 0° longitude), although his calculations better approximated the Cape Verde Islands (22°–25° west). This location was simply the westernmost point of land of the known world, as opposed to Alexandria which had earlier served as his prime meridian. Indeed, it was a convention adopted by al-Idrisi. It is also true that in medieval times European maps often used Jerusalem as the primary line of longitude, although this convention changed again with the emergence of Ptolemy-revival maps.

As A. R. T. Jonkers (2005: 1–7) points out, with their westerly location off the coast of Africa, the Canaries (plus Madeira), the Azores (discovered 1427–52), and the Cape Verdes (discovered 1455–61) all served as springboards for voyages of discovery, including sources of supplies. With their prominent volcanic peaks, the islands also offered a positional fix for navigators. A variety of locations from within these island groups featured in ship logs as prime meridian; St. Miguel, Corvo, Flores, Santa Maria in the Azores; Boavista and St. Jago in the Cape Verdes; and Ferro and Tenerife in the Canaries. Each was different, with the Azores more western or further from the coast of Africa than Cape Verde, in turn, more westerly than the Canaries. Each or any of these islands served as a standard and, as discussed below, with Ferro Island coming to be favored by the French, and also adopted by Jesuit surveyors in Asia as their prime meridian. On the other hand, Dutch cartographers like Gerardus Mercator, Jodocus Hondius, and Willem Janszoon Blaeu deviated from the use of the Ferro meridian. Mercator came under the influence of the idea that there could be a nature-given reference line where no difference would be observed between geographic and magnetic poles. His lead was followed by cartographers like Ortelius and Hondius. From the early 1640s, some Dutch cartographers were using Tenerife as a prime meridian in maps, with a significant increase in use after 1662.[1] After 1675, Tenerife was the predominant meridian on Dutch maps and in 1787 the Amsterdam Admiralty declared it the official prime meridian of the Dutch Republic. The towering 3,718 meter Teide volcano of Tenerife also achieved prominence as the prime meridian with Dutch cartographers (after abandoning the idea of a magnetic meridian).

A natural or magnetic meridian was suggested by Columbus who observed that, at a point in the mid-Atlantic, the compass pointed due north. Although the theory of magnetic declination had already been known in Europe and much earlier in China (Smith and Needham 1967: 1213–1214), practical navigation in the Atlantic gave new importance to this understanding. As determined by João de Lisboa in 1508, magnetic declination was viewed as low in waters near the Atlantic islands (Cape Verde; the Canaries; the Azores), suggesting a natural meridian (the agonic meridian) and with the idea catching on in a number of countries, Holland included. Several locations in the Americas (Acapulco), and in Asia (Pedro Branca off Singapore and Canton in southern China) also came to be associated with zero magnetic declination, although without dislodging the Atlantic longitudinal reference. But the theory lost favor with Blaue becoming a champion of a Dutch national meridian. When, in 1635, Englishman Henry Gellibrand proved the invalidity of time-invariant geomagnetic longitude schemes, and revealed that the Earth's magnetic field changes with time, the theory was virtually abandoned. Nevertheless, as explained below, the cartographic legacy remained, as with the maps of Mercator, Ortelius, and Blaue (see Jonkers 2005).

It is important as well to register certain map conventions observed in France. French cartographers, such as Jean Baptiste Bourguignon d'Anville, would observe the Paris meridian actually passing through the Church of Saint-Sulpice, as their prime meridian and this would endure for the next 200 years. It should also be recalled that, in line with classical precedence, longitude was long counted from 0° to 360° (as indeed with the Treaty of Tordesillas) as opposed to the modern convention of numbering 0° to 180°, respectively, east (E) or west (W).

Pioneer Astronomical Observations in Asia and the New World

By the fourteenth century, astronomers and navigators could measure latitude with just a small margin of error, vastly improving upon the crude measurements dating back to the age of Ptolemy. As explained, this was done by observing the Pole Star or the Sun with a nautical astrolabe. The determination of longitude, as alluded, was much more difficult. In practical terms, as Durand and Curtis (2013: 35) summarize, "it required knowledge of the precise time not only at a ship's present location—which was not too difficult to discern—but also the time at the reference location in Europe." In other words, longitude corresponds with time.

According to Pimentel (1712: iii), because longitude was so notoriously difficult to calculate many errors crept into the *roteiros* or sailing guides, and different nations were in disagreement as to many of the facts. He also noted that mariner's charts of many nations falsified the distance from the (western) tip of Europe until the end of Asia in line with Ptolemy. As he further pointed out, Pedro Nunes refuted these arguments in a treatise on navigation dedicated to Infante Luís of Portugal, the Duke of Beja (1506–55). As stated, he used words to the effect that those world itineraries which separated the Canaries from China by 180° were so ignorant that they granted greater credit to Ptolemy than to the recent navigations that acknowledged change. As he registered, it also had to be understood that the measurement of places through polar (star) observation was still so tentative that consensus on the truth remained elusive.

As discussed earlier, clocks were not sufficiently accurate to establish such calculations through the sixteenth century. As Jay Lieske (1983: 51) confirms, in the early days clocks were the principal source of error and with observations made in local apparent time. In the wake of Galileo many of the early geographical charts were developed by using satellite eclipses to determine longitude. However, the development in 1656 of the pendulum clock by Dutch scientist-astronomer Christiaan Huygens eventually permitted more accurate measurements on land. Refinements upon Galileo's telescope also added improvements. Besides bringing telescopes, the first-arriving Jesuits in China were armed with time pieces including sun dials, important as the measurement of time went hand-in-hand with the measurement of longitude or relativized space.

Beginning with Spanish and Portuguese voyaging in the Atlantic Ocean, it is hard to conceive the European expansion project without the measurement of longitude, whether as a marker of territorial conquests or as a guide to navigation. As mentioned in chapter 5, the Iberian seafarers were also schooled in Ptolemaic theory. According to Randles (1984: 15), although lists of longitudes for European cities were already in existence in the sixteenth century, it is not altogether clear how these numbers were used in the construction of maps at that time in any systematic way. Commencing with the observation of eclipses in Mexico City in the 1540s, referenced against Toledo, and at Peru in the 1580s, improvements in the use of longitude for cartographic purposes were achieved outside of Europe in the Spanish empire (Randles 1984: 13–15). On the other side of the globe, arriving Jesuits commenced to make astronomical observations in India in 1593, albeit with no great precision, and with Goa continuing to be an observation post for Jesuits transiting through to Macau and China, at least those sanctioned by Portugal.

THE JESUIT ADVANCE ON MING CHINA AND THE DISAMBIGUATION OF CATHAY/CHINA

As Joanna Waley-Cohen (1999: 63–64) summarizes, some 900 Jesuits worked in China in the seventeenth and eighteenth centuries, more numerous and influential than other such orders as the Franciscans, Dominicans, Augustinians, and the secular French Society of Foreign Missions. For the Jesuits, this would be a “top-down strategy” targeting the court not the masses. Taking cues from Francis Xavier’s earlier experience in Japan, and a general curiosity about the sphere, celestial bodies, eclipses, phases of the Moon, and so on, in Asian courts, astronomy was seen as a winning way in making progress on the missionary front (Sasaki 2003: 32). Arriving in Macau in 1579, Michelle Ruggieri was joined by Matteo Ricci in 1582 and with both of them entering China (Zhaojing) together in 1583. Reaching Nanjing in 1599, Ricci observed firsthand an observatory equipped with special buildings and brass instruments that had survived over 200 years. Some years later he observed a parallel observatory in Beijing with similar instruments, also dating to the Yuan dynasty. Ricci himself was impressed with the bronze-cast instruments at Nanjing, including a (celestial) globe with parallels and meridians marked out degree by degree, an armillary sphere, and a gnomon. According to Joseph Needham (1959: 369), Ricci was more impressed with the Yuan-era instruments than the Ming-era astronomers whom he sought to supplant with newer knowledge.

Rigorously schooled at the Jesuit Collegio Romano under leading Jesuit mathematician Christoph Clavius in such practical sciences as globe and mapmaking and the functioning mechanism of the astrolabe, Ricci put his knowledge to practical use in China as with the construction of clocks, armillary devices or, rather, newer devices, especially as horology and astronomical measurement had a long pedigree in China. Clocks also served Jesuit diplomacy (Tang 2015: 257–281). As widely commented upon, Ricci’s techniques of memorization also attracted interest among Chinese elites (Spence 1985). In initiating a truly two-way exchange, he also distinguished himself as a translator into Chinese of Latin works on mathematics, astronomy, and geography. Among his translated works gaining favor in China was a description of Ptolemy’s *Astrolabium* and Euclid’s *Elements* (Sasaki 2003: 2, 33, 41). Famously, Ricci became the first Jesuit to serve the Chinese imperial court as astronomer.

From 1584, as explained below, Ricci began to publish his maps, gaining influence in China, Korea, and Japan even after his death (and notwithstanding Tokugawa censorship from the 1630s). Ricci also pioneered the construction of globes in China, producing three in 1586, based upon models he studied in the Collegio Romano (Halseburghe 1994: 221). This was advanced in 1623 by Jesuits Manuel Dias and Nicolo Longobardi who produced a Chinese terrestrial globe. Even so, as Helen Wallis and E. D. Grinstead (1962: 83) explain, the first terrestrial globes had been introduced into China from Arab sources during the Mongol-Yuan dynasty. On the Chinese part, Jesuit mathematical science was needed to reform the calendrical system. As Chikara Sasaki (2003: 40) points out, coinciding with Ricci's arrival in Macau in 1582, with Clavius as a key adviser, the Catholic Church had just adjusted the calendar under Pope Gregory XIII, indicating that the Jesuits were equal to the task (see Figure 6.1).

Cathay or China?

In constant letter communication back to Europe, Ricci may also have been the first to signal that the China he visited (Nanjing) was actually the Catai/ Cathay of Marco Polo. This was a deduction he made in October 1596 in a letter to Claudio Acquaviva, Jesuit Superior General in Rome. But he was not alone if we consider discussions on the issue entering Jesuit letter documentation of the times.[2] In a letter dated September 8, 1584, to Juan Bautista Román, the Spanish Crown agent in Macau, Ricci placed the coordinates

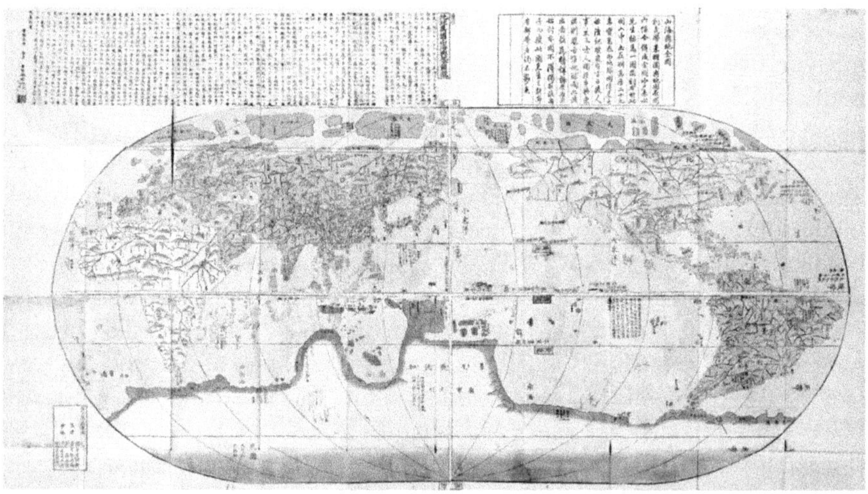

Figure 6.1 ***Sankai yochi zenzu,*** **山海輿地全圖説 Nagakubo Sekisui; Matteo Ricci, Published: Naniwa, Akitaya Taemon [1785?] Scale ca. 1:27,000,000 at the Equator.** Wood block print, hand colored. Shows continents and oceans of the world at about 1600. Based on Matteo Ricci's World map which was published in China in 1602. *Source*: Retrieved from Library of Congress, Geography and Map Division. Digital Id http://hdl.loc.gov/loc.gmd/g3200.ct001813.

for Ming-era China at large within 120° and 136° or 137° to the east of the Canaries (actually from Cochinchina in the west to Chincheo (Xiamen in Fujian) ("the coastal province of China facing Japan" in the east). From Juaquin (Zhaojing), then the residence of the Viceroy in Guangdong where he sojourned along with Ruggieri, as Ricci stated, "as I have clearly seen by two lunar eclipses, one observed in Macau, with great diligence, and the other here in Juaquín the meridian of which is not very different from Macau 120 degrees shortly more or less." He placed the coast of China between 20° and 28° north pole height, dividing China from Tartary China at 44 or 45 "degrees of the pole." As he qualified, "I say more or less because they do not use height, but I have surmised by the number of days on the road by the distance reduced to our leagues." As he surmised, the kingdom called China since the time of Ptolemy existed because of the potency of the king of China who set up the great wall of many hundreds of leagues in length dividing off the realm from the Tartars.[3]

Ricci made several observations of lunar eclipses, that of June 5, 1583 (undoubtedly in Macau) and those of September 15 and on November 29, 1584, at Zhaojing to which he refers. According to Sardo (2013: 66 n. 1), Ricci was successful in "radically correcting beliefs held until then," especially as to the geographical construction of Ming China as perceived by Europe. However, it is not that Ricci corrected his own erroneous Aristotelian conceptualization of the universe—Galileo had yet to publish his masterpiece—but, from his own calculations, the Jesuit significantly revised the Ptolemaic reckoning of the place of China within the Eurasian landmass. Specifically, to quote Sardo:

> The known world, the vast Eurasia extending for more than 200 meridians from the Fortunate Islands (the present-day Canaries) to the shores of the Pacific, suddenly shrinks by roughly a third of its size and moves dramatically away from the American continent.

Now "a vast ocean divides the continents." As Sardo muses upon the revelations of Ricci and Ruggieri who as discussed below produced a pioneering atlas of China, not even Columbus or Magellan had challenged the ideas of the ancients in so scientific a manner (Sardo 2013: 59).

To be sure, the notion that the country called *Serica* by Ptolemy and Catai (Cathay)/*Mangi* (Manzi) by Polo, following the Persian, Turkic, and Mongolian convention of dividing the realm into northern and southern kingdoms was actually one entity known as China had long puzzled Europe. As Piergiorgio Odifreddi (2010: 107) points out, with the eclipse of the Mongols in 1368, Europe was literally left in the dark as to the geopolitics of Ming China. It mattered as well and with Columbus erroneously taking on board a letter to be delivered to the Great Khan, well over one hundred years since the advent of the Great Ming.

With the reality eventually acknowledged by the Iberians, important texts in this respect include Athanasius Kircher's *China monumentis* (1667) in turn informed by ideas circulated by the Polish Jesuit Michał Piotr Boym. According to Liu Jing Jing (2013: 217–18), such ideas are carried in a manuscript found in the Vatican library titled *Magni Cathay quod olim Serica et modo Sinarumest Monarchia Quindecim Regnorum Octodecim Tabulae.* This is undated but the context is the 1640s. Setting aside intellectual debates, the revelation of Ming China to Europe was actually settled on the ground half a century earlier. The role of Jesuit adventurer and Ricci confidant from his time in Goa, Bento de Góis in positively demonstrating that Cathay was indeed China cannot be ignored. The first European to travel overland from Agra in India to western China via Afghanistan and the Pamirs, he died in April 1607 before making direct contact with Ricci in Beijing.

THE CHINESE ASTRONOMICAL TRADITION AND THE EUROPEAN CHALLENGE

With an astronomical tradition at least contemporary with the ancient Greeks, as mentioned in chapter 2, under the Mongol-Yuan dynasty China gained new astronomical knowledge and techniques from Persian and other Islamic sources. Eventually this exchange led to the construction of an observatory in 1271, long predating the first observatories set up in Europe. With its millennium-long bureaucratic tradition and sophisticated proto-scientific-industrial base, well established under the Song dynasty (960–1279), we wonder as to how the incoming Ming dynasty (1368–1644) responded to challenges stemming from the arrival of European interlopers on the coast of China in the early decades of the sixteenth century? To be sure, this question has focused many minds, especially as the establishment of a virtual Portuguese city-state in Macau adjacent the Pearl River estuary leading upriver to Canton. Not only did Macau serve as a beachhead for the Jesuit missionary (and scientific) advance on China, but also kicked off an East-West knowledge exchange.

French Jesuit Du Halde's Take on Chinese Astronomy

Nowhere else was the state of art of Chinese astronomy of the early eighteenth century better expressed contemporaneously than in Jean-Baptiste Du Halde's *Description Geographique, Historique, Chronologique, Politique, et Physique de l'Empire de la Chine et de la Tartarie Chinoise* (first appearing in Paris in four volumes in 1735), in turn derived from letters sent by the first-arriving Jesuit missionaries in China in the late sixteenth century. Du Halde had never been to China but, from 1711 to 1743, oversaw the publication of *Lettres édifiantes et curiieuses* written by Jesuit missionaries in China, in thirty-four volumes ranged between 1703 and 1776. In turn, as discussed in a separate chapter, the project led to the publication of a first map of Korea by Jean-Baptiste Régis along with forty-two maps of the Chinese provinces executed by d'Anville.

In his history, Du Halde upheld a triple concern: first, to encapsulate the astonishing antiquity of Chinese astronomical observations; second, to produce a snapshot of Chinese astronomical knowledge at the point of contact with the first-arriving Jesuit missionaries in the late 1500s, such as embodied by the Nanjing observatory; and third, to relate the achievements of the Jesuit scientists in constructing a new observatory in Beijing and in successfully demonstrating to the emperor errors in Chinese calendrical calculations.

As Du Halde (1741: 81–83) explains, of the thirty-six eclipses recorded by Confucius (as confirmed by Jesuit experts in the late seventeenth century), "there are two but false, and two doubtful; all the rest are certain." Notably, German missionary Johann Adam Schall von Bell (1592–1666) calculated and verified the eclipse of (Emperor) Tchong Kang (Zhong Kang) (2155 BCE). This was written in the *Chou King (Shu Jing)* (100 BCE). The eclipse of 776 BCE was recorded in the *Chi King* (*Shih Ching*) [Book of Odes] of the Han dynasty. French Jesuit Antoine Gaubil (1688–1759), author of *Traité de l'Astronomie Chinoise* (1739), also conducted a major study of Chinese texts. As cited by Du Halde, he remarked, "I have found a great many Observations corresponding with those of Europe and Asia." In confirming the veracity of these ancient Chinese observations, Gaubil wondered whether Herodotus, Thucydides, Plutarch, and Dion had achieved (as much) or more. Taken together, this mix of Jesuit sources demonstrates both the antiquity and general accuracy of Chinese astronomy.

As Du Halde (1741: 81–83) embroiders upon Gaubil's letter of 1729, reaching back 120 years before Christ, "you have the Number and Extent of their Constellations, and what stars answered the Solstices and Equinoxes, and this by Observation; you have also the Declination of the Stars, the Distance of the Tropics and the two Poles." As he continued, "The Chinese were acquainted with the Motion of the Sun and Moon from West to East, and likewise of the Planets and fixed Stars, tho' they did not determine the motion of the latter till 400 Years after Christ; they also had a Knowledge of the Solar and Lunar months, and have given Saturn, Jupiter, Mars, Venus and Mercury Revolutions very near ours, but they have not a thorough Knowledge of the Rules of Retrogradation and the Stations, yet as well in China as Europe, some have made the Heavens and Planets revolve around the Earth, and others have supposed that they move around the Sun, but the number of the latter is small, and this opinion is not found in the Writings of some particular person." Here, Gaubil is confirming that the Ptolemaic convention of a geocentric universe was not universally accepted in China, just as the Jesuits had to adjust their own position on heliocentricity. At the time when Gaubil was writing, the scientific revelations of Copernicus, Brahe, Kepler, and Newton were eroding church dogma but it was only in 1758 that the Catholic Church dropped the general prohibition of books advocating heliocentrism from the *Librorum Prohibitorum* or Index of Forbidden Books.

As Du Halde (1741: 83–87) summarizes, since the Han dynasty (206 BCE–220 CE) the Chinese had produced several treatises on astronomy and, from a reading of these texts, one could judge that the Chinese "have been pretty well acquainted, for above 2,000 years, with the length of the Solar year of 365 days, and almost six hours, that they have known even the diurnal motion of the Sun and Moon, have understood the method of taking the meridian altitudes of the Sun by the shadow of a gnomon, and have well enough calculated, by the length of the shadow, the height of the Pole and the declination of the Sun; they have likewise known sufficiently exact the right ascension of the stars and the time of their passing by the meridian, how the same stars in the same year rise or set with the Sun, and how they pass by the meridian sometimes at the rising and sometimes at the setting of the Sun; they have also given names to the stars, and have divided the heavens into different constellations, by which they could trace the place of the planets, and distinguish them from the fixed stars."

Trumping all these observations, as Du Halde continued, in the Chinese province of Henan, there was the city of Teng Fong (Dengfeng) deemed "in the middle of the world, because it was in the middle of their empire." There, was a tower, still extant at the time of the Jesuit missions, along with an observatory deploying a brass table and style to record shadows to measure retrogradation of the Sun. Here, Tcheo Cong (Zhou Gong), acclaimed as the most highly skilled mathematician in China of all time, made his astrological observations, "1200 years before Ptolemy was born." According to classical Chinese texts, Emperor Yu the Great, founder of the Xia dynasty, the first dynasty in Chinese history, made Yangcheng his capital in 21 BCE. In Chinese lore, as expressed by Du Halde, this was deemed the center of Heaven and Earth. A still extant stele dating from the Tang dynasty (618–960) marks the site of an eighth-century geodetic survey that established the length of a degree of the meridian. Zhou Gong, commemorated by a temple, is believed to have carried the earliest measurement of the Sun's shadow at Yangcheng. Generations of astronomers measured the Sun's shadows at Yangcheng or used the results to make calendars. The tower as observed by the Jesuits, and known today as the Gaocheng or Dengfeng Observatory, was originally built in 1279 by Guo Shoujing (1231–1316) during the Mongol dynasty, and repaired in 1542 during the Ming Dynasty (1368–1644). Observations of the lengths of the Sun's shadow were used to determine the latitude in order to fix provincial and other territorial boundaries.[4]

The basic conundrum for Chinese calendrical reform, according to Oxford historian Allen Chapman (2006: 271), was the "geometrical limitations" of the 360 and one quarter digit circle or solar cycle adopted by the Chinese, effectively disallowing the striking of neat multiples or fractions of the radius. The Chinese had never gone on to develop a mathematical theory-based model for the movement of celestial objects but based their imperial calendars upon runs of practical observations going back thousands of years. Their method stood outside of the Western system of predictive models such as prevailed by 1600. Such knowledge forearmed arriving Jesuit astronomers, as with Schall who, with accuracy, could always predict the exact altitude of the Sun for a given day of the year or accurately predict a forthcoming eclipse (as he did on September 1, 1644). Notably, the Cologne-born Schall became Director of the Imperial Observatory and the Tribunal of Mathematics under the Shunzhi emperor (r. 1644–61), the third under the Qing, at least until losing his influence following the death of his patron in 1661.

By 1668, the Belgian Ferdinand Verbeist (1623–88) had become the leading Jesuit astronomer at the Qing court. At this time, the Kangxi emperor (r. 1661–1722) gave Verbeist the resources to reequip the old imperial observatory. Verbeist, in turn, was well informed of the contributions to scientific astronomy made by Tycho Brahe, including his *Astronomaie Instauratae Mechanica* (1598), the illustrations of which provided the prototypes for the re-engineering of the Beijing observatory, practically replicating Brahe's observatory as it appeared in Denmark in 1597. Still, Verbeist was obliged to entirely innovate as with the casting process in fireclay molds, as well as adapting to Chinese aesthetic-geometric conventions (Chapman 2006: 273). He also had to impress because the precedent in China of great celestial globes, astrolabes, and sextants had already been established thanks to the contribution made by the Muslim astronomers (see Figure 6.2).

As described by Du Halde, the Nanjing observatory such as visited by Ricci in 1599, hosted three main instruments. The first was described as a "great globe, with all the parallel circles and meridians engraved and divided into degrees; it was so large that three men could not encompass it, and supported by a large brazen cube, which opened on one side to give entrance to a man that he might turn the globe about when the observer required it; there were neither stars nor countries designed on the globe, so that it served equally for terrestrial and celestial observations." A second instrument was a sphere four yards in diameter, "instead of circles were double rings, representing the usual circles of the sphere, which were divided into three hundred and sixty-five, and every degree into the same number of minutes; in the

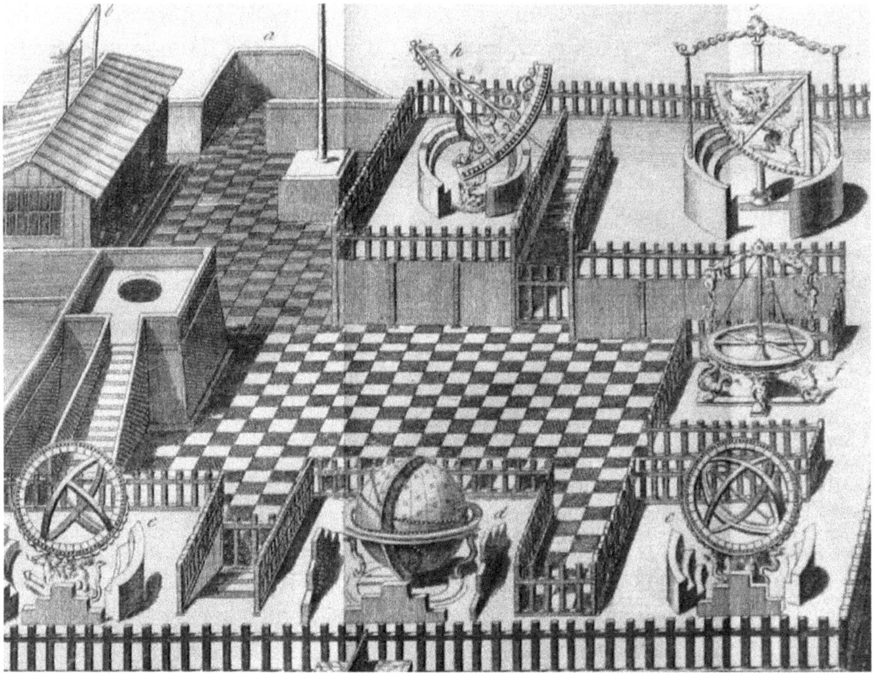

Figure 6.2 Imperial Observatory, Beijing: Equinox Sphere (c); Celestial Globe (d); Armillary Sphere (e); Azimuthal Horizon (f); Quadrant (g); Sextant (h). *Source*: From Jean-Baptiste Du Halde, *The General History of China* (London, 1741) (author's collection).

middle of the globe of the earth was a kind of tube, which turned every way at the pleasure of the observer. To view the stars, and to mark the place on the degrees by the situation of the tube." The third instrument was a quadrant, "raised eight or ten yards upon a stone-table, pointing to the north" used to measure through observation of shadows, the true points of the solstices and equinoxes. But the greatest of these instruments was composed of three or four astrolabes "joined together with movable rulers and sights to observe with; one, inclined to the south, represented the equinoctial, and the other that crossed it, the meridian" (Du Halde 1741: 84–85).

Du Halde (1741: 84) fails to mention the Mongol observatory erected in 1272 just north of the Ming-era observatory in Beijing built in 1492. Neither does he offer a detailed description of the Beijing observatory such as also visited by Ricci. Rather, he alludes to the Muslim inheritance, as with the Ming adoption of "equation tables" devised by the Arabs "who entered with the Tartars into China." The question of Mongol borrowing from Arab science is certainly worthy of careful consideration but, it is also true as Du Halde continues, that the Jesuits sought to demonstrate their superior science to the

Chinese court to discredit the Mongol inheritance, just as the older Nanjing and Beijing observatories observed by Ricci fell into disuse under the Ming. The fall of the Ming dynasty (1644) and the conquest of China by the Manchu Qing was a major setback for the Jesuits, depending upon location and circumstances, but the Shunzhi emperor recognized the Jesuit's skills.

As Du Halde (1741: 100–105) explains, in the service of the emperor or "son of heaven," the Bureau of Rites required detailed astronomical information as with the longitudes and latitudes of eclipses, just as court ritual demanded that these were observed ceremonially. But the principal work of the Tribunal was the production of a calendar printed in millions of editions and distributed throughout the empire. Bearing the emperor's seal, its production and distribution was accompanied with great ceremony. Tribunals of mathematics served the same purpose. But with the accuracy of the calendar coming under official scrutiny, the Jesuit group in Beijing headed by Verbeist was personally summoned by the Shunzhi emperor to offer an opinion and, through three experiments involving the measurement of the length of a shadow, convinced the court of his superior method. In doing so, Verbeist was opposed by certain mandarins and, with some tenacity, by a Muslim court astronomer who had been engaged to reform the calendar and armed with little more than ancient Arab tables. In the event, the emperor charged Verbeist to examine the Muslim astronomers' findings. In this contest of wills and science, and over the opposition of conservative (neo-Confucian) opinion, Verbeist prevailed and, accordingly, was loaded with honors and granted the authority to reform the calendar, otherwise as much "an Arabian as Chinese calendar." On his advice, accepted by the emperor, an intercalary month in the then calendrical year was removed. Verbeist not only reformed the mathematics tribunal but, from 1673, rebuilt the observatory with numerous instruments, still extant in modern Beijing today. Elaborately crafted, the instruments are supported on a pedestal signifying Chinese mythological elements as with cast dragons representing the "emperor's arms."

THE CASSINI DYNASTY, THE SUN KING, AND THE MEASUREMENT OF THE GLOBE

As with Britain, France also sought to harness its scientific establishment to the cause of astronomy, geography, and navigation. France under the "Sun King" Louis XIV, backed by his powerful chief minister Jean-Baptiste Colbert (1643–1715), played especially significant roles in launching French scientific cartography (Brown 1965: 780). But national rivalry and confusion also reigned over the adoption of a prime meridian.

As explained by Harris (1764: iv), starting with King Louis XIII (r. 1610–43), and with Cardinal Richelieu serving as his chief minister (1624–42), France sought to establish its priority on matters geographical astronomical. Under an edict of April 25, 1634, "in order to keep as close as possible to Ptolemy's choice of the Canary Islands as the prime meridian," Louis XIII "directed all French geographers to place their first meridian in the island of Ferro, the most western island of the Canaries. . . . The Dutch maps generally place the first meridian at the peak of Tenerife. The Arabians fix themselves at the Strait of Gibraltar. The old Spaniards would need have it at Toledo. . . . Here in England, we generally reckon the first meridian that of London, and instead of counting as the old method was, from the west to the east." With the Ferro Island reckoned at 20° 30' west of Paris, this was later rounded to 20° disguising it as the Paris meridian. In doing so, Cardinal Richelieu rejected the notion of a line consistent with zero magnetic variation such as had been widely mooted since the time of Columbus's discovery of this phenomenon on his voyage (Ariel and Ariel Berger 2005: 85–86). As Harris (1764: iv) observed, Venice, Bologna, Cracow, Lisbon, Toledo, Nuremberg, Frankfurt, Copenhagen, St. Petersburg, and Ulm all offered their locations as national meridians, and with Great Britain establishing its own meridian at the Royal Greenwich Observatory established under King Charles II in 1675. Harris of course was writing before an international conference of October 1854 established Greenwich as the internationally recognized prime meridian, and with some nations including France still maintaining their national meridians until eventually falling in line.

In 1666, backed by Colbert and with Louis XIV lending his endorsement, an Académie Royale des Sciences was established in Paris as an organ of government. The following year, under pressure to solve the longitude problem, the king approved the building of a new observatory in Paris. As Jacob Soll (2009: 99) explains, Colbert recruited international scholars and scientists. Among them was Giovanni Domenico Cassini (1625–1712) (Cassini I), author of a landmark work published in 1667 on observations of Jupiter and Saturn. Arriving in France in 1669, Cassini served the rest of his career as astronomer to Louis XIV. Other members of the family, as with son Jacques Cassini (Cassini II), carried on the tradition from 1712 to 1756, and with César-François Cassini de Thury (1714–84) (Cassini III) becoming director of the Observatoire de Paris in 1771. Completed in 1671, the Paris Observatory thus predated its Greenwich counterpart by a few years. From 1679 the Paris Observatory published the world's first national almanac, the *Connaissance des temps,* using eclipses in Jupiter's satellites to aid seafarers in establishing longitude. It also hosted the Meridian or Cassini Room with the Paris meridian traced on the floor.

As Lloyd A. Brown (1956: 780) points out, while the ideal of Hipparchus and Ptolemy to locate each place on the earth scientifically according to longitude and latitude was still current, two new pieces of apparatus, namely the telescope and a timekeeper would lead to both a revolution in mapmaking *and* efforts to measure the earth. The French project was also boosted by the collaboration of Colbert with Christiaan Huygens whose pendulum clock and tables based on measuring the Sun's passage offered a new method of accurately measuring longitude. Besides sponsoring the pendulum clock, Colbert, Huygens, Cassini, and other scientists collaborated on the manufacture of a machine to measure longitude (Sol 2009: 8, 99).

As discussed below, Colbert would also dispatch Jesuit mathematicians and astronomers to China via Siam to "deepen knowledge of astronomy and to measure the earth" (Landry-Derron 2001: 424). To achieve this goal, as Theodore N. Foss (1994: 140–141) explains, Jesuit colleges became training grounds for geographers including scholarly elites, mapmaker dynasties, and missionaries, and with a "bond of union" developing between the Jesuits and French cartographers.

The French Astronomical Project in Siam and China

Even prior to arriving in China, the French Jesuit mission was directed at Ayutthaya, the capital of Siam during this epoch with diplomatic alliance in mind. On their part, the Jesuits saw advantage in obtaining access to Asian courts, as in Siam and China, through demonstrations of astronomical prowess as with the prediction of eclipses. King Narai (r. 1656–88) of Siam, who had opened up commercial and diplomatic activities with foreign nations including the Portuguese, Dutch, English, and Persians had earlier dispatched a mission to France in 1680 though ending disastrously in shipwreck in Africa, followed by a successful mission in 1684 (see Smithies 2003: 1–4).

Backed by Giovanni Cassini and endorsed by Louis XIV, the fourteen Jesuits chosen for the mission set sail on March 1, 1685. Dubbed the Chaumont Embassy, the two-ship voyage was led by Chevalier de Chaumont and François-Timoléon de Choisy. The journey to Siam took seven months and nineteen days. Under this initiative, France would vie to enter China whether or not with Portuguese permission to stage through Macau, and with French Jesuits spearheading France's push into the China area. In line with the *padroado* convention between the Vatican and Portugal relating to missionary activity in Brazil, India, and China, the Portuguese Crown jealously guarded its rights. It is important then to see the Chaumont Embassy to Siam as a preliminary to future events in and around Macau and in Qing China, especially under Emperor Kangxi.

As minutely detailed by one of the participants, the missionary-mathematician Guy Tachard (1688), the expedition made several celestial observations at Cape Town (having convinced the wary Dutch). They also brought with them an array of astronomical instruments, telescopes, quadrants, pendulum clocks, and tables of the satellites of Jupiter. On December 11, 1685, at Lopburi, north of Ayutthaya, the Jesuits observed the lunar eclipse in the company of King Narai. The following year, on February 20, they measured the longitude of Lopburi, also observing a lunar eclipse on April 18, and a solar eclipse on April 30. Narai also requested the missionaries to build an observatory at Lopburi. In the event, King Narai was not in the least swayed by the Jesuit demonstration of Christian faith. Rivaled by Phra Horathibodi, the court astronomer during Narai's reign, the Jesuits learned that local astronomical knowledge (possibly informed by Persian) was not to be underestimated. Nevertheless, according to Lars Gislén (2004: 143), both the king and court ambassador Kosa Pan collected and understood European maps and globes. Enticed by palace insider, the Greek Constantine Phaulkon, Tachard returned to France to suggest an alliance between the Siamese court and Louis XIV. A second French mission of 1687 under Simon de la Loubère joined by Tachard and organized by Colbert also brought a military force charged to occupy strategic points (see Smithies 2003: 1–4) (see Figure 6.3).

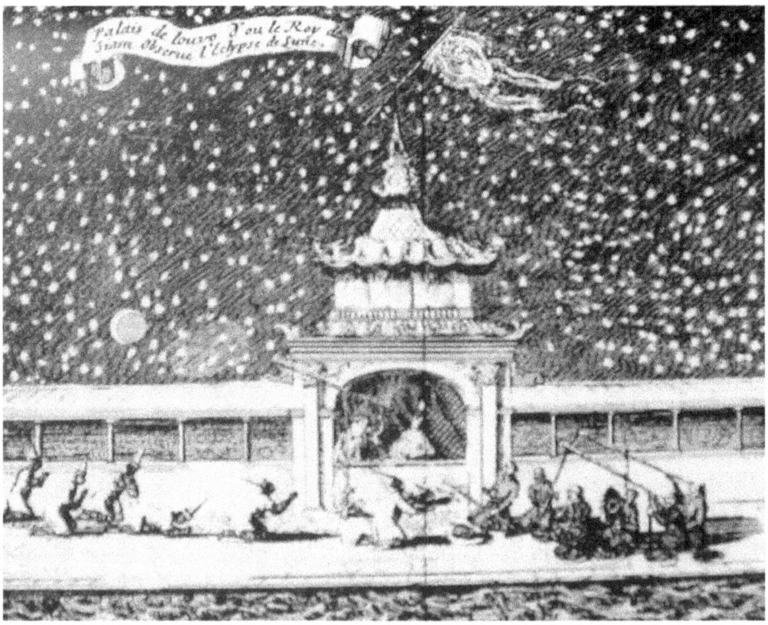

Figure 6.3 King Narai Observes a Lunar Eclipse with French Jesuits at Lopburi, 1685. *Source*: Guy Tachard, *A Relation of the Voyage to Siam*, London, 1688 (author's collection).

As Cassini commented upon the mission, and as published in the *Mémoires de l'Academie des Sciences* (1725: 38), all the observations of eclipses taken in Siam in February, March, and April 1686 were extremely important because they could be used to locate all the others taken (globally) in the same month, allowing comparisons to be made with observations taken in Paris and so revealing the difference between the meridians in the same cities. Here, he echoed the sentiments of mathematician/astronomer Philippe de la Hire, part of a circle including Cassini and Descartes. "All the observations given us to know the position of Louveu [Lopburi] in Siam with regard to Paris, as exactly as if the towns were even in France, and without doubt of which time observations are made of the principal towns and places in China and Tartary. As he summed up, "We will have perfect knowledge of the great countries that up until now we only know imperfectly."

Political events soured in Siam with a palace coup of May 1668 against the king. One group headed back to France via India while another group comprising Superior Jean de Fontaney, Joachim Bouvet, Jean-François Gerbillon, Louis Le Comte, Claude de Visdelou headed to China. After a disastrous false start forcing a return, and by means of passage aboard a Chinese junk, the Jesuit group sailed direct to Ningbo (in Zhejiang) arriving in February 1668, having by-passed Portuguese Macau. Summoned to Beijing by the emperor, de Fontaney moved on to Nanjing making astronomical observations as he also did in Siam. Bouvet and Gerbillon went on to serve the Kangxi emperor in the mapping of China. Making several return trips to Europe, de Fontaney subsequently became an informant for Du Halde on matters Chinese (Michaud 1816: 207).

Besides bearing presents from King Louis XIV to the Chinese emperor, the Jesuit party brought with them vital astronomical measuring instruments including telescopes and improved timepieces. While telescopes introduced by the Jesuits had been used in the observation of eclipses in Beijing since 1631, mechanical clocks brought into China from 1583 were also recognized as important in calculating the timing of eclipses, indeed deemed essential in court ceremony alongside traditional devices and methods. By 1718 the pendulum clock had become the main timing device used by Jesuit astronomers in China (Yunli 2000: 135). Not only would the mission scientifically link France with China but, as explained below, would also help to rectify the Ptolemaic misrepresentation of the easternmost extent of the Eurasian landmass through the geometrical mapping of China. As Han Qi (2001: 139) points out, Jesuit astronomical calculation would test the reliability of Chinese chronology and also help to clarify in Europe the antiquity of Chinese civilization.

Three other Jesuits making good their exit from Siam, Jean Richaud, Jean-Vevant Bouchet, and Charles Francois Dolu, reached Pondicherry, the French enclave on the southeastern seaboard of India established in 1673. There, commencing on April 26, 1690, Richaud carried out systematic observations of binary stars (Alpha Centauri), comets, zodiacal light, dark clouds, and other phenomena, bringing with him the twelve-foot focus telescope that was used in Siam (Kameswara-Rao et al.; 1984). In commenting upon observations of eclipses of satellites of Jupiter made by Richaud in Pondicherry as well as other Jesuit observations conducted in China, Cassini found confirmation of his own findings advancing and correcting Galileo such as appeared in his 1610 publication *Sidereus Nuncius* (Auzout 1736: 151).

Qing-era Celestial Observations in Macau

Having survived the turbulent Ming-Qing transition, the Jesuit mission in Macau and China carried on. Preceding the arrival of the French Jesuit astronomical mission in Beijing, two Jesuits of Flemish origin literally helped to put Macau on the world map. Although not strictly part of the official French mission, their astronomical observations in Macau and China nevertheless fed back to Cassini in Paris. As such, they were active collaborators with the French state project. One, François Noël, is better remembered as a translator of Chinese classics. According to a short biographical note constructed by Paul Rule (2003: 189), Noël arrived in Macau on August 9, 1685, before crossing over into China in 1687. The other, mathematician and geographer Antoine Thomas (1644–1709), departed Europe via Lisbon and was therefore sanctioned by the Portuguese Crown. This was four years prior to Noël arriving in Macau via Goa and Siam in 1682 (and conducting observations in both these places). He was ten years senior to Noël in the Jesuit order, then holding the rank of Vice Provincial. Both would make astronomical observations in Macau, with Thomas preceding Noël.

As discussed below, Thomas observed an eclipse of the Sun in Macau in 1683 prior to being summoned to Beijing in 1685 by Emperor Kangxi where he would serve as secretary to an ailing Ferdinand Verbeist before taking up the office of vice president of the Bureau of Mathematics. He would also become a close adviser to the emperor and would participate in the Jesuit mapping project as well as the Qing negotiations with Russia at Nerchinsk in 1689 leading to the first treaty between Russia and China inter alia touching on boundaries and territory. Among other accomplishments, Thomas produced two maps, one *Duplex Iter Terrestre in Chinam ex Persia & ex Mogor iuxta descriptionem, 1690,* and the other *Tabula Geographica Orientalis* (1690), inter alia, marking the distance between Beijing and Moscow, part of a project to establish a land route for missionaries to reach China (Florovsky 1951; Roque de Oliveira 2012: 496–498).

The Eclipse of the Sun and Moon as Observed by Antoine Thomas in Macau

As revealed in a letter written by Thomas and published in *Mémoires* (1729, Vol. 7: 696), on July 24, 1683 he observed an eclipse of the Sun from Monte Fortress in Macau. Armed with a quarter circle (or quadrant) and a *lunette* (optical telescope) of seven Roman feet length, he then observed images of the Sun on a strategically placed sheet of paper. The space occupied by the image was divided into twelve points surrounded by concentric circles. "The time was good, the air very still. The edge of the Moon commenced to cover the edge of the Sun." He measured the height of the Sun above the horizon at 31° 2'. Nevertheless, his observation did not accord with the Riccioli tables.[5]

Some twelve years later (June 16, 1695), Thomas observed an eclipse of the Moon in Macau from Jesuit Company premises on a "small island," a reference to the then bucolic Ilha Verde (Green Island), "located one minute north of the Monte fortress." He had earlier observed the height of the Pole Star in 1682 at 22° 15'. In so determining Macau's latitude, he was reasonably accurate and with the modern measurement at 22° 19' N. Starting with an observation of stars, Thomas then observed the initial stage of the eclipse—"the veritable shadow of the earth"—and then, the total immersion. Although the commencement of the eclipse was partly obscured by clouds, he was nevertheless able to observe the conclusion. As footnoted in *Mémoires* (1729, Vol. 7: 699), the same eclipse was simultaneously observed by the Jesuit fathers who had traveled to Siam on board the French royal vessels.

Lest it not be forgotten, these were missionary-astronomers. Addressing the astronomers of Europe, members of the Royal Academy of Sciences in France, as Thomas advised:

> I am persuaded Monsieurs . . . it will be a great pleasure to contribute thus to the conversion of China, where without Astronomy, we would not perhaps have the liberty to preach Jesus Christ. And if we were exact. After all in the calculation of the eclipses, the enemies of Europe would not miss out in drawing great advantage against us to the prejudice of the Christian religion. (author's translation, *Mémoires* 1729, Vol. 7: 696)

Such is a telling remark, not only in stating the well-known motivation of the Jesuit mission in China to win "top-down" conversions from the Ming and Qing courts, but his willingness to collaborate with the French Royal Academy and, willy-nilly, the French state. As Thomas promised in a letter of September 13, 1689, he would send a detailed description of Korea (eight provinces) along with a description of Tartary "between China and the frontier with Moscow" (*Mémoires* 1729 Vol. 7: 743).

Observations by *François Noël* and Comparative Analysis

En route to Macau on June 16, 1685, Noël had observed an eclipse of the Moon while anchored on a ship off the coast of Sumatra in the Malacca Straits and with Thomas observing the same eclipse in Macau. Earlier, according to Rule (2003: 142–145), Noël had collaborated with Thomas on a project to upgrade maps of Siam, Macau, and Japan. As explained in the report, "Observations de la Hauteur du Pole en plusieurs Villes de la Chine, par la Pere Noël," published in *Mémoires* (1725), between October 1687 and November 1689, Noël observed the mid-day Sun with the quarter circle (quadrant) in various parts of China, as with Hoai-Ngan on the Grand Canal, Canton, Macau, and Nanjing, calculating both longitude and latitude. He also carried a timepiece. Specifically, on November 30, 1685, Noël observed the commencement of the lunar eclipse in Macau having corrected for time. Simultaneous observations made in Paris (November 29) allowed calculations of differences in the meridians between Paris and Macau (7° 25' 45). Meantime, observations made by Jesuit astronomers in Avignon allowed calculation of difference between Avignon and Macau (7° 16' 22'). The difference between Madrid and Macau was calculated at 7° 51' 9''. Differences in the meridians between Macau and Nagasaki (as measured by Jesuit Charles Spinola) were also recorded. "Following our hypothesis, the longitude of Paris was taken at 22° 30', therefore the longitude of Macau was determined at 133° 56' E," although not in agreement with de la Hire who determined Macau at 136° 15' E. (*Mémoires* 1727, Vol. 7: 706). Measured against Ferro, this compares accurately with the modern longitude of Macau which is 113° 55' E. Eventually, both Noël and Thomas were caught up with the Rites Controversy—the dispute besetting the Vatican over Jesuit accommodation to Confucianism and Chinese rituals—and with Noël charged by the latter to make the arduous sea and land voyage to Rome to plead their case.

From Malacca, in 1689, Jesuit fathers Patrisius Comilh and Claude de Bèze who were taken captive by the Dutch while en route to Macau, also took secret observations from a prison tower. According to their calculations, the latitude of Malacca as determined on September 21 was 2° 12', which they corrected to 2° 15' in a second observation of September 28. Notwithstanding difficult conditions imposed by their prison environment, including the inability to set up a clock, their observations to calculate longitude also enabled comparisons to be made between Paris and Macau. Transferred by the Dutch from Malacca to Batavia (present-day Jakarta) on Java and from there to Holland (along with a number of French prisoners), they only gained their liberty in early 1691, whereupon they (miraculously) arrived in Paris (Auzout 1736: 142). According to Lieske (1982: 54), when reduced from apparent time to universal time, "the observations are quite good."

For their efforts, as with seaborne navigators, Jesuit astronomical calculations in China were far more accurate for latitude than longitude (near impossible on board ship). However, from fixed observation stations as with Macau's Monte Fortress or offshore Ilha Verde, repeated observations feeding back to Paris obviously led to improvements over the years. No doubt, as well, the Jesuits mapped Macau in their time although the existence of a reported seventeenth-century "Jesuit map" of Macau remains to be discovered.

French Astronomer Cassini's Sizing Up of China

In this regard, it is worth taking note of Giovanni Cassini's statement relating to Jesuit astronomical observations made in India and China (*Mémoires* 1725: 176–179). Even so the French astronomer appears not to have acknowledged Abel Tasman's discoveries of Tasmania and New Zealand (1642–43), much less those of Semyon Dezhnev (1648) to the Arctic Ocean-Pacific Ocean region of what is now the Russian Far East. Taking observation of Jupiter and Saturn in India, Jesuit Richaud also offered corrections to the French polymath. But Cassini also had matters of state as well as science at heart. As servant of the Sun King, it could hardly be otherwise. While acute as to the scientific observations made in Macau, he was more interested in how Macau could be used as a datum point to gain broader insights into the true geography of China and Eurasia. As an astronomer, that meant the accurate measurement of longitude and latitude. As a classicist he also knew Ptolemy but not with any particular respect. In particular he wished to know the furthest extremity of Asia which, as he well knew, Ptolemy misrepresented. Surprisingly for a person of his position he also misjudged the true extension eastwards of the Eurasian continent and this assertion can be verified by a glance at a modern map of China, Korea, and the Russian Far East.

As derived from Jesuit sources, Cassini believed that this point was at an island "between China and Japan" located at the mouth of the Kiam or Yam River.[6] Ambiguously named "çummin" from a Portuguese phonetic rendering, the name and description fits Chongming Island (崇明島) formerly known as Chungming, an alluvial island at the mouth of the Yangtze River. Although described as hosting a single sprawling village, it was then a site of intense Catholic missionary activity going on to become an established community. Here, an eclipse of the Moon was observed on October 8, 1688, with a view to measuring its longitude (Auzout 1736: 176). As explained: "The position of çummin island located at the eastern extremity of the continent of Asia needs to be determined with the greatest accuracy possible by measuring its precise longitude." As noted, certain inconsistencies in Father Noël's measurements had to be taken into account especially with respect to the difference between the longitude of Macau and Hoai-Ngau or Hoi Nganfu.[7] Based upon observation of the satellites of Jupiter, according to Noël, the longitude of çummin Island was 142° 16' 45. "This compared to 'our 141° 15' 14', the best we can establish at present" (Auzout 1736: 177). Given that the modern longitude of Chongming is 121° 23' 50 E, and duly adjusting the Noël-Cassini readings as referenced to the Ferro meridian, then the Jesuit measurements might be considered more or less accurate.

As Cassini wishfully stated:

> If the missionaries visit the island in the future then we can anticipate observations of eclipses, the satellites of Jupiter, to determine longitude with more accuracy, because it is of great importance insofar as the island is very close to the eastern-most coast of China, which terminates the continent of Asia.
>
> And because we have observations of satellites made by astronomers sent expressly by order of the king to the island of Gorée, which is the closest point of Cape Verde, the western-most part of Africa and all the continent of our world, we would have the total longitude of the continent that encompasses Asia, Europe and Africa. (author's translation, Auzout 1736: 178)

First visited by Portuguese navigator Dinis Dias in 1444, Gorée Island is located just south of Cape Verde off the coast of Senegambia near modern Dakar. A fortified island, it was also linked to the "French Atlantic" slave and sugar trade.

Matched by observations made in Guadeloupe and Martinique, under astronomers Varin, Hayes, and Glos, Gorée thus became the focus of the French king's efforts to measure the Atlantic (and the world), as well as to establish a precise measurement of its longitude in relation to Paris. This was in 1682. The method of determining longitude and latitude of places through observations of eclipses of the satellites of Jupiter had been tested on the coast of France and some other places in Europe. Satisfied with the results, the Royal Academy of Science reckoned that they now had the means to achieve a "perfection de la géographie" (*Mémoires* 1727, Vol. 7: 431). Pendulum clocks were part of the instrumentation and crucial for success of the mission (Brown 1956: 796). In this light, Cassini's logic was reasonable if only the actual physical exploration of the coast of China-Siberia-Sakhalin could actually be accomplished (and that would await the French navigator La Pérouse in the mid-eighteenth century).

As Cassini continued with reference to Ptolemy:

> We can consider the progress geography has made in Asia over the last century, since Ptolemy demonstrated that the capital of Sines (Sinae) to be at 180°, beyond which he placed unknown continent, instead of the eastern coast of China, of which the longitude must be greater than that of all the Continents, whether at 141° or 143° longitude.
>
> We cannot believe that the part of Asia which Ptolemy called Sines, be that which we call China. It comprises the present-day kingdoms of Siam,

> Cambodia, with part of the island of Borneo, and half of Java, that were not distinguished from the Continent, relative to modern charts. (author's translation, Auzout 1736: 178)

It can be inferred that Cassini is referring to the tendency of Ptolemy to blur mainland and island Southeast Asia with the Eurasian continent, unless he was referring to these polities as tributaries of China, including the Sultanate of Brunei and Mataram on Java? But Cassini was also privy to the astronomical observations made by the Jesuits in Siam and knew the coordinates, at least, of the mouth of the Menam Chao Phraya River.

> First, Ptolemy gave to Sines as borders the eastern and middle coasts of an "unknown land," instead of what Chinese recognize as two coasts of the Ocean.
>
> Secondly, Ptolemy gave to Sine as borders on the western side, India Extra Gangem, which is correctly the countries west of the Kingdom of Siam.
>
> Third, Ptolemy gave to Sines a great Gulf from 16° in the north enclosed by a great peninsula, terminating at the Golden Peninsula (the Auroa Chersonese) at 8° to the south and an eastern land (Continent) which advances beyond the Equator up to 8° 30' latitude south. (author's translation, Auzout 1736: 178)

Referring to Ptolemy's *Sinus Magnus*, such as discussed in the following chapter, Cassini continued:

> If we consider this last by degrees of latitude, we will find that the Great Gulf is nothing but the Gulf of Siam, which at the entrance to the Menam River is 13° north; that the great western peninsula is none other than Malacca joined to the island of Sumatra, which is totally separated from the Continent, the Strait which is between Malacca and Sumatra is estimated as a Gulf called Ferinus of which Pliny attributed the latitude 2° north, as well as Malacca which doesn't appear strange, even in this century diverse supposed islands as with Tierra del Fuego, California, Korea, etc.
>
> There are no other lands having southerly latitudes that Ptolemy attributed to oriental towns of Sines than the island of Borneo and Java and others adjacent which should pass as part of the eastern Continent where otherwise the capital city of the Sines that Ptolemy placed at 3° south latitude and 180° longitude. (author's translation, Auzout 1736: 178)

"For sure," he continues, the ancients upheld only a "very confused knowledge" of those countries they called Sines derived from various land and sea voyages.

> From the ancients, he learned nothing except the length of the journey and perhaps the length of the day of the year in different places. That Ptolemy placed at the head of his Table and for which he drew up latitudes of as the principle laws of his discipline. We can't accept the longitudes given by Ptolemy where

> he offers an excess of 45° attributed to the median part of Sine, of which the longitude surpasses 135°. In any case, the excellent geographer Marin[us] de Tyr[e] showed it to be 225° and did not fall into the absurdity of Strabo who made India the antipodes of Spain. (author's translation, Auzout 1736: 179)

Finally, alluding to the Ptolemaic trap arising from the Alexandrian's exaggeration of the size of the Eurasian continent (and a trap to which Cassini did not avoid), he continued:

> All the countries which comprise European Asia and Africa are found by modern observations to be a quarter less extended from west to east than the ancients supposed. There remains between Asia and America an unknown part opposite Europe in the same zone, of which the Jesuit fathers who have been sent to the capital of the oriental's king's mathematicians by land and seas will one day give news. (author's translation, Auzout 1736: 180)

Interpreting Cassini

Macau may appear peripheral in this story but, to reiterate, along with three or four other global New World cities, the Portuguese-controlled enclave in China was a privileged site for trained astronomers of the late sixteenth and seventeenth centuries to operate. With the Dutch occupation of Portuguese Malacca in 1641, that city lost its status. Especially given that skill in the use and assembly of instruments required much backup, this science was practically a monopoly of the Jesuit order. The other fact which comes to mind is that the French Crown knew this and, as demonstrated by the French diplomatic-military push on Siam, sought to use the French Jesuits as diplomatic tools, even by-passing Macau and the *padroado* convention in their push on China. Because Macau was the site of intense and repeated astronomical observations, it still fell into Cassini's reckoning as a datum point for all other measurements taken east of the Portuguese-controlled enclave. But in all of this, almost 200 years on from Columbus, we are amazed with the critical attention granted Ptolemy in Cassini's thinking as he put in place his grand design to master geography scientifically.

THE RUSSIAN MAPPING OF "TARTARY" AND NEW QING/CZARIST VALIDATIONS OF TERRITORY

The Russian mapping of "Tartary" offers yet another tangent upon the European framing of Eurasia (see Gunn 2003: 123–124). Maps and atlases produced by the St. Petersburg Academy of Sciences, along with those drawing upon intensive expeditionary and survey activity, were as much concerned with setting down boundaries as with framing Russian space in the sprawling Asian Far East, not excepting Russian America, a reference to Alaska and other claimed territory. By the eighteenth century, the northern and eastern borders of the Asian continent were defined and mapped, the correlation between the Asian continent and North America had been established, and the sea route to Okhotsk, Kamchatka, and Japan had been established (see Frolova 1999).

According to Johannes Keuning (1954: 99), "As the Russians penetrated deeper further into Siberia their geographical interest in these regions grew, and the want of an outline map of the whole of the Siberia region was also more keenly felt." Eventually new evocations of continental Asia including "Tartary" broadly defined as central Asia including Siberia and Manchuria, began to appear as with Anthony Jenkins's (Jenkinson) *Russaiai Moroviae, et Tartarie Descripto* (ca. 1562–67), a "wall map" arising out of the author's travels as far east as Bokhara as agent of the Muscovy or British-Russian Trade Company. According to Krystyna Szykuła (2012) who rediscovered a copy of the map in Poland, Jenkins's production "had no geometrical web" (and therefore was not Ptolemaic in its design). He recorded latitude on the map as measured by astrolabe, but ignored longitude which he could not measure accurately. Rather he adopted an innovative projection based upon ten triangular points using cities, river mouths, and so on, as vortexes.

Jenkins was also a major source for Abraham Ortelius's *Tartariae Sive Magni Chami Regni Typu,* improving upon Tartary-Siberia. Appearing in the *Teatrum Orbis Terrarum* altas of 1570, it was the first printed map of the region. *Tartariae* (1600) by Hondius assimilated results of the expedition Willem Barents to Novaya Zemla (1595–97). Even so, this map still reflected Ptolemy's conception of a north shore of Asia. Another version of Tartary appeared in Jansonius's *Tartaria Siva Magni Chami Imperium*, published in Amsterdam in 1647. In any case, according to Bagrow (1952: 88), the atlases of Mercator, Ortelius, Hondius, and so on represented legendary countries, for which Pliny, Strabo, the Bible, Marco Polo, and Plano Carpini had served as material" (and with the later individual a medieval Franciscan visitor to the court of the Great Khan). But now Western Europe dreamed of reaching China through the Arctic circle.

Still, it would take a century before the old conception of Tartary would be radically revised. This was accomplished in 1687 when Nicolas Witsen (1641–1717), the enterprising burgomaster of Amsterdam, produced his six-sheet map of Tartary offering a new interpretation of the northeast of the continent and thus remaining the standard model for the mapping of this region until the expeditions of Vitus Bering (1725–49). In 1692, Witsen published *Noord en Oost Tartaryen* (North and East Tartary), in part a text-guide to his maps. A second edition followed in 1705 with additional maps, and with final edition produced in 1785. A number of smaller derivative maps were soon published.

To be sure, as Keuning (1954) alerted, Witsen's role as geographer and cartographer deserves more attention. Even though his sojourn in Russia (Moscow) as part of a States General diplomatic mission was short, he prodigiously collected materials as with local conversations with Tartars and Persians, as well as plying his political connections with concerned literati, and even a supposed life-long association with Peter the Great. His direct sources are mainly Russian and Cossack. He possessed and used Japanese maps, including a Jesuit account of Yezo (Hokkaido) (1622). He also produced a map based on the explorations of Hokkaido and the Kurile Islands by Maarten Gerritsz de Vries in 1643. Although Witsen's map had graduation, it lacked for want of accurate measurement of latitude much less for determination of longitude (see Kizilov 2012: 183).

For Witsen, north and east Tartary were squarely within the Russian empire. Nevertheless, as Bagrow (1952: 90) interprets, Witsen still kept to the old sixteenth-century configuration of northern Asia as suggested by Ortelius, Mercator, and Hondius. Allowing precursor Russian mapping of the Siberian northeast, Bagrow concludes that European maps did not adopt the rectangular shape of the first Russian maps of Siberia. At the Nerchinsk negotiations with China in 1689 between Czar Peter the Great (r. 1672–1725) and the Kangxi emperor, Russian maps were taken as a basis. Even so, Emperor Kangxi was served by the Jesuit linguist Jean-François Gerbillion (1654–1707) who made numerous journeys into Tartary (Manchuria and Mongolia). Other Jesuits also served the mission, including mathematician and mapmaker Thomas. European cartographers gradually altered the shape of Northeastern Siberia through different projections and through adding Russian materials. The work eventually acquired shape under the German cartographer Johann Homann (1702–07), as with his *Geographica Nova ex Oriente Gratiosissima*, or Map of the Caspian Sea and Kamchatka, from 1725. Overall, as Li Narangoa and Robert Cribb (2014: 2) remark in a more recent interpretation, Witsen identified the northeastern part of Asia as a distinct region owing to its history at the center of great empires, especially the Mongols under Genghis Khan (r. 1206–1227) and his successors.

CONCLUSION

In this chapter we wondered as to the Jesuit challenge to Chinese astronomical traditions, alert as well to new awareness of the size and shape of China versus the Ptolemaic stereotypes. As noted, Ricci became the first Jesuit serving the Chinese imperial court as astronomer. In his turn, Adam Schall von Bell would take up a leading position under the Shunzhi emperor in the Imperial Observatory and the Tribunal of Mathematics and with Jesuit Ferdinand Verbeist subsequently stepping in. Notwithstanding its astronomical demonstrations, the Jesuit mission to Siam was obviously a failure, even a "clash of civilizations." Chinese resistance or even rejection of European science in this era may be inferred from our narrative, although mapping and calendrical reform was undoubtedly an exception.

As this chapter has exposed, the broad outlines of the scientific revolution stemming from the Copernican revolution led to the triumph of astronomical investigation, such as vested in national scientific projects. As revealed, the results yielded by the French state project backed by Cassini and with French Jesuits in tow produced some surprising spin-offs as well as setbacks. The understanding that simultaneously conducted astronomical observations taken from strategic sites globally as well as inside China could facilitate the measurement of the true size of the Earth including the Eurasian landmass was actually part of a quiet revolution in geo-science. As mentioned, the French project was also boosted by Colbert's collaboration with Huygens whose pendulum clock and tables based on measuring the Sun's passage offered a new method of accurately measuring longitude (Soll 2009: 99). Though unstated, such critical geographic information would service France's late arriving push in the wake of the Portuguese, Spanish, and Dutch into Asia, especially as it happened in the mid- to late nineteenth century in relationship to late Qing China and Vietnam under the Nguyen. Macau indeed was a crossroads for scientific communication between East and West, but a fuller discussion of this issue is outside the scope of this brief chapter.

Obviously the Jesuits were intrigued to know and map Chinese space both mentally and in line with Cartesian principles. This they did as they traveled across China from Macau in stages to the Chinese capital. On the way they also collected local lore, gazetteers, and examples of local mapping. Taking just Ricci and Ruggieri as examples, as Song Liming (2013: 156) summarizes, "if Ricci is the first person who compiled world maps in Chinese, Ruggieri is the first person who compiled maps of China in Western languages." Arriving back in Europe the atlases produced by the Jesuits under the Ming offered a new image of China's spatiality outside of the Ptolemaic template. Not only was the Cathay of Polo exposed but China was framed province by province. The basic topography and features, including towns, the Great Wall, rivers, and even the approximate shape of the coast line of China was revealed, hitherto drawn on even post-Ptolemaic maps as virtually a straight line.

It is important to note that Emperor Kangxi commanded the Jesuits to chart the entire length of the Great Wall from the Gobi Desert to the sea. Manchuria and eastern Siberia were mapped, followed by Shanxi and Shaanxi. Other missionaries were dispatched to Mongolia. Mapping the Amur River was also charged to the Jesuits, part of a strategy to consolidate diplomatic gains with Russia and with the longitude and latitude of hundreds of points determined as well as measuring distances between key points. The results of the "Kangxi" or Jesuit atlas reached the French King and, as analyzed in a following chapter, would feed into Du Halde's and d'Anville's atlas, as well as broader European knowledge. No less important from our perspective is that Russian mapping, imperfectly accomplished by Witsen and carried through by the St. Petersburg Academy and others, had finally done away with "Tartary" and come up with far more convincing renditions of Tibet, Mongolia, Korea, the Russian Far East and, ipso facto, China's vast land borders, practically a concept outside of imperial history.

NOTES

1. While the Dutch in particular referenced Tenerife in the Canaries, as Jonkers (2005: 6) clarifies, this was not meant to honor Ptolemy but was out of practical considerations. French and English trading companies adopted different practices, generally avoiding the meridian of Tenerife. Common English practice up to the 1750s was to "renounce the supremacy of any prime meridian," and to rely upon way stations and hundreds of meridians. Also, in sailing east, the English measured in leagues or physical distance, rather than degrees of arc or longitude.

2. Notably, the Jesuítasna Ásia record (Codice 49-IV-53 505–513) offers tantalizing references to Jesuit debates on China from within a Ptolemaic framing of Asia. See Leão (1998: 66).

3. Relación de Juan Bautista Román, Macán, 8 de setiembre 1584, Archivo General de Indias AGI Filipinas 29, ANM Colección Fernández Navarrete XVIII, fol. 146, dto. 31 y Archivo de la Real Academia de la Historia, Colección Juan Baptista Muñoz, 9–4797, vol. 18, ff. 249–258 Localización y transcripción: Manel Ollé Rodríguez Universitat Pompeu Fabra, Barcelona <http://www.upf.edu/asia/projectes/che/autors/autors.htm#obrasnopub>

4. Becoming a World Heritage listed site in 2010, UNESCO offers a detailed description of the cultural and symbolic dimension of the observatory. http://www2.astronomicalheritage.net/index.php/show-entity?idunescowhc=1305

5. The Rome-based astronomer Giovanni Battista Riccioli (1598–1671) composed a table of 2,700 entries based upon information received from former Jesuit pupils in China and India concerning lunar and solar eclipses as well as transits of Venus, in turn widely consulted by Jesuit astronomers in the field.

6. Dubbed Quian by Marco Polo, the Yangtze is indicated as Kian on the earliest English maps as with Herman Moll, "The Empire of China and the island of Japan, agreeable to modern history," 1736.

7. Hoai Ngau (or Hoai Ngan-fou) is described in a British traveler accounts as a large walled populous southern city astride the Huang Ho administered by a mandarin in charge of navigation and canals.

Chapter 7

The European Framing of Ptolemy's *India Extra Gangem*

Having examined the Jesuit mapping of Qing China under the Kangxi emperor, this chapter offers a reprise on the European framing of Asian regions in part arising from that project. In particular I reference French mapmaker and geographer Jean Baptiste Bourguignon d'Anville (1697–1782), one of the more influential mappers of Asia of this period and one of the more reflective through his writings. This was the age of the French Enlightenment, as with the writings of Jean-Jacques Rousseau, Denis Diderot, Abbé Henri Grégoire, and other *philosophes* taking reason as their measure and, through their secular approach to life, often falling foul of religious establishments. That certainly fitted the case of d'Anville's contemporary, Abbé Raynal (1711–96) who, teamed up with Diderot, wrote an iconoclastic history of the world, with one part focused upon Asia. With Raynal-Diderot as a foil, what we seek in this chapter is to determine how d'Anville actually drafted his image-making representations of Asia, as with boundaries, scale, dimensions, latitude and longitude, toponymy, and boundaries, familiar to historians of cartography reaching back to Ptolemy, and how they contributed to the European framing and/or redefinition of Asian regions as with India, mainland Southeast Asia, Qing China, and the Russian Far East.

First, this chapter positions d'Anville's scientific mapmaking with reference to classical renditions of Asian space. As a foil to d'Anville's neoclassical pretensions, a second section discusses the new relativist thinking on global geography such as presented by the French encyclopedists and *philosophes*. Then follows an examination of Priest Henri Grégoire's intervention on longitude to the French revolutionary parliament, a literal wake-up call for France alongside English rivals.

JEAN-BAPTISTE D'ANVILLE'S NEW ASIAN WORLD REGION

The last half of the eighteenth century might be taken as a transitional age in the mapping of India, China, and Japan, especially in revising Ptolemaic stereotypes through scientific measurement and direct observation. In the case of India, the British launched their trigonometric survey of the subcontinent starting in 1802. Between 1800 and 1822, the Tokugawa launched their own mapping of the Japanese archipelago using European methods. In 1737, as discussed below, d'Anville produced his landmark *Nouvel Atlas de la Chine, de la Tartarie Chinoise et du Thibet*, a sweeping province-by-province cartographic representation of the Qing empire, incorporating such data as the measurement of hundreds of points of longitude and latitude as well as the distances between points collected by a French Jesuit survey commanded by the Kangxi emperor. Literally, as argued below, European elites came to embrace a new Asian world region, one firmly embedded in Cartesian space representing a major revision to the old die-hard Ptolemaic-Polean image.[1] Even so, this new spatialization of Asia at the hands of d'Anville would be a long intellectual journey of discovery, involving interrogation of Ptolemaic texts, the assimilation of newly minted data arriving in Paris, and the incorporation of the fruits of Jesuit surveys conducted inside China itself.

Becoming leading mapmaker to King Louis XV (r. 1715–74) and member of the academy of sciences in 1773, d'Anville obviously worked in the employ of the state. All his maps and works carried a dedication to his patron the Duke of Orleans. According to Christine Peto (2007: 45), he joined such scholar-geographers as Guillaume Delisle, Philippe Bauche, Jacques Nicolas Bellin, and other members of the French Academie in serving "the interests of those in power and influence." Part of an "epistomological shift" away from humanistic exegesis of earlier mapmakers, as Peto (2007: 45) exposes, this group were "less concerned with the 'cult of image' as with the Sun King (Louis XIV) and more directed by scientific authority and rational thought in service of well-reasoned government," suggesting a new "alignment of mapping and power for a state."

By definition, as Peto (2007: 45) points out, such *géographes de cabinet* (or armchair geographers) rarely left their offices for the field, although they delegated out local research as with the mapping of France. Still it is obvious that d'Anville never traveled East. Another feature of d'Anville's work, as explored below, is that he favored elaborate discussion on maps, especially including Ptolemaic and other classical references, alongside text or *mémoire instructif.* Nevertheless, his map productions were not just intellectual exercises, especially as they were put to use. An example was his 1752 map produced for the French East India Company referencing Pondicherry, along with Chandannagar in West Bengal.

Yet, d'Anville had a free hand intellectually, as with combining antiquarian study of cartography and sources with original research. This was the case with his *États formés en Europe après la chute de l'empire romain en occident* (1771) and *Mémoire et abrégé de géographie ancienne et générale* (1834), achieved in later life. But unlike the ancients whom he critiqued, d'Anville sought to verify place names and other features. Such critical attention shows up on many of his maps, simply by leaving blank space where doubt existed (Peto 2007: 80), and avoiding the invention and fantasies of predecessors. In one map of Arabia he drew upon Arab sources as with *Carte d'Arabie dressée uniquement sur la description qu'en a fait Abdulfeda Ismaël, sultan de Hamah, en son livre de géographie intitulé Takouin-al-buldan, qu'il acheva d'écrire vers l'an de l'ère chrétiéne 1321* [*sic*].

Yet, d'Anville was more reflective about his map of India actually executed in two parts, one "greater India" and the other "greater China," including the Southeast Asia region as it is understood today. In doing so, the two maps assimilated and displayed an enormous amount of geographical knowledge, although also ignoring much local knowledge as well (as with the case of Japan) and even neglecting anterior Portuguese and Spanish knowledge on the peripheries of the better known trade routes as with such places as Timor and New Guinea. A northern coast of Australia is registered but detail in French mapping would await later Napoleonic voyages.

We should not neglect the second of d'Anville's grand Asia maps, namely, *Seconde partie de la carte d'Asie contenant la Chine et partie de la Tratarie, l'Inde . . . Sumatra, Java, Borneo, Moluques, Philippines et du Japon* (1752). With respect to the eastern mainland of Southeast Asia, this map displays an enlarged and bounded Kingdom of Cambodia, in turn bordered to the north by a bounded Kingdom of Laos and with Cochinchina and a misplaced Champa facing Vietnam's eastern seaboard. Tonkin (northern Vietnam) is partially bounded by present-day Guangxi province of China to the east. In the western mainland, the kingdom of Siam is bounded by Laos to the east, albeit west of the Mekong River, and with the Kingdom of Pegu, the Kingdom of Mien [Mien-koeu as known to the Chinese and Marco Polo], and the kingdom of Ava to the northwest and west, and with Malaya (so-named) to the south. Other states as with Arakan, Tipora, Bengal, and Tibet are noted. The interiors of Cambodia and Laos are particularly scarce in detail, just white spaces. The Chinese empire dominates this map and with Japan, Formosa, Ryukyu (Luchu), and Korea named. Island Southeast Asia, by contrast, appears only as geographical representations of island groups (with an added note on the "city of Brunei"). Ancient (charter) states and proto-states on the mainland are thus well identified. Some (Patani), are not recorded. Some, as with Champa located in the Mekong delta, are obviously misplaced though the Champa heartland in central Vietnam had virtually disappeared by this date. But, almost everywhere, the bounds or "geo-bodies" (Thongchai 1994), of the Asian Buddhist kingdoms are awarded fixed boundaries where none before existed.

D'Anville, Ptolemy, and the Construction of India Extra Gangem

Notably, in 1752, d'Anville produced a large format map of India, titled *Carte de l'Inde Dressée pour la Compagnie des Indes par le Sr. d'Anville Secretaire de S.A.S. Mgr. le Duc d'Orleans,* drawn from numerous sources, including Ptolemaic, Turkish, Indian, along with Jesuit surveys, as with the work of Jean Venant Bouchet of 1719 and Claude Stanislaus Boudier of 1734. Reflecting the state of knowledge of this area at mid-century, the map displays fine detail of south India and coastlines. It includes in miniature detailed maps of the mouths of the Ganges, the Arakan, and Sirian (Rangoon) Rivers, Goa and surroundings, Junk Selon (Phuket), the Mergui archipelago, and Aceh with Pulao Way. For a large size map, there was a conspicuous absence of place names, especially in the northern interior.

In 1753, d'Anville published *Éclaircissemens géographiques sur la carte de l'Inde,* an elaborate description of the 1752 map of India. In 1775, by stark contrast, he published *Antiquitégéo graphique de l'Indeet de plusieurs autres contrées de la haute Asie,* a work which deliberately replays classical references to India and China. I have chosen to highlight these two works because of its discussion, however convoluted, on Ptolemy's *Aurea Chersonese*, the elongated promontory which, in its classical version folds round the *Sinus Magnus*, separating India from China, but also obscuring the basic configuration of mainland Southeast Asia as with a clear delineation of the Indochinese and Malay Peninsulas.

Ptolemy's Golden Peninsula Reprised

As mentioned in chapter 4, the configuration by Ptolemy of a so-called land bridge linking Africa and the Southeast Asia region had long deadened accurate European renditions of an open ocean. Even when Polo, not to mention the Arab voyages, demonstrated a communicable sea passage between India and China via the Malacca Straits, the exaggerated Golden Peninsula stubbornly remained in European imagination and, in cartographic representation, variously bisected, fragmented, or dislocated its true location north of the Equator. As geographer Victor Savage (1984: 28–33) points out, the notion of a Golden Peninsula (sometimes island) with Southeast Asia as a source of gold was an enduring trope in Western discovery literature down until the nineteenth century, reinforced by the myth that Soloman's Ophir shared this location. The Ptolemaic world-in-motion also found its place in Camões's *Lusiadas* where, in one stanza, he makes the association between the Chersonese and Ophir, also identifying Sumatra and the Malay Peninsula as its true location.

Essentially, d'Anville's *Éclaircissemens géographiques* divides into five parts; first, the part of India traversed by the Indus and its tributaries; second, the part of India traversed by the Ganges, third, maritime India, extending from the mouths of the Indus until Cape Comorin; fourth, maritime India from Cape Comorin to the Ganges; and fifth, the extension of the coast of India from the Ganges until the entrance of the Malacca Straits. The book is introduced with an essay on the *Itineraire* of India. On the Golden Peninsula, d'Anville wrote:

> I pass now to the Golden Chersonese, or Malayan peninsula. Do not expect that Ptolemy had no fault in its representation. Is it to be better informed of a part of India which is less distant [than China, or] how to determine positions of this peninsula beyond India and the Ganges? Let us give him the grace to have known the name of Malaya for the promontory of the Malacca peninsula which he named Malae-colon. The deep gulf marked to the east of the peninsula, under the name of Perinniclus is drawn from Perimula, a maritime town [that] doesn't exist. Seemingly, in making a correction, of which the *Geography* of Ptolemy is susceptible, to take this gulf as the depression of the Gulf of Malacca, the opening of which was a small island named Pera(k), facing the coast and town of the Perak River [doubtless Pangkor]. This gulf terminates in Ptolemy with a very pronounced promontory named [in Greek] Great, which no better extends than the cape of Romania, and of which the coast turns suddenly north to form the Gulf of Siam, very deep and as recognizable as in Ptolemy as with [in Greek] Great. (author's translation, d'Anville 1753: 137–138)

As mentioned in chapter 4, the Ptolemaic construct of the Golden or Great Peninsula and the true location of Cattigara also confounded Columbus. This was not just an academic discussion because, as discussed below, the Cattigara location even confused the Magellan-Delcano circumnavigation. Map renditions of Ptolemy placing Cape Cattigara at the extremity of Asia are also understood to have lured Vespucci, Columbus, and others into believing that, by sailing westward, they could enter the Sinus Magnus via the strait of Cattigara (as Polo had done) and the route to Malacca. As mentioned in chapter 4, Columbus fell into the trap set by Ptolemaic-revival atlases of exaggerating the eastward extent of the Eurasian landmass, thus deluding him as to a quick westward-sailing passage to the Indies.

Ptolemy's Cattigara Reprised

In *Éclaircissemens géographiques* (1753: 137–138) d'Anville also seeks to locate Ptolemy's Cattigara with specific reference to the Arab geographies:

> According to Edrisi [al-Idrisi], in the first part of the third climate, the first port of Sin [China] is Lucqin. Abdulseda [Abu'l-Fida], Edmadeddin Ismael [1273–1331] adds that this port of Sin is next to a very great river; and I have discovered that Loukin is still the name of a branch of the great river of Cambodia [Mekong] of which the mouth faces Poulo Condore. Thus, the Arab geographers are in accord with Ptolemy to give the same name of Sin to the country that we have judged it conforms. Besides Lucqin, in the second climate, ninth part, corresponding with the same part of the first climate, Idrisi mentions a maritime town, comprising, he says, in the country of Sin under the name of Caitaghora. Then in Ptolemy we find, Cattigara as a port of Sines (author's translation).

By locating Cattigara in the Mekong delta (as in his 1763 map *Orbis Veteribus Notus*), d'Anville would become part of an extended discussion reignited by the "rediscovery" in 1942 of the ancient Hindu trading port city of Oc Eo by Louis Malleret, an archaeologist working with the Ecole Français d'Extrême-Orient. Today, because of siltation, the Oc Eo site is located some 30 kilometers inland from Ha Tien in southern An Giang province of present-day Vietnam or, precisely, 10° 15' 17' N, 105° 9' 6' E. Archaeology has revealed a wealth of Roman relics, evidence of the strategic location astride the long-distance maritime trade linking east and west. By contrast, Malacca was a young emporium not existing in Ptolemy's time. For some, Oc Eo equates with Ptolemy's Cattigara, plausibly derivative of Kota-Nagara (lit. city-state), fitting with the Hinduized *negara* states of the Malayanized Sanskrit world. But why the location of Cattigara even matters is a question addressed by Suárez (1999: 79; 91).

On China

On China, as d'Anville (1753: 160–161) wrote in *Éclaircissemens géographiques,* "It is perhaps sufficient to demonstrate the great displacement of the capital of China made by Ptolemy, without rigorously determining its true location. I have researched this." D'Anville (1573) then turns to a discussion as to the true location of Polo's Zaiton. "I am persuaded to have read in

Marco Polo that the name of Zaiton placed it between the middle provinces of China and Champa."

> Abultieda cites Zaiten as a port in the country of Cin, and on the reports of people who have been there, attests that it is distant a half-day journey from the sea along a freshwater canal, such as verified by the situation of Sin-hoa. Marco Polo calls it one of the richest markets in the world comparing its commerce to that of Alexandria. And of whom Emperor Coblai [Kublai Khan], ruling at this time, drew great revenue. (author's translation, d'Anville 1753: 160–161)

Obviously this kind of discussion added little to the facts of navigation and hard data accumulating on China in the age of the Canton trade.

D'Anville's Treatise on Classical Geography

As d'Anville explains in an opening page of his *Antiquité géographique de l'Inde* (1775: iii), a virtual treatise on classical geography, in 1752 he was solicited by persons on high to draft a map of India of some scope. In first resisting this idea, he cited lack of knowledge of vast interior spaces across the subcontinent. But, as his superiors answered back, his task was to illustrate what was known. It also occurred to him that, entering print in multiple copies and widely distributed, such a map would feed the rivalry of great nations competing in commerce, and even those maneuvering for war. In fact, as he acknowledges, French and British military officials were already exploring these parts, and mapping these empty spaces including even the upper Ganges and parts of Tibet. He also expressed repugnance at reprinting *Éclaircissemens géographiques*. Rather, he sought to treat on "ancient geography." Rather than just enumerating the names of towns, he argued, it was infinitely more vital or interesting to investigate the expeditions made by Alexander the Great from Lahore to Kabul to Kandahar and the lower Indus. The study of the ancient Western world and the Orient were different tasks.

> The islands known to us only from Ptolemy, as enclosed in the gulf formed by the Ganges, would take their place in the Oriental Ocean having been displaced by more than 600 leagues. Can we conceive that three little islands, under a single position after Ptolemy, without any other question in the *Geographie*, and understanding a more distant existence in the Gulf of Siam, having been applied to all that we know of Japan, between six and seven hundred twenty degree places beyond the point where it was necessary to stop? (author's translation, d'Anville 1775: xj–xij)

Further noting of China in *Antiquité géographique de l'Inde,* as d'Anville explained:

> The name of Sina (Sinae) of which there are monuments from antiquity, only reveals that Ptolemy no longer holds up as before in designating China and the Chinese nation. *La Serique* [or Seres, the classical Greek name for China] no longer occupies the northern part of China until the sea, neither is it limited in all its extent in Tartary, but it is contiguous with Tibet, between the north and the west (*le couchant*) with regard to China. (author's translation, d'Anville 1775 xij)

Certainly d'Anville is correct in deriding the false divisioning of China between *Sinae* and *Seres*, dating back to the ancients, though it was actually the first-arriving Jesuits in China who made this revelation.

By way of illustrating d'Anville's method, *Antiquité géographique de l'Inde* included a separate map *Ad Antiquam Indiae Geographiam Tabula* (1765), displaying India and part of Southeast Asia, with an inset in the lower part entitled *Ptolemaicae Tabulae* showing the island of Ceylon also labeled Taprobane. The Bay of Bengal is denoted as the *Gangeticus Sinus*; the Malay Peninsula is denoted *Aurea Chersonese*; its southern tip Magnil Prom; the Gulf of Siam is still denoted *Magnus Sinus*. The Andamans zone (still host to pre-contact tribal peoples) is marked anthropophagi. What he seeks is a "simple exposition of the facts and their consequences." What then emerges in this map is a hybrid modern map matrix with coordinates added, but with Ptolemaic and/or Latinized names embedded.

In a first section, *Antiquité géographique de l'Inde* deals with the "Part of India enclosed by the Indus," including the etymology of India, Hind, and so on as comprehended from classical times. In particular, he exposes Ptolemy's disregard for facts as revealed by history and places as confirmed by geography. Writing of the descent made by Alexander down the Indus, d'Anville (1775: 33) writes, "We are not surprised from this article and other adjacent objects that similar detail in Ptolemy, does not account for the historic circumstances of Alexander's expedition." As he concludes this section, "Alexander's expedition was the single largest morsel (*morceau*) informing ancient history with regard to India" (d'Anville 1775: 47).

Matching the unparalleled collections of the Royal Library (the future BNF), d'Anville's sources are vast and recondite, and give us some idea as to an approach to this subject. Even so, few local Indian sources are brought into play. Taking a count of indexed entries as a guide, they include classical Greco-Roman scholars, Aristotle, 6; Strabo, 23; Pliny, 37; Eratosthène, 5; Pomponius Mela (*De situ orbislibri III.*), 6; the Periplus, 34; Arrien (*Indica, Expédition d'Alexandre*, 36; alongside Ptolemy. On the Muslim side, the

Turkish Geography of Abidindan; the Persian author Sheriffudin, 9; al-Idrisi (a Maronite translation), 18; Abu-rihan, 5; Ebn Said, 3; Gezirat al-Arab (Gezirat Andalos); Abu Rayhan Al-Biruni on latitude; and Abdulseda (1320), 14; Marco Polo, 20; are consulted. On the Portuguese side, João de Barros, 14; Garcia de Horta, 1; and Manoel Pimentel, *Arte de Navigar* (1699; 1712). Of the French (and English) travelers, François Bernier, *Voyage*, (1710), 2; Jean de Thévenot, *Le Voyageur*, 8; John Ovington, *A Voyage to Surat in the Year 1689,* 1; and T. S. Bayer citing a Tamil source, 5. French collections include Renaudot's *Anciennes Relations . . . des voyages* (1718), 8; French mapmakers, Guillaume Delisle, 15; Edward Terry, *A Voyage to East-India* (1655); 3. Jesuit sources include Jean Hardouin, 8; Herbelot Molaninville, *Bibliothèque Orientale*, Pierre du Jarric, *Historie de Inde* (1610–1614), 4; François Sanson, *The Present State of Persia* (1695) 11; Jerome Osorio, *History of the Portuguese* (1752) 1; Etienne Souciet *Mémoires*, 1; and Charles Le Beau, *Liban Historie du bas empire* (1770).

In a second section of the book entitled *The Part of India Traversed by the Ganges*, d'Anville offers a complement to the discussion on the Indus. Here he draws upon works of antiquity, but is also alert to new information as to positioning with regard to longitude, such as revealed from the 1730s by the Jesuit missionary Claude Boudier of the French establishment at Chandannagar (comparing meridians with Jesuit Antoine Gaubil in Beijing), and with separate calculations relating to Fetipor and Agra. Ptolemy, he allows, improved upon Strabo by giving place to five mouths of the Ganges, each named (though on most other matters offbeat) (d'Anville 1775: 73).

In a third section of the book titled, "The Part of India prolonged towards the middle," Gujarat, the Gulf of Cambay, Surat, the two coasts of India, Cape Comorin, and Ceylon are all discussed. With respect to the Gulf of Cambay, a site of intense trade activity, d'Anville (1775: 85) finds the author of *Periplus of the Erythraean Sea* better informed than Ptolemy on this region. The longitude of Surat was crucial for map construction. He also notes that the determination of longitude for Goa, namely that written in the tables of cosmographer Manoel Pimentel (1712), was way off compared to the measurements conducted by Jesuit mathematician François Noël, such as discussed in chapter 6.[2] Of the Malabar coast as far as Cape Comorin, d'Anville again compares Ptolemy unfavorably to the author of the *Periplus* and with a major distortion of the extremity of the peninsula. Determining latitude of the promontory—around 8°—was also contentious (d'Anville 1775: 119).

D'Anville then enters a long detour on Taprobane, observing the Latinization of its form via Pliny, Pomponius Mela, and Ptolemy. As he understood, it was Alexander's expedition that brought knowledge of the island to classical learning. For Hipparchus and Mela, the locale could have been closer to Ethiopia. This was truly Ceylon, he confirms, but also transported to a longitude closer to Sumatra. At least its pear shape is correct in Ptolemy, but (in reality) proportionally four times smaller. Ptolemy correctly marked the Mowil-Ganga River entering the Bay of Trincomalee. Ptolemy traced three rivers and mountains (plausibly Adam's Peak, well known in Arab lore). Elephants are attested; but Ptolemy exaggerated size. Ptolemy also called the island Simundi (also Salice) —Serendib—Sailon. Obviously no vestige of the name Taprobane exists, nor Simundi. Ptolemy also located two towns on Taprobane, both in the heights of the source of the major river (Sindcanda/Canda-uda). We should not neglect the Maldives and Laccadives (as mapped and measured for latitude and longitude by English navigator Henry Cornwall, also informing d'Anville's maps). Notably, Ptolemy placed a multitude of islands facing Taprobane (actually numbering 1,378), possibly a misplaced Maldives (actually numbering 1,900 islands). But he also offers closer detail on nineteen islands, a number precisely matching the nineteen Laccadives (d'Anville 1775: 157). D'Anville reinforces this view by appending a map *Insulaenumero xix apus Ptoleaem.* He also cites al-Idrisi to add information broadly confirming the 1,900 island hypothesis, albeit separated by two seas, suggesting the inclusion of the Andamans in the Bay of Bengal (on the route to Java). Summing up, as d'Anville (1775: 159–60) states, in treating these questions he seeks to establish "a correspondence of positions in different ages, more or less posterior to that of antiquity." After all, he points out, Ptolemy's maps of this side of the Ganges, not including Tabrobane, fill ten pages in folio of his Atlas in two columns "yet revealing a great drought of knowledge." He reprints the Ptolemy map "to show up its imperfection."

In a fourth Part or Supplement (First Memoire), "Limits to the Known World of the Ancients beyond the Ganges," d'Anville (1775: 161) writes that "Ptolemy is the first of the ancients to bring us beyond the Ganges, and, accordingly, marks the mouths of the Ganges as his point of departure." But any discussion on Ptolemy also raises issues of longitude and its measurement. D'Anville agrees with Ptolemy to mark the Canaries or Islands of Fortune as zero meridian). However, what Ptolemy indicated as 130° was determined by observations at Chandannagar by Boudier in 1732–32 as 86°4' or 5' (actually 88°3800' E. by modern reckoning). Obviously this is a big discrepancy and was an early confirmation of Ptolemy's exaggeration of the eastern extent of the Eurasian landmass.

On Cingapura/Singapore

Relating a lecture he delivered at the Royal Academy of Belle Arts in 1763, d'Anville claims not to be preceded by any other writer with regard to a fitting location of the *Magnum promontorium* or Great Promontory and "Zaba," as it relates to Ptolemy's *Geographia.* However, following a careful reexamination of de Barros's *Decadas*, he modified his opinion, accordingly:

> I was surprised to see in this historian, that the *Geography* is in no way indifferent but rather truthful on this position. He states (Book V of Second Decade, chap. 1.), that in ancient times and before the foundation of Malacca, the most considerable place in this country was Cingapura, on the most distant point of land on the continent of Asia, so close to the Equator as to be less than a fraction of a degree. He adds that this point must be the tip of the great promontory marked on the map by Ptolemy, upon which he places the great city of Zaba, *ondesiuou a grandecidade Zaba [where the great city of Zaba is located],* the general rendez-vous and meeting place for navigators from the Occident and Orient. (author's translation, Anville 1775: 171)

According to Thomas John Newbold (1839: 44), *Political and Statistical Account of the British Settlements in ... Volume 2* (citing Wilford), Zaba (from the Sanskrit Zabaja) was an emporium back to even Ptolemy's time, remaining so through to the period described by the two Muslim travelers entering Renaudot's account or even longer. In Newbold's time it was called Batu Sabor on the Johor River. By contrast, without dismissing the Johor location, T. Braddell (1887: 239) contended that it was more probably "only a variation of the Jaba or Java." While de Barros would associate Zaba with Singapore, modern scholars place Zaba as an island off the coast of Indochina (Lach 1965: 507). Zaba obviously did not survive but Cingapura gained a longer life, as with Ortelius (1570); Lambert Andreas/J. Metellus, *Regnum Sian* (1596), but rendered Singapura with Barent Langenes (1598), and Theodore de Bry (1603).[3] Modern archaeology and modern Singapore's search for historical roots actually reinforce the classical view.

On Siam

Seeking to interpret Siam against Ptolemaic points of reference, as well as ruminating upon nomenclature, as d'Anville states:

> Deep in the Great Gulf, Ptolemy show us a river between the principal, under the name of Serus. In fact, the Siamese call this the Menam. This is not a proper

> noun but an appellative used to describe rivers, the same as the Ganges or the Sind with respect to different parts of India. Nevertheless, I should remark that the true name of Serus should read Sri, which I found in an ancient chart, just like the entry of the Menam or Siam river into the sea. I admit that it was with a certain repugnance that I wrote Siam here in my map of Asia, instead of writing Sian as it should be, after the Portuguese orthography Sião. (author's translation, d'Anville: 1775: 173)

Here, d'Anville reflects that Simon de la Loubère, head of a French mission dispatched to Siam by Louis XIV and arriving in 1687, preceded him in writing "Siamois" and so, he didn't challenge the new convention "despite attachments I owe to original forms of denomination."[4] He also acknowledged that it was the Portuguese who bequeathed (to Europe) the first rendition of names of these (Asian) polities. Sian was indeed the early standard, as with Ortelius (1570), and Andreas/Metellus, *Regnum Sian* (1596), actually using Sian to designate the Southeast Asian mainland in general (Suárez 1999: 187). The Italians Giacomo Cantelli da Vignola/Giacomo Giovanni Rossi (1688) were also phonetically closer to the Portuguese, with Sian, as was Petrius Bertius (1616).

D'Anville's China Mapping

Map Representations of China as Mediated by Du Halde

As introduced in chapter 6, Du Halde's *Description* (1735) was visually enhanced by forty-three maps drawn by d'Anville. Commissioned by the Kangxi emperor from Jesuit surveys, as mentioned, the maps were xylographically printed in Beijing and then sent to Paris to be engraved by d'Anville. They comprise the second European atlas of China, the first of Tibet and Korea, and included the first published map and account of Vitus Bering's expedition of 1733–43 to the North Pacific—this being the first map to depict any part of Alaska. Besides the maps, Du Halde's book also offers sixty-five plates after Antoine Humblot, including eight double plates (costumes, village views, hunting, fishing, etc.), ten city plans, one plate showing the Beijing observatory, a portrait of Confucius, a numismatic plate, and a music plate. It is noteworthy that Ryukyu was not included in the Jesuit surveys or in the Du Halde-commissioned maps. This lapse was observed by Jesuit Antoine Gaubil who, in 1752, sent his own rough maps of the island chain to the French government with a suggestion to establish relations with the Ryukyu kingdom, also feeding into the cartographic productions of French astronomer and Siberia explorer, Joseph-Nicolas Delisle (1644–1720) (Szcześniak 1955).

D'Anville's China mapping stands apart from his deep and abiding concern to intellectually connect up or, at least, ground his cartographic research especially on India and *India Extra Gangem* within the classical tradition. In any case, as highlighted in the Introduction, China hardly figured in the Ptolemaic template. Yet, there is another reason for d'Anville's switch of interest, namely the wealth of empirical data entering Europe through the agency of the Jesuits, via the so-called Jesuit letters, and also first news of the bold new China-wide map survey being conducted by the Kangxi emperor of China. This would be a decade-long process. In France, the Duke of Orleans was acting as regent, and with Louis XV becoming the king.

On April 15, 1728, Du Halde initiated contact with d'Anville with a view to commissioning maps for his forthcoming *Description.* Under the agreement, d'Anville was to miniaturize and design maps of the Chinese provinces as well as to create three general maps. The first volume contained two general maps and twenty-five maps and plans of China, the fourth volume contained twenty-five maps. It was Du Halde who supplied him the source material, notably those maps commissioned by Emperor Kangxi and executed by the Jesuit missionaries in-country in collaboration with Chinese officials. Commenced in 1707, the work was completed in 1715 and published in Chinese and Manchu. On their part, the Jesuits sent to Europe traces of these maps, and with toponyms transcribed into Roman letters. D'Anville was charged to produce a miniature. In fact, d'Anville produced seventeen provincial maps each titled *Carte préparatoire* and four general maps, namely, *Mongolie, Mandchourie, Corée, Japon et N. de la Chine*; *Carte générale du Tibet ou Bouttan et des pays de Kashgar*; *Carte générale de la Chine*; and *Carte générale de la Tartarie chinoise.*

Adding to his fame, d'Anville's map of China and Central Asia (1734) was included in the first edition of Du Halde's *Description*, a by-product of the Jesuit mapping of China (comprising forty-two separate Chinese province-level maps), in turn based on the systematic geographic survey of the entire Chinese realm (ca. 1700). As with Du Halde's book, d'Anville's maps of China became standard sources throughout the eighteenth century, and his maps went through many reprints as with the English language rendition of *Description.* Faithful to the original and the centerpiece of d'Anville's China oeuvre in-sofar as it captured the vast expanse of the Qing empire, it encompasses Mongolia in the north, Korea in the east, while diplomatically leaving white space in the southwest adjoining Vietnam, Laos, and Pegu (present-day Burma). Macau is indicated. A large island off the coast of Fujian, reasonably represented, carries the dual label Formosa-Taiwan. Encased in a longitude/latitude grid, it is significant that the top of the map represents longitude according to the "meridian of Peking," ranging from around 15° W to 20° E. However, the base of the map is calibrated "longitude from London" placing Beijing at 114° E.

To be sure, as Marcel Destombes (1976) demonstrates, the original source materials used by the Jesuits were not Occidental but derived from local (Ming-era) maps. D'Anville owned an original woodcut from a 1721 atlas published in Chinese and Manchu. The Orientalist Julius von Klaproth claims that he bought copies of the Jesuit maps used by d'Anville later sold to the Russian government in 1818. It seems likely that the city plans were copied in the palace by the Chinese for Gaubil during the period 1727 or earlier until 1730 and sent to Jesuit mathematician and astronomer Etienne Souciet, his Paris correspondent. In turn, the copies were finally handed over to Du Halde, at a time when d'Anville began working on the maps for these volumes (see Haguet 2011).

D'Anville's *Nouvel atlas de la Chine, de la Tartarie chinoise, et du Thibet*

In 1737, as alluded, d'Anville published his *Nouvel atlas de la Chine, de la Tartarie chinoise, et du Thibet*. In Europe this was acknowledged as the second major atlas of China produced since the Martinus Martini's *Novus Atlas Sinensis* published by Blaeu in 1655. Launched in Holland, d'Anville's *Nouvel atlas* was a companion work to Du Halde's *Description.* As mentioned, d'Anville partnered with Du Halde in supplying the maps and with his work remaining a standard Western source for the geography of China and adjacent regions until well into the nineteenth century, when it was finally superseded by more accurate maps.

In addition to the maps on China and Korea, Du Halde's work included the highly important map, *Carte des Pays traversés par le Capne Beerings,* containing the first printed configuration of any part of Alaska. Both the report and map, were based on manuscripts that the Russian explorer Bering had presented to the King of Poland who, in turn, presented them to Du Halde for inclusion in this work. Since the Jesuits were not directly sanctioned to work in Korea, the "Tartar lord" (Mukedeng), a reference to the Manchu official, Mu Kedeng, referred to as a "troubleshooter and trusted assistant for the Kangxi emperor," made measurements and observations in the Manchurian and Korean regions in 1709 and 1710. The French Jesuits Jean-Baptiste Régis, Pierre Jartoux, and the Austrian Jesuit Xavier-Ehrenbert Fridelli, were also involved albeit under constant surveillance while in Korea. The "Tartar lord" was, however, given a map used by Régis to produce a map that came out in the Kangxi atlas. This, in turn, was used and edited by d'Anville and published in Du Halde's *Description* under the name, "Royaume de Corée" (atlas de la Chine 1737).[5]

All in all, the atlas contained the following forty-two maps comprising both general and province maps. To adhere to the original, we maintain the French nomenclature and numbering. The general maps comprised, respectively: 1. *Carte le plus générale de toutes comprenant la Chine, La Tartarie& le Thibet*; 2. *Carte générale de la Chine.* The Chinese province maps comprised 3–18. Then followed, *Carte générale de la Tartarie chinoise*; 19, and a series of twelve separate "*Feuille particulière,*" respectively, numbers 20–31, including "Feuille particulière Korea; 31." Then followed, *La Royaume de Corée Thibe*; 32, *Carte générale du Thibet*; 33, and nine additional "Feuille particulière," respectively, numbers 32–41, including "*Siberia/Beering Street*" and, finally, 42, "*Voyage du Capitaine Beerings.*"

According to map specialist, Mario Cams (2013b), alongside material supplied by the missionaries in China, d'Anville consulted a range of non-Jesuit sources. They included his correspondence with Joseph-Nicholas Delisle (1688–1768), Guillaume Delisle's brother and head of the School of Astronomy in St. Petersburg, as well as a manuscript Dutch chart depicting the seas adjacent to Japan. Both inspired d'Anville to redraw the islands off the coast of Tartary and to include Japan on the general map of the region. All the Jesuit and non-Jesuit sources eventually allowed the cartographer to complete the *General Map of China* (1730), the *General Map of Chinese Tartary* (1732), and the *General Map of Tibet* (1733).

It is true that later-arriving colonialists—French especially—would indeed construct boundaries through gunboat diplomacy but, in his age, d'Anville was both celebratory and mocking of Ptolemy's follies at the same time. He also assimilated vast new sources of cartographic data. Little wonder then that, building upon Jesuit knowledge, d'Anville mapped the provinces and provincial boundaries of China like no one else before him. But for a powerful centralized tax collecting-conscript-raising bureaucratic empire like China, the Chinese were more than thankful. As a final note and, as taken up in the following chapter, one Southeast Asian country that does appreciate d'Anville today is Vietnam, especially his rendition of Qing China with its southern boundaries drawn at Hainan Island thus missing out the South China Seas, including the Paracels and Spratly Islands, adding cartographic support to Vietnam's dynastic and current territorial claims over its "Eastern Sea" (see Figure 7.1).

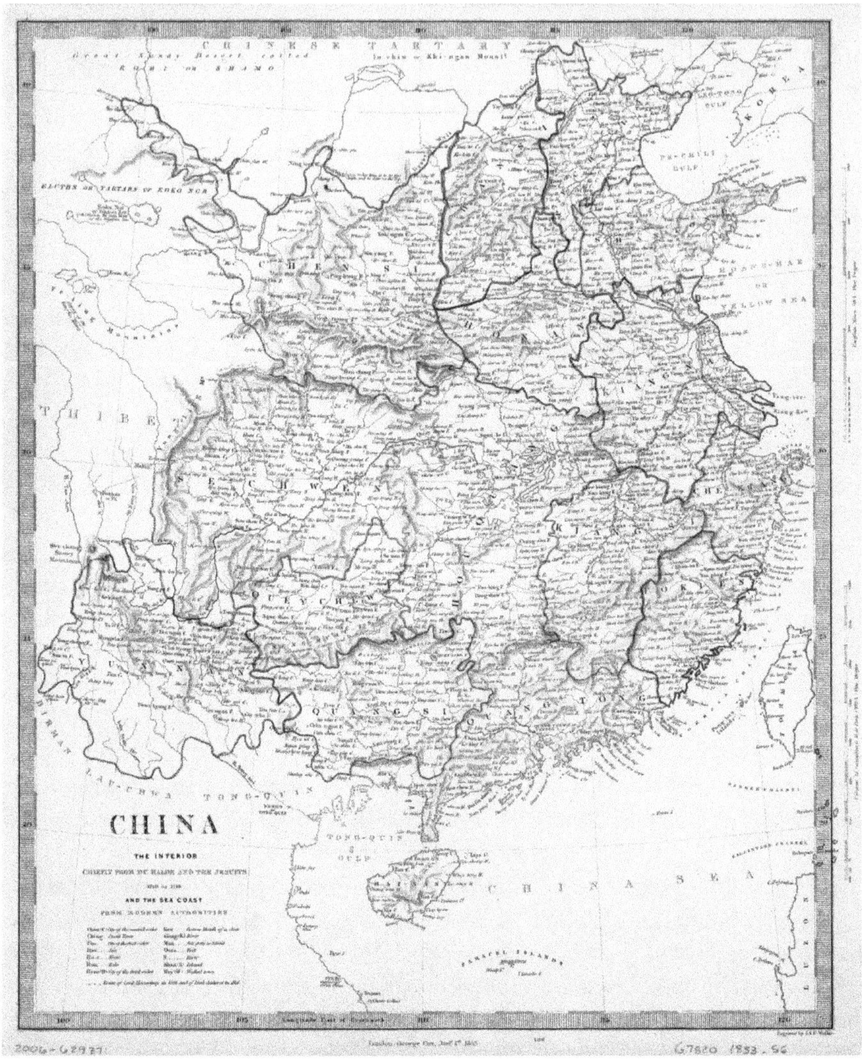

Figure 7.1 ***China: The Interior, Chiefly from du Halde and the Jesuits 1710 to 1718, and the Sea Coast from Modern Authorities*, London: George Cox, January 1, 1853.** One of a number of maps of China of this age derivative of the Jesuit-d'Anville template. *Source*: Retrieved from Library of Congress, Geography and Map Division. Digital Id http://hdl.loc.gov/loc.gmd/g7820.ct005224.

FRENCH ENCYCLOPEDISTS: A NEW RELATIVISM

European determinism and triumphalism over colonial expansion did not go unchallenged even in its times, although it took rare individuals and circumstances to challenge state power, church hierarchies, prejudice, and powerful vested interests in the colonial project, slavery included. Figures associated with the French Enlightenment led this double-barreled critique against both the established order (the *ancien regime*) and the colonial-structured framing of the globe especially by injecting a new relativism into the "debate."

The dominant and unlikely figure in this movement was, as introduced, the French cleric Abbé Raynal, author of the multivolume *L'Histoire philosophique et politique des établissements et du commerce des Européens dans les deux Indes* (A Philosophical and Political History of the Settlements and Trade of the Europeans in the East and West Indies). First published anonymously in French in 1770, it was followed by several expanded versions, especially including contributions by fellow *philosophe* and author of the landmark *Encyclopédie, ou dictionnaire raisonné des sciences, des arts et des métiers*, Denis Diderot (1713–84). Running to a million words, and with Diderot possibly contributing a large percentage, the scope of Raynal's *Histoire* is global, especially relating to Spanish conquests in the Americas and the West Indies; the Portuguese conquest of Brazil; and the English and French colonies in North America, including the institution of slavery which he condemns. However, as discussed below, in one dedicated volume he expands upon European trade with Arabia and India and the sequential conquests of the Portuguese, Dutch, and English in the East Indies and Asia. In a final chapter running to hundreds of pages, Raynal also examines commerce, religion, slavery, among other topics, all from a French Enlightenment perspective, alternated with tirades on political and social questions. Much of the *Histoire's* success, and subsequent reputation, was based upon its attacks on tyranny, slavery, and colonial exploitation.

Not surprisingly, the French crown condemned the book as "dangerous" and "contrary to the principles of religion." A decree of Parliament of 1781 ordered Raynal's arrest and the confiscation of his property. The Catholic Church officially banned the History in 1774 and ordered copies to be burnt. As a result, Raynal was forced into exile. Nevertheless, the book gained widespread attraction for its peons to liberty, including in the North American colonies. As discussed below, he offered a laudatory view of China, comparing it favorably to European barbarism, as evidenced by complicity in the slave trade and colonialism, and so on. More than twenty editions appeared by the time of Raynal's death in 1796, including the translation into English by J. A. Justamond (1784) to which I refer below.

The Asia World Region in Enlightenment Vision

The message in Raynal-Diderot, as developed in Volume I, is one of an increasingly "globalized" (not his term) space knotted together by the combined and sequential actions of the Portuguese and Dutch, English (and French) trading companies, literally linking the Persian Gulf to the Moluccas in a vast trading-commodity exchange system mediated only by local dynasties and despots (enlightened in the case of China). With the exception of a passing aside on the mythologized Atlantis Island, this work is devoid of classical references and thus completely outside of a Ptolemaic frame of reference.

The very first line of this tome sets the tone, "No event has been so interesting to mankind in general; and to the inhabitants of Europe in particular, as the discovery of the New World, and the passage to India by the Cape of Good Hope" (Raynal 1784: 1). Notably, the first volume of *Histoire* incorporates a general introduction to the entire oeuvre, the sequence is "Discoveries, wars and conquests of the Portuguese in the East Indies"; "The settlements, wars, Policy and Trade of the Dutch in the East Indies"; and "Settlement Trade and Conquest of the English in the East Indies."

But, as evoked by the title of his oeuvre, Raynal divides the world into West (Indies) and (East) Indies. So, in the case of Asia, all the ancient nations and civilizations are equally swept or subject to Portuguese, Dutch, and English depredations. Possibly for the first time in European letters, European colonialism as a system of political and social domination is foregrounded. Civilizations, nations, and regions certainly enter the picture with lengthy portrayals and asides drawn from contemporary accounts, as with India and China. However, their peoples and cultures are also profiled as subject to new forces unleashed by the European encroachment. Breaking from the othering implicit in a range of travel collections from Polo down, the Enlightenment literature subjectifies as well as objectifies the colonial subjects-in-making. Raynal foreshadows the eclipse of the Mughal empire at the hands of the English in India; he prefigures the eclipse of the "Sanskrit complex" of Southeast Asian polities—variously, Burma, Siam, Cambodia, and Java, to Dutch and English commerce.

On the two coasts of India, Portuguese, Dutch, English, and French invade, trade, nibble away, fortify, plunder, attack, and wheel and deal with local tyrants. This is a narrative of creeping colonialism before full-blown imperialism, but the seeds are there. While slavery does not enter the picture of European commerce in the East Indies as it does in the West, plunder (the Moluccas) and unequal exchange certainly does. Portuguese wars were religious wars. Following Afonso de Albuquerque's subjection of Malacca, "the kings of Siam, Pegu, and several others, alarmed at a conquest so fatal to their independence, sent Ambassadors to congratulate Albuquerque, to make him an offer of their trade and to desire an alliance with Portugal" (Raynal 1784: 124). "They supposed that the Pope in bestowing the kingdoms of Asia upon the Portuguese monarchs, had not withholden the property of individuals from their subjects. Being absolute masters of the eastern seas, they extorted a tribute from the ships of every country; they ravaged the coasts, insulted the princes, and became in a short time, the terror and scourge of all nations."

Otherwise, "A mixture of avarice, debauchery, cruelty and devotion, prevailed everywhere in their manners" (Raynal 1784: 184–145). However, as the narrative continued, "the decline of the maritime power of the Portuguese, encouraged the Dutch to attack the settlements belonging to that nation, and greatly facilitated the conquest of them." Holland stepped into the breach, mainly though small trading posts, but also taking the monopoly of spices. Dutch highhandedness and avariciousness is not spared from Raynal's pen, although mindful of the superior organizational skills of the Dutch East India Company even when weighed up against losses in Formosa, Tonkin, and Siam. Local tyranny and complicity is not neglected, and Raynal does not shy away from condemning the role of English and Dutch in procuring Indian opium for the Southeast Asia-China trade (Raynal 1784: 482–484).

In a long chapter on the English on the coast of India, he wrote, "The conquest of Bengal, the boundaries of which have since that period been extended, as far as that heap of mountains which separate the Thibet and Tartary from Indostan, without making any essential alteration to the external form of the English company, has produced a material change in the object of it. They are no longer a commercial society, they are a territorial power" (Raynal 1784: 507). As Raynal concluded this book and volume, "Unfortunate Indians! Endeavour to reconcile yourselves to your chains. In vain have your supplications been carried to the ministry, to the senate and the people." "Do you think that the nation, when roused, as it will be, from its present delirium and intoxication, will not call upon you to answer for your oppression?" As Raynal teases the reader—and lectures the natives (doubtless also outraging the authorities)—if justice is not forthcoming, "What are you waiting for" (Raynal 1784: 434–435). These were incendiary words indeed against colonial tyrants.

Additional versions of the book included maps of the so-named regions engraved by Thomas Kitchin (1718–84), "hydrographer" to the English king. Mapmaker and author, Kitchin is also known for "The World From the Best Authorities," published in William Guthrie's *New Geographical Grammar*, 1777. Specific to the Justamond translation of 1784, he added his, "*Map of the European Settlements in the East Indies and on the Coast of Africa Including Part of Europe*," undoubtedly incorporating the broad state of knowledge of Asia at that time. Siberia makes an entry and with Tartary divided three-fold; independent, Russian, and Chinese, respectively. Kamchatka, the Kuriles, and Yezo or Hokkaido are indicated although, together with Honshu and Kyushu, are not well formed. Papua is an island separate from New Guinea. The eastern coast of Australia is roughly indicated as per the Cook discovery of 1770. The Indian Ocean littoral is well developed, as is the archipelago, Indochina, and the coast of China, with rivers well demarcated and named. An inset map displays "European settlements" on the two coasts of India. No evident Ptolemaic elements remain on the map, although longitude, as indicated, is measured east of Ferro (following Ptolemy), and with London situated at 18° E.

Tartary

The vast region of Tartary came to be defined and redefined, not only on European maps but in text through the seventeenth and eighteenth centuries. As Raynal (vol. 2) (1784: 246–249) pronounced, there was a vast difference separating Russia and China, also summarizing shifting conceptions and nomenclature over the ages:

> Between these two empires, the greatness of which astonishes the imagination, there is a great space, known in the earliest times by the name of Scythia, and since by that of Tartary. This region, taken to its fullest extent, is bounded on the west by the Caspian Sea and Persia, on the south by Persia, Hindustan, the kingdoms of Arracan and Ava, China, and Corea; on the east by the Eastern sea; and on the north by the Frozen ocean. One part of these vast deserts is subject to the Chinese empire; another is under the dominion of Russia; the third is independent, and is called Kharism [the region south of the Aral Sea], and greater and lesser Bucharia [Bokharia or "Chinese Turkestan"].

As described, the inhabitants living by fishing and hunting were nomadic and had an aversion to the settled life. Tartars of the present bear striking resemblance to their forefathers. More generally, the people were followers of the Great Lama residing at Putali, a location partly in Tartary and partly in India, "called Bhutan by the inhabitants of Hindustan, Tangut by Tartars, Tsangli by Chinese, and Thibet by the Europeans." The religion of Lama, he continued, was adopted "all over Thibet, and Mongolia; almost universal in lesser and greater Bucharia, and several provinces of Tartary and has some followers in the kingdom of Cashmere, in India, and in China." Noting the passion of the Tartars for war and plunder, "it was to check the inroads of these robbers into China that three hundred years before the Christian era, that famous wall was built, which extends from the river Huambo to Kamchatka." He also expounds upon the Portuguese priority at Macau as conceded by Ming China alongside the rejection (defeat actually) of the Dutch who backed off to the island of Taiwan, at least until their expulsion.

REVOLUTIONARY PRIEST HENRI GRÉGOIRE'S INTERVENTION ON LONGITUDE

Coming into being in 1795, a Bureau des Longitudes was established in Paris under the responsibility of the Paris Observatory, with Giovanni Cassini as one of its founding members. More broadly conceived with the goal of wresting back "*la maîtrise des mers aux Anglais*" (mastery of the seas from the English), it was vested with the improvement of nautical navigation, the standardization of timekeeping, geodesy, astronomical observation, and determination of longitudes at sea (Bureau des Longitudes, 2014). In other words, harnessed to a national project, the Bureau sought to advance France's own national and colonial project in the face of intra-European competition such as playing out in Africa, India and the Asia-Pacific. Undoubtedly the events surrounding the French Revolution of 1789 slowed the French project but it also had its supporters from within.

In this respect it is of interest to refer to the report made to the French National Convention meeting of June 25, 1795 on the establishment of the Bureau des Longitudes. This was Year II of the French Revolutionary calendar. Author of this report, Henri Grégoire (1750–1831), also known as the Abbé Grégoire, was a French Catholic priest, revolutionary leader, abolitionist, progressive reformist, engaged "politician," and litterateur. After Napoléon Bonaparte took over, he became a member of the legislature and senate.[6]

In referring to the "war of 1761" (or the Seven Years' War between France and Great Britain commencing in 1756), as Grégoire adverted, English "success" proved that naval superiority often determines the outcome of wars. Accordingly, one of the most effective ways to stifle "British tyranny" would be to compete on their terms, namely in harnessing the state to the service of science. The English, he observed, had spared no expense in "pushing this science to the point of perfection," convinced as they were that "without Astronomy they had neither trade nor navy." Alongside astrology, he pointed out, superstition still existed in France as with popular belief in meteors, auroras, and comets. Astronomy, however, was on a different plane. "Finally, without Astronomy, Geography is still in the cradle." "However, it is through celestial observations and experience in knowing the meridians that we have determined the shape of the Earth and revealed the true system of the world" (Grégoire 1795: 1).[7]

Looking back to antiquity, as the priest-turned revolutionary observed, the legacy of astronomy's impact on the marine and trade is revealed in "the glory and richness of Phoenicia, Carthage and Rhodes." To make this point, he also struck a biblical reference to the ships of Soloman and Hiram making Ophir the main port for commerce with the countries bordering on the Red Sea and Indian Ocean, although Ophir also has African and Asian points of reference, as with India, Ceylon, or even further east ("With its help, the combined merchant fleets of Asiongaber lashed to Ophyr)." He also referred to the semi-mythologized account of Hanno the Navigator, a Carthaginian explorer ca. 500 BCE celebrated for his naval exploration beyond the Straits of Gibraltar and south down the coast of Morocco ("Hannon, a race in 26 days until pushed to Senegal, and consigned his journey in the Periplus, of which we have the abstract") (Grégoire 1795: 3).

But, as Grégoire acknowledged, this and other classical and medieval voyages were accomplished by coastal navigation. Given the limitations of the magnetic compass, the navigator could not dispense with astronomy. In a reference to Galileo, he continued, "The discovery of Jupiter's satellites, the perfection of marine charts, sufficed to produce a revolution in the human mind and in trade and diplomatic relations" (Grégoire 1795: 4). However, he declaimed, "The most important discovery, which at first was seen as a chimera, and one that has much exercised mathematicians over the last two centuries, is the determination of longitude at sea." "The problem is this: knowing time on the ship, namely time at the first selected meridian or place of departure. The difference in reduction of hours on parts of the Equator gives the longitude of the ship, relating it to the meridian selected for comparison. "/*la différence des heures réduites en parties de l'équateur donne la longitude du navire, en la rapportant au méridien choisi pour terme de comparaison.* /We count on the Equator fifteen degrees for an hour, and consequently four minutes to a degree" (Grégoire 1795: 5).

Turning to the question of political and economic rivals in the race for trade and colonial possessions, the National Convention assemblyman remarked, "Almost all the nations frequenting the sea, staged contests relating to longitudes, but nothing beats what England did in this respect." Referring to a committee established in London in 1714 bringing together "the greatest men of the country, with Newton among their number," he observed that prizes were proposed for those who could determine longitude within a half degree. "While Astronomy perfected methods for measuring distances from the Moon to the Sun and the stars, which gives the difference of meridians, watchmaking facilitated the manufacture of marine time pieces, of which the idea was not new, but of which the application was." He named Englishmen John Bird, Jesse Ramsden, and especially John Harrison (who went on to invent the marine chronometer) as individuals whose clocks were successfully tested in various voyages to Barbados and Jamaica. As he pointed out, in France two illustrious rivals had entered the race, namely, Pierre Leroi (Leroy), son of Royal clockmaker Julien Leroi, and Swiss-born horologist Ferdinand Berthoud, who adopted France as his second home. At great expense, France had also fitted out armed corvettes and frigates to examine new methods for determining longitude at sea, tested in the course of long voyages. Of these experiments, as Grégoire (1795: 5) summarized, the marine timepieces of Berthoud especially "retained an accuracy that has not hitherto been reached."

Noting that the English had been publishing their Nautical Almanac since 1767, as Grégoire commented, the idea was actually French. As explained, when British Astronomer Royal Nevil Maskelyne returned from St. Helena (where he established an observatory), he adopted an idea first presented in 1755 by Nicolas Louis de Lacaille based on his meticulous research in and around Cape Town, and with de Lacaille the author of a southern star catalogue titled *Coelum Australe Stelliferum* (1763). However, as Grégoire lamented, the English Nautical Almanac became the operating manual of their sailors, published five and even six years in advance, while at home the *Connaissance des Temps* was printed only for the current year, hardly an essential book for sailors undergoing a voyage (Grégoire 1795: 9).

As the priest-revolutionary further deplored, French hydrographic charts, especially those of the Mediterranean relating to trade with the Levant, were particularly wanting. Referring to the erudite *Voyage du Jeune Anacharsis en Grece/Travels of the Younger Anarcharsis in Greece* (1788) by Jean Jacques Bartholomew (1716–95), he decried that the maps were made from plans using a compass rose. As he continued, "The Caspian Sea, with the adjacent countries, is still so little known that some geographers have varied five degrees on its position." He held out hopes that [Josephe de] Beauchamp, formerly of the observatory in Baghdad [built in 1784, at the expense of Louis XVI] and consul in Muscat would "hopefully rectify these errors in the course of a trip, including trade and Astronomy to promise great benefits" (Grégoire 1795: 9).

As Grégoire pontificated, doubtless looking ahead the elevation of the Paris meridian, "the time is probably not far distant for nations, abjuring childish pride, to adopt a common meridian, that Ptolemy had otherwise set at the westernmost of the Canary Islands." As he further pointed out, his fellows had failed to recall the efforts made by other people in the erection of "magnificent observatories," not excluding the Chinese. As the priest-apostle of the French revolution also allowed, the "two tyrants of France" had actually worked for the advancement of astronomy. "Under Louis XIV, the meridian and perpendicular were drawn. Under Louis XV, coteries of scholars somehow divided the globe to observe the transit of Venus to measure terrestrial degrees; one on Rodrigues Island [Mauritius], at the Cape of Good Hope, in California, and others in Lapland, and Peru" (Grégoire 1795: 12). He might also have mentioned, as discussed below, the French-sponsored Jesuit astronomical missions to China

Even though the French *Connaissance de Temps* still held to Paris as the prime meridian, Maskelyne's calculations also went into French translation (Sobel 1996: 167). Under a decree of January 30, 1854, the Bureau des Longitudes was charged with a special mission, leading to the realization of *éphémérides* or astronomical tables, along with the formal establishment in 1881 of a *Service des Calculs* or special government department vested with the organization of grand scientific expeditions, inter alia involving geodesic measurements, observations of solar eclipses, and observations of the transit of Venus across the Sun. True to its mission, all the findings entered publications of the Annales of the Bureau between 1877 and 1949 (Bureau des Longitudes, 2014).

CONCLUSION

Specifically this chapter sought to determine how French Royal mapmaker Jean Baptiste Bourguignon d'Anville actually drafted his image-making representations of Asia, literally transforming regional world maps into global maps complete with map graticule. Primarily a cartographic geographer, we found that d'Anville also fell into the classical tradition of describing geography through his elaborate published commentaries. We can see in this approach both the cosmography and choreography of the classical Ptolemy. Yet, paradoxically, d'Anville was also at the forefront of assimilating new cartographic data such as would render Ptolemy irrelevant as far as his conceptions of China were concerned.

With Raynal-Diderot as a foil, the contrast in approach to that of the official mapmaker to the King of France could not have been greater. Raynal-Diderot promoted a virtually borderless world under assault from rapacious European colonial capitalism with slavery one of its less savory methods and land-grabbing, monopoly control over valuable resources and rent-taking as its prime motive. In the hands of the Portuguese, Dutch, English, and French trading companies, this was a zero sum world of wheeler-dealing mediated locally only by collaborating despots where they did not resist. Whereas d'Anville looked to Ptolemy as a point of reference, to the Enlightenment *philosophes* the world created by colonial intervention simply rendered the old geography redundant. Needless to say, such an iconoclastic work along with its principal author was driven out of the kingdom, albeit receiving an even greater European audience through translations and reprints. How Cartesian framings of Asia such as achieved by d'Anville contributed to or fed into imperialist or even post-colonial imaginings of nation-states is reserved for the conclusion to this work.

NOTES

1. As defined by Websters, a Cartesian coordinate is "either of two coordinates that locate a point on a plane and measure its distance from either of two intersecting straight-line axes along a line parallel to the other axis."

2. Pimentel (1712) offers the latitude of Goa as 15° 25' N and longitude 99° 45' E. He also tabulated coordinates for all major ports from Lisbon, the Mediterranean, the African coast, Arabia, and the Indian coast, including Malacca (2° 24' N 124°40' E Long.), Macau (22°15' N 136°10' E Long), and Nagasaki (32° 50' N 151° 10' E Long). Essentially, Pimentel's work was a veritable navigation guide in the era of celestial observation, not excluding the Pacific Ocean, and the Atlantic including Brazil.

3. The modern archaeology of ancient Singapore (Temasek) only kicked off in the 1980s. See Miksic (1984). A description along with his publications can be found on this government website http://eresources.nlb.gov.sg/history/events/2ebfebad-a4d5–4bbb-bf43-c7db6e30eb7d

4. In fact, La Loubière (1668) was preceded by a number of European authors and mapmakers in rendering the name of this country, Siam, as with Heylyn (1652), as described.

5. Description in http://www.swaen.com/Anville-China.php. According to Lim Jongtae (2013: 304), Mu Kedeng's mission to Korea was specifically to gather geographical information on the country. In this, he was accompanied by Fridelli when crossing the Qing-Korean border.

6. An icon of anti-racism and emancipatory ideas, in the twentieth century Grégoire became a hero to such individuals as Ho Chi Minh, along with French Jews (Sepinwall 2005).

7. My translation and rendering of Grégoire's address.

Chapter 8

Reimagining Ptolemy's *Sinus Magnus* and the Historical Geography of the China Seas

Known to the ancients as the *Sinus Magnus,* bordered by the Golden Peninsula and identified as integral with the *India Extra Gangem* region, it became part of common lore for eastward-sailing navigators that this was the gateway sea to the shores of China, the source of silk and ceramics and sites of major trading ports linked with the Indian Ocean trade. On the cusp of the European age of discovery, as with the Columbus arrival in the Caribbean Sea, the *Sinus Magnus* was even conflated with the Indian Ocean. This we termed the Ptolemaic "trap" deceiving early westward-sailing navigators. Still, in the Ptolemaic template, this was a closed sea and this was an assumption embedded in the world maps drawn by Münster, Ortelius, and others. In the wake of Magellan, we seek to ascertain exactly what was the Spanish priority in the Pacific Ocean. However, as discussed below, for westward bound sailors the true location of key Ptolemaic markers, also led to fundamental and consequential errors as to the general coordinates of China and surrounding seas. But, with the advent of modern Anglo-Dutch-French colonialism, contest over "ownership" of the Asian seas and islands also raised more general questions as to statecraft in Asia, and especially in the Sinic world. With China inscribed as the central kingdom/world under heaven, we cannot ignore the codes and protocols regulating the way that tributary states interacted especially when it related to maritime commerce. Neither can we ignore China's own maritime forays into the so-named Nan Hai or South Seas, spectacular in the case of the Ming-era voyages of Zheng He conducted between 1405 and 1433, but also preceded by Mongol-Yuan fleets ranging as far as Java (1293). Outside of state-centric narratives and claims of "ownership," my own approach is to view the "China seas" as civilizational crossroads through which Austronesians (Cham), Vietnamese, Arabs, Persians, and others passed (and into which Europeans intruded).

To recapitulate, Ptolemy's *Sinus Magnus* (sometimes *Magnus Sinus*) came to be associated in Ptolemy recovery atlases with some expanse of ocean east of the Golden Chersonese and bounding *Sinae* (*regio Sinarum*) or China itself, in other words a Great Gulf. Still, not one geographer had a monopoly on truth when it came to framing the *Sinus Magnus* with the mythologized Cattigara as an identifier, such as entering Greek and Roman texts. The question often turned on whether or not Cattigara (and the *Sinus Magnus*) could be identified with a generalized Gulf of Siam, or a site near the mouths of the Mekong River, or whether located in the Gulf of Tonkin and/or the South China Sea or, even the Pacific Ocean at large.

As we have seen, Ptolemy's *Geographia* posited only a three-continent view of the world and with measurements misrepresenting its size by at least one quarter. Obviously the New World stood outside of such a conception (as indeed did an Austral continent disambiguated from a Great South Land). The Spanish Crown demanded of its discoverers to chart and measure new lands. It then remained the task of royal cosmographers to fit the new discoveries into the map graticule in line with Ptolemaic prescription, although there would be considerable trial and error in this process and with British, Dutch, French, and navigators of other nationalities hot on the trail.[1]

As such, this chapter offers a multiple optic upon the political geography of the "China Sea" or western Pacific Ocean as it came to be known to European navigators and cartographers. Setting political geography aside, a first section brings to attention the first apprehensions of the seas washing the coast of Vietnam and China since the age of Marco Polo by early arriving Iberians, once again demonstrating that Ptolemaic and classical conceptions of the eastern hemisphere died hard. As discussed in a second section, the belief that the Indian Ocean was enclosed on the south by land proved to be the most enduring element of the Ptolemaic canon, and not unmasked until the voyages of James Cook and others to the South Pacific in the last decades of the eighteenth century. A third section offers a recapitulation of classical Western understandings over the framing or construction of this maritime space, in part arguments turning upon the legal concepts of *Mare clausum* (closed sea) versus *Mare liberum* (open sea). A fourth section discusses French/English mapping and surveying of the South China Sea pending "ownership" claims. A final section discusses what I view as a clash of concepts besetting understandings of the South China Sea, such as persist down until the present.

SPANISH CONSTRUCTIONS OF CHINA AND THE CARTOGRAPHIC "CAPTURE" OF THE CHINA SEAS

Until this point we have not examined the real effects of the Spanish irruption into the Pacific Ocean mere decades after the completion of the first circumnavigation. Not all repeat Spanish expeditions across the Pacific following Magellan were successful, indeed disaster followed disaster, but the voyage of Miguel López de Legazpi of 1564–65 was significant. Departing Mexico in 1563 riding the westerly blowing trade wind approximately 13° N, he also touched at the archipelago known to Magellan as, variously, Islas de Larrons and Islas de las Velas Latinas. As explained by the Jesuit Charles Le Gobien (1700), with cross in hand Legazpi took "possession" for Spain, naming them the Marianas after Marie-Anne of Austria. Continuing on to Luzon and Mindanao, he named the archipelago earlier dubbed Saint Lazarre by Magellan, *Las Philippinas*. This was to honor Philip II (although the voyage of Ruy Lopez de Villabos had already done so in 1543).[2] Moving on to the Manila Bay area and having expelled the outliers of the Brunei-Sulu Sultanate, Legazpi founded Manila as the Spanish capital and center of power in the western Pacific. With the first Franciscan missionaries arriving with the Legazpi expedition, Jesuits, Dominicans, and other Catholic orders soon followed.[3]

Establishing a new territorialization based upon sea power, long-distance trade, and with Catholic Christian conversion of pagan natives at a premium, the Philippines entered the Spanish fold as a vice-royalty of New Spain (Mexico), adding the Marianas (including Saipan and Guam). No less epochal was the advent in 1571 of the Manila-Acapulco or Manila Galleon system connecting up the Fujian coast of China with the Americas, a system surviving down until 1813 (Giraldez 2015), but also announcing a new age of Spanish (and future American) imperialism in the western Pacific (see Figure 8.1).

The Spanish Longitude Quest in the Pacific/ Juan de Herrera/ Jamie Juan

As with the Portuguese entering the Atlantic and Indian Oceans, navigational science was at a premium in the even vaster Pacific Ocean and we detour to examine the Spanish priority in this area.[4] As Portuondo (2009: 86–91) explains, during a sojourn in Portugal in 1582 Juan de Herrera (1530–97) serving at the court of Philip II designed a project to establish the extent of the Spanish empire and to plot its boundaries. He also sought to reconcile Portuguese with Spanish charts to the advantage of the latter. Himself an instrument maker and focused upon mathematical solutions to determining longitude, he chose the young mathematician Jamie Juan with a brief to sail to Mexico and the Philippines and to conduct astronomical observations and to fix coordinates. Juan was also instructed to use instruments designed by Herrera. Along the way he was instructed to make observations at sea and at all ports of debarkation, including the Pacific crossing. The observation of eclipses and tidal regimes was added to his tasks. This was an eight-year commission and he was well provided. Having measured an eclipse of the Moon

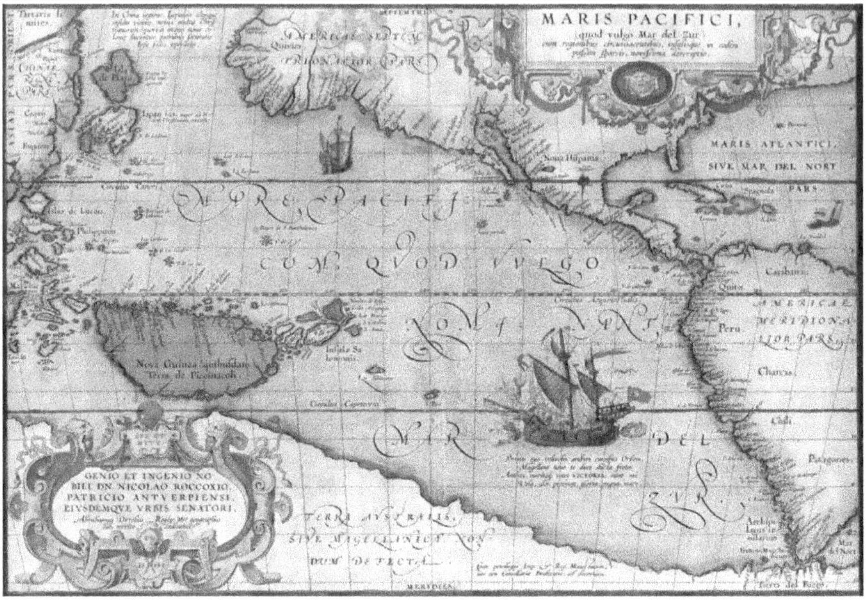

Figure 8.1 ***Maris Pacifici (quod vulgo Mar del Zur) cum regionibus circumiacentibus, insulisque in eodem passim sparsis, novissima descriptio,* Abraham Ortelius.** Map of the Pacific Ocean in large part reflecting the Spanish discoveries as with the Philippines and an oversized New Guinea. Magellan's ship Victoria is prominent. Terra Australis or a Great South Land is conflated with Antarctica in the Ptolemaic mode. Plate 6 from Theatro d'elorbe de la tierra / de Abrahamo Ortelio. In Anversa: Libraria Plantiniana, 1612. Scale [ca. 1:42,000,000]. *Source*: Map reproduction courtesy of the Norman B. Leventhal Map & Education Center at the Boston Public Library.

in Mexico City in 1584, Juan was sent to the Philippines to accomplish the same, but died soon after arrival. From Manila, another official stepped in and measured the eclipse of March 23, 1587.

The risks to navigation involved in not paying attention to the longitude question were also pointed out in an undated letter written by cartographer Lopo Homem to King John III of Portugal. As revealed, between 1529 and 1557 with the first measurements of longitude taken in India (Chaul) adapted to Portuguese sea charts (as proposed by Pedro Nunes), shipwrecks and accidents were frequent owing to inaccuracies in the measurement of longitude. Accordingly, Homem advised the Portuguese monarch to rescind the obligation of Portuguese pilots to use Nunes's chart for navigation.[5]

Spanish Constructions of China/ Hernando de los Ríos Coronel Maps

In a memorial addressed to the Spanish king from Manila dated June 27, 1597, the Spanish navigator and adventurer Hernando de los Ríos Coronel gave an account of a book he was writing on the use of the astrolabe and the art of navigation and the importance of Hermosa (Formosa) as an access port to mainland China. This was accompanied by a map. He also advertised safe sailing directions from the Philippines back to Mexico at the latitude of 45°.[6] Believed to have studied navigation at the Casa de Contratación in Seville, de los Rios was known for his mathematical prowess going on to produce more than 200 *derroteros* (or sea logs), also entering the service of the King of Spain in order to test a sea compass (Crossley 2011: 3, 27).[7]

As John Newsome Crossley (2011: 47) points out, de los Ríos's *Discrepción de la Isla Hermosa* is the first Spanish map of Formosa/Hermosa and the island of Luzon in color and with Formosa depicted as one island, not multiple. The Lequios (Ryukyus) are included, although not Japan proper. The coast of China is well depicted with river mouths, promontories, and other realistic coastal features. Macau is named and positioned with respect to Canton although not accurately, indeed on the wrong side of the Pearl River estuary. The crowding of islands around the estuary possibly confuses with the Ladrones (or Wanshan archipelago), then serving as an important navigational landmark. Formosa is not well aligned vis-à-vis the coast of China or Luzon, although Keelung on the northern part of the island is realistically drawn. Longitude is marked and counted from the island of Corvo in the Azores. An oversized scale in leagues is placed diagonally. Manila is located at 167° 30' E. and with Macau at 161° E, both inaccurate by far (but depending upon the prime meridian adopted).

With its pivotal location and fortifications, Manila would emerge as a new front line for Spanish adventures in the western Pacific. For example, de los Ríos boasted that Spain could take Cambodia and Champa with 300 men, though his first target was Formosa. A Spanish intervention in Cambodia followed but with little result (San Antonio 1914 passim; Crossley 2011: 43–45). According to Charlotte Pham Minh Hà (2016), having mastered the passage across the South China Sea south of the Paracels, Spanish trade between Manila and Vietnam reached its peak in the 1660s and such crossings became common at the end of the seventeenth century. Besides China, Japan also entered Spanish expansionist plans, however unrealistic. Trade relations on the part of the Spanish with various Japanese domains were always tenuous, until they were completely expelled by a wary Shogunate, itself with designs on Manila. The establishment by the Spanish of strategic fortresses on Formosa between 1626 and 1642 in an attempt to intercept the China trade at the source came to naught with defeat by rival Dutch power.

The dynastic union between Spain and Portugal (1580–1640) also gave impetus to Manila-based missionaries seeking to parlay Macau's special status to enhance their own entrée into China (San Antonio 1914: 137–138). While it lasted, Manila offered another tangent for Portuguese voyages from Macau and with stakeholders seeking to profit from the porcelain-for-silver trade with the Americas (Pires 1994). Setting aside crucial navigational priorities relating to the Manila Galleon trade, the Spanish project of measuring coordinates in the Asia-Pacific region is less well documented, at least until the nineteenth century.[8]

The Herrera-Velasco Maps

The maps reproduced by Herrera such as described in chapter 6 and distinguished here as Herrera-Velasco maps (after Juan López de Velasco, cosmographer under Philip II) include two on the Pacific, earlier produced by Velasco in manuscript form. Velasco relied on primary sources without ever having been to the Americas in order to make his maps. The first of these is *Descripción de las Yndias Ocidentales*, appearing in Herrera y Tordesillas's *Descripción de las Indias* (1601). As Christina Findlay Connett (2014: 183–184) explains, this was the first of fourteen maps illustrating Spanish possessions in the Atlantic and Pacific. Prominently marked is the Papal line of demarcation in the Atlantic established by the Treaty of Tordesillas along with its antemeridian, the later Pacific line of demarcation defined by the Treaty of Saragossa (or Zaragoza). While Herrera is faithful to Velasco's calculations in displaying the Asian line, the text miscalculates the longitude of Malacca by 34° (40°) too far west and to Spain's obvious advantage (see Suárez 1999: 163).

The second Herrera-Velasco map to illustrate the Pacific, *Descripción de las Indias del Poniente* (1570), is of a different scale. Encompassing Bengal and the East Indies at one extremity, it includes the Moluccas, the Philippines, and the Solomon Islands, at the other. Japan is better represented here as is the coast of China, Vietnam, the Gulf of Siam, the floating island of Borneo and with the South China Sea inferred but not labeled. Lequoi mayor (Ryukyu) and Lequiomenore (Formosa) are indicated. Canton is named at the head of a misnamed "Rio Ganges." Macau is absent. The designation Las Philippinas (the first on a map) stands at the center and with major islands identified by a number keyed to a word list (a first on a map). Manila, also named, is represented by a church. Similarly, the various Ladrones (Marianas) as with Saipan are numbered and keyed, as are the important Moluccas. A false island named San Juan east of Mindanao also made its appearance enduring on European maps for the next two centuries (Suárez 1999: 172–173).

Descripción (the book) is the major locus of Herrera's evocation of the Spanish conquest of the Philippines as well as Portuguese contacts with China-Macau and Japan, albeit comprising several pages only of the entire text. The Magellan and Legazpi voyages are referenced. The Portuguese loss of the Moluccas is noted. Some shrewd information is given as to the spread of Islam in the archipelago. China is apprised and sized up with especial reference to Tomé Pires, the Portuguese ambassador to the court in Beijing (also noting its latitude though not with accuracy). Macau, only permanently settled by the Portuguese in 1557, is mentioned as a place from where they engaged in commerce with China. Lequoi mayor and Lequoiminore are referenced with the latter also named Isla Hemosa (Formosa). Described "as big as Sicily," this is closer to the truth relative to the smaller and scattered Ryukyus.

As Connett (2014: 331) explains, the Herrera-Velasco maps illustrate the vastness of the space waiting to be occupied by the Spaniards, while also revealing the numerous cities already established by the empire. Importantly, these were the first printed maps to show the Spanish Indies all the way to the Asian coast and therefore the entirety of the Spanish territories. Rhumb lines and compass roses are absent. Bare lines of delineation are restricted to the tropics and the Equator (highly accurate with respect to the location of Malacca), and with Toledo marked as the prime meridian of zero. The Manila Galleon track is indicated (on the first not the second map). Velasco includes the demarcation of the Occidental and Oriental boundaries of the Spanish empire as black lines, but these are not labeled as such. The Spanish conquest priority is clearly described in a duly embellished cartouche.

The Spanish advance is further revealed in *"Map Taboas geraes de toda a navegação Mapas generales originales y universales des todo el orue con los puertos principales y fortalezas de Ambas Indias y una descripcion topographica de la region Austral Magallonica año de 1692."* Magellan is cited and the knowledge advances upon his navigation. An inscription at the bottom hails the "discovery" in 1602 by Pedro Fernandes de Queirós of a southern continent. In the north the Russian Far East separated from North America by the Straits of Anian is well referenced as is the general shape of Japan including Ezo (Hokkaido). Nagasaki is named. In fact, the Pacific Ocean is framed, though China is out of the picture.

A Ming Map of China Known to Spain

Ming China had long been producing cartographic renditions of the empire building upon Song and Mongol antecedents, just as they began to assimilate Arabic and, tentatively, European knowledge and concepts. The Ming also drew upon the maritime knowledge collected by the Zheng He expeditions. It was inevitable that copies of such maps would fall into the hands of the first-arriving Europeans. As it turned out, it was Spanish Manila and probably not Portuguese Macau that became the first point of transaction. An important instance was that of the *Ku Chin Hsing Sheng ChihTu* (古今形勝之圖研究) or *Carta de la configuración Antigua y moderna* first published in 1555 by Yushi, a Ming dynasty official, falling into Spanish hands in 1575. According to Lee Yu-Chung and José Luis Caño Ortigosa (2015: 36), the acquisition of the map by the Spanish amounted to a major coup, allowing the Castilian Crown to considerably expand its knowledge of China, its territory, and the broad trajectory of the geography of the Far East. More than that, it facilitated the Spaniards to take the lead among other European nations in a spectacular way, in grasping knowledge of the Sinic world, its culture and its language. Removed to Spain the map obviously became a major Spanish geographical source but, over the centuries becoming a virtual archival oddity in the Archivo General de Indias de Seville until its recovery and study in the modern period by a range of Spanish, Chinese, Japanese, and other scholars. From the geographical point of view, the map includes, in modern terms, Northeast and Southeast Asia, including the Indochina peninsula. Japan appears (microscopically) in the east along with the Ryukyu islands. Java, Sumatra, and Borneo are represented. The Indian Ocean appears suggestively. The Philippines, already occupied by the Spanish, were not well known, or rather, understood as islands of little importance, without a precise outline. However, as Lee and Ortigosa (2015: 36) acknowledge, the accuracy of the map in many places is called into question.

Far more important from our perspective is the uses of the map by the Spanish Crown. As Lee and Ortigosa (2015: 40) note, of particular importance to the Spanish was the recorded data in the upper right corner of the map where the terms and names of the Chinese bureaucracy appear. Similarly, data registered in the upper left corner of the map reveals much of the Chinese perception of the world. The Spanish authorities would also have perceived that the country was divided between two capitals and thirteen provinces, in which cities were hierarchized. Their names are framed in boxes of one or two lines. A dotted line extending from the Korean peninsula to the west represents the Great Wall of China, among other place names. In particular, the different cultural and ethnic groups of the empire are also noted (see Figure 8.2).

Citing Chinese historian, Jin Guoping (2004: 318–319), Lee and Ortigosa (2015: 40) adjudge that the map could well be one of the antecedents or sources used by Luiz Jorge de Barbuda to produce what, as indicated in chapter 4, was the first European map of China, *Chinea, olim Sinarum Reginis, Noua descriotion.* This is credible as the Barbuda (update) was made in 1584, several years subsequent to the arrival of the *"Carta de la configuración Antigua y moderna"* in Madrid in 1575. As Lee and Ortigosa (2015: 41) point out, within a year of acquisition of the map, Martín de Rada and his Chinese companions, Wang Wanggao and the translators Lin Bixiu and Chen Huiran, entered Fujian province for the purpose of establishing the first official contact between China and Spain. This is a reference to the de Rada/de Luarca diplomatic mission to Xiamen mounted from Manila (see Boxer 1953).

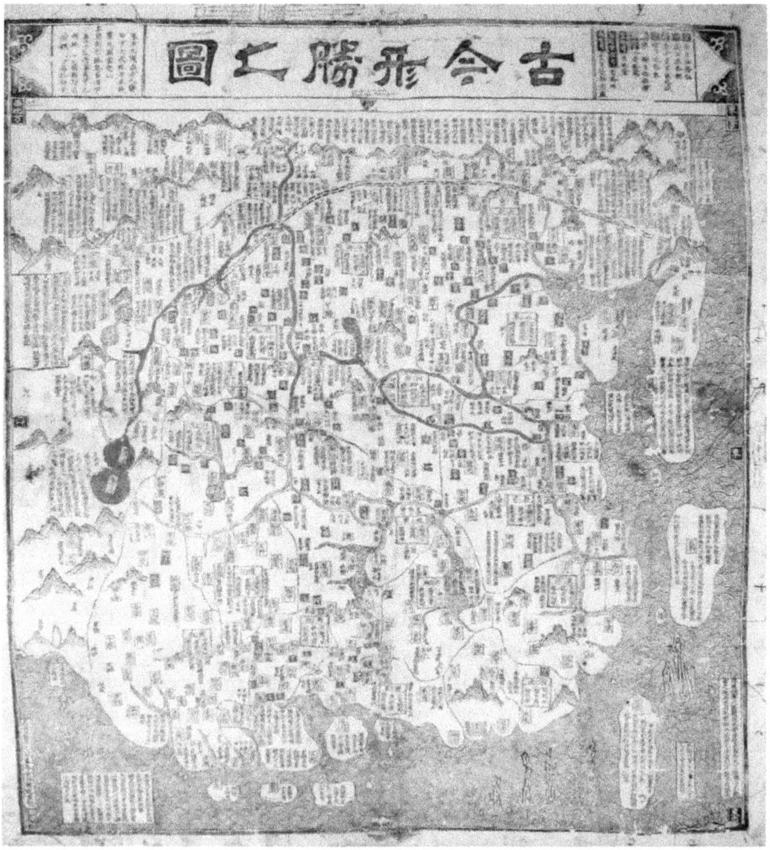

Figure 8.2 ***Ku Chin Hsing Sheng ChihTu/ Carta de la configuaraciôn antiga y moderna.*** *Source*: Complements of Research Center for Humanities and Social Sciences, National Tsing Hua University, Taiwan. Author's numbered facsimile copy.

In turn, the reports written by Martín de Rada along with documents collected in Fujian served as a source for Juan Gonzalez de Mendoza, author of a major Spanish text on China. Both Mexico City and Manila would emerge as printing centers and with woodblock books including translations of Chinese classics struck at Manila by Chinese printers (see Gunn 2003: 93).

UNPACKING PTOLEMY'S GREAT SOUTH LAND

The classical belief that the lands of the Northern Hemisphere should be balanced by land in the south was particularly enduring. It also found resonance in medieval Christian thinking and, as mentioned in chapter 5, led directly to a veritable Great South Land or *Terra Australis Incognita* imaginary. As Paul Mapp (2011: 105–106) points out in his study of Spanish exploration of the New World, "One reason Europeans continued to entertain notions of a southern continent was because they could rest their surmises not only on Ptolemy's ancient authority but also on more contemporary reason and faith." Besides credence in a symmetrical planet with a southern landmass balancing out a northern, such as entering world maps by Mercator and others, further credence in a Great South Land came from the failure of sailing far enough south to confirm this belief.

The Spanish Search for a South Land

Spanish exploration of the Pacific did not stop with the occupation of the Philippines and their advance on the Moluccas but found significant sequels in the South Pacific voyages of Queirós between 1605 and 1606 in search of continental land.[9] According to Sheehan (2008: 354; 371), following a first voyage in 1568 as pilot on an ill-fated voyage of Álvaro de Mendaña y Neira to the western Pacific, in 1598 the Portugal-born Queirós petitioned King Philip III (r. 1598–1621) of Spain to support another voyage into the Pacific. It was Mendaña who named the Islas Salomon believing them to be the mythical Ophir, as alluded, and with the Spanish holding out for other candidate sites in Asia. In presenting his theories in Manila, Mexico, Lima, Rome, and Madrid, Queirós also talked up the islands that had acquired an association with the biblical Solomon as a source of gold. Visiting Rome in 1600, Queirós obtained the support of the Pope. He was also marked out by the Crown to test compasses and calculations of longitude as perfected in Rome. Departing Callao on December 21, 1605, and accompanied by Luis Váez de Torres, the Queirós voyage was not just a search for the Solomon Islands but a quest to discover the continental land (Sheehan 2008: 396). While the Queirós-Torres voyages were mixed as to results, Torres did establish the insularity of New Guinea, virtually marking it out from the Australian continent as with the straits which bears his name.

The Search for a Terra Australis Incognita

The Search for a *Terra Australis Incognita* or Great South Land such as entering the classical template was not for want of trying on the part of discoverers and cartographers. Treating the question as imaginary geography, as Avan Judd Stallard (2016: 73) writes, "Once Tierra del Fuego had been appropriated by cosmographers and cartographers as a part of the southern continent, the notion of a sprawling southern landmass lying just beneath the tip of South America became a fundamental tenet of cosmography." Mapmakers such as Ortelius and Mercator further exaggerated the presence of a massive Antarctica or *Terra Australis Incognita* waiting to be discovered. As Stallard (2016: 57–58) alerts, the Francesco Rosselli worldmap of 1508 is likely the first map revealing a southern (not South American) continent. Located south of the Cape of Good Hope and situated from within a longitude-latitude grid and labeled *Antarcticus*, it stands outside of classical conceptions. Versions of the map reflect modifications to the northern and southern boundaries of the island continent, known later as *Terra Australis*. By the age of Rosselli and others, "the idea of an Antipodes had crystallized into more than just a theory." For example, in his 1570 map Ortelius depicts "*Terra Australis Nondum Cognita*" as a large southern continent on the bottom of the map practically balancing out the northern landmass matched by a northern Arctic continent. An encircling southern ocean is allowed. He appears to acknowledge an Australia located south of Java but merged into the southern supercontinent.

It is certainly intriguing as Stallard (2016: xvi) questions, why did people keep believing in a great southern continent in the face of overwhelming proof of its non-existence? Indeed, how did the possible existence of a southern continent such as portrayed from ancient through to medieval times transform into the positing of the probable existence all in the course of just a few decades? Further, why did belief in the existence of *Terra Australis* become so widespread, and why did it evoke such conviction in so many of its advocates? Additionally, why did people continue to believe in the southern continent in the face of repeated acts of empirical falsification.

While the history of navigation and first settlement across the Pacific belongs to ancient Austronesian navigators, we are of course concerned in this text with European impacts. In any case, with the Spanish blazing their way across the Pacific, the English and Dutch were not far behind. As with the Spanish, the discovery of an Austral continent was also a preoccupation of future arriving navigators. Francis Drake was the first Englishman to complete a circumnavigation in (1577–80), second after Magellan. A two-ship venture under VOC merchant Isaac Le Maire and with Willem Corneliszoon Schouten as navigator departed Holland in June 1615 and rounded Cape Hoorn on January 31, 1616. This was a navigational first revealing that Tierra del Fiego was but a small island and not part of a southern continent. Yet, in that regard they held to the belief. By revealing a passage from the Atlantic to the Pacific to the south of Tierra del Fuego, they also broke the presumption of Spanish "ownership" and control. In their Pacific Ocean crossing, Le Maire and Schouten found and named a number of South Pacific islands including some of today's Tonga and Futuna Islands. Coasting north of New Guinea they arrived in the Dutch-controlled Moluccas in mid-August.

In actual fact, even with the voyages of Drake, Schouten, and Le Maire, joined by Abel Tasman (1642), William Dampier (1688, 1670), and Jacob Roggeveen (1721–22), a vast blank or unknown remained on period maps representing the ocean between South America and New Zealand. Famously, British Captain James Cook also entered the search for a Great South Land in the late eighteenth century (see Suárez 2004: 128–139). On his second voyage (1772–75), Cook's ships *HMS Resolution* and *Adventure* crossed the Antarctic Circle on January 17, 1773, in December 1773, and again in January 1774 (coming to within 120 miles of Antarctica at the closest). Even when the insularity of New Zealand and Australia were determined in the late seventeenth century, geographers continued to believe that the icy continent was much larger than its actual size. Antarctica thus remained the last region on Earth to be discovered. Without ignoring a number of early Antarctic voyages in the decades following Cook, the Russian expedition of Fabian Gottlieb von Bellingshausen and Mikhail Lazarev of 1820 was the second and closer circumnavigation of Antarctic continent. As Robert Clancy et al. (2014: 104) explain, this was part of a determined 180-year Russian venture to measure the easternmost extent of its empire and to discover new lands in the Pacific and southern ocean. To be sure, as map history reveals, Antarctica was by now decisively cut down to its actual size.

Willy-nilly, Cook's voyage to the east coast of Australia led to the British settlement on New South Wales. It also added a clearer definition of the northern and western part of the land known to the Dutch and early cartographers as New Holland (see Suárez 2004: 78–105). Nevertheless, it would await the exploration of British navigator Mathew Flinders to actually circumnavigate and survey the coastline of Australia in 1801–02. Flinders, in turn, has been credited with popularizing the transfer of the name *Terra Australis* (the fabled Great South Land) to Australia. Concluding that the *Terra Australis* as hypothesized by Aristotle and Ptolemy did not exist, he applied the name Australia to the continent which he had circumnavigated although, as we have seen in chapter 6, English clergyman Heylyn much earlier applied this name to his own imagined southern continent. As Flinders wrote in his 1814 book, *Voyage to Terra Australis,* "There is no probability, that any other detached body of land, of nearly equal extent, will ever be found in a more southern latitude; the name Terra Australis will, therefore, remain descriptive of the geographical importance of this country, and of its situation on the globe: it has antiquity to recommend it" (Flinders 1814: Vol. I, iii). As Stallard (2010: 309), comments, even after Australia was officially accepted as the name of the continent by British governors, Terra Australia, Great South Land, and Antipodes all remained in common usage for some time.

GROTIUS'S *MARE CLOSUM* (CLOSED SEA), *MARE LIBERUM* (OPEN SEA), AND THE LAW OF THE SEA

The publication in 1609 by Dutch jurist Hugo Grocius (1583–1645) of his landmark work of international legal jurisprudence *Mare liberum* (open sea) called down a powerful challenge to the doctrine of exclusive ownership of the seas, reaching back to the Romans with their notion of *Mare nostrum* in reference to the Mediterranean. Preceded by his *De Indis* (Concerning the Indies) more commonly known as *De iurepraedae* (On the Law of Prize and Booty) (ca. 1604), it is generally understood that both works are forerunners of the modern law of the sea. But it was in *Mare liberum* that he fully articulated the principle of the freedom of the seas, and that the sea should be free and open to use by all countries, essentially limiting national rights and jurisdiction over the oceans to a narrow belt of sea surrounding a nation's coastline. Specifically, Grotius sought equal rights for all nations to participate in the lucrative spice trade in the Indies. The remainder of the seas was proclaimed to be free to all and belonging to none. He also asserted that the Portuguese monopoly based upon claims to ownership "could be considered 'worthless' when viewed in the context of historic trade relations between the Arabs and the Chinese" (Morton 2016). Grotius's theory of "just war" also supported the notion of seizing Iberian ships. Translated to the East Asia area, Chinese trading junks were fair game as well, and with VOC acting upon the "law" with impunity. Taking *De Indis* as a point of reference, and viewed in terms of world-systems analysis, as Eric Wilson (2009: 230) declares, the Grotius's project sought to make the world "safe for Holland."

Needless to say, many other Europeans nations challenged this doctrine as an assault upon their own rights (as with the English assertion of closed seas). In the Latin edition of his *Mare Clausum* (1635), legal theorist John Selden first publicly articulated the argument that a sovereign could establish dominion over or appropriate the seas. The much-delayed English version appeared as *Mare clausum: The Right and Dominion of the Sea* (1663). Initially the Atlantic was uppermost in this closed sea argument at least until the overturn of the Navigation Acts during the American Revolution and the active pursuit by Great Britain of free trade policies in East Asia such as signaled by the Opium Wars (Batchelor 2013: 58). By this time, however, Holland "the nascent hegemon of the seventeenth century" had gone global and with the Grotius heritage established as part of the new normative and intellectual foundation of international public order (Wilson 2009: 239).

Making the argument that the Portuguese were not the first to sail these seas, Grotius also looked to classic sources while mentioning Muslim, Ethiopian, Arab, Persian, and Indian navigators, albeit forgetting Chinese:

> That also in the flourishing estate of Rome they were wont to sail from the Arabian gulf to India and the islands of the Indian Ocean, even to Golden Chersonesus (which most men suppose to be Japan), the voyage described of Pliny, the embassages from the Indies to Augustus, to Claudius also from the island Taprobane, besides the worthy acts of Trajan and Ptolemy's tables, sufficiently declare. Strabo witnesseth that even in his time a fleet of Alexandrian merchants out of the Arabian Gulf sailed to the furthest parts of Aethiopia, and so of India, when in times past few ships durst attempt it. Thereby the people of Rome had great revenues. Pliny addeth that they sailed having shipped bands of archers for fear of pirates, and that India alone took away yearly from the empire of Rome 500 sestertia (if you add Arabia and Seres they took 1,000), and that the merchandise were sold for a hundredfold more. (Grotius 1609: chap 5: That the sea or right of sailing on it is not proper to the Portugals by title of possession)

By writing his book, according to David Armitage (2004), Grotius intervened in two political debates of pivotal significance for his own country. The first was the relationship between the United Provinces and the Spanish monarchy, from which the Dutch had broken away in 1581. The second, and of greater importance for this essay, was the Dutch right to commercial penetration in the Asian Seas all the way to Japan. As Peter Borschberg (2002; 2011) has outlined in a number of texts, the background to the writing of *mare liberum* had been the dispute arising from the Dutch seizure of a Portuguese vessel *Santa Catarina* in the Straits of Singapore in February 1603. Especially as the contents of the carrack sold in Amsterdam grossed a sum equivalent to just less than the annual revenue of the English government at the time and more than double the capital of the English East India Company, a debate immediately emerged as to the legitimacy of the Dutch capture of a Portuguese vessel in the distant seas of the East Indies.

Dutch "free sea" ambitions in East Asia would eventually be dealt a death blow, especially with their 1662 defeat on Formosa/Taiwan at the hands of the Southern Ming loyalist Zheng dynasty. By the middle of the twentieth century, however, there was an impetus to extend national claims over offshore zones, often rich in resources. Such claims subsequently led to the United Nations Convention on the Law of the Sea (UNCLOS), with negotiations taking place between 1973 and 1982, and coming into force in 1994. In effect, UNCLOS which sets limits on navigation, archipelagic status, transit regimes, exclusive economic zones, and so on, overturned the Grotian doctrine of *Mare liberum*. Returning to Wilson's (2009) argument, it is also true that "the world-system governs the historical development of international law, and international law is a primary means of the ideological legitimation of the world-system." As developed below, disputes over "ownership" of the South China Sea would test this assumption over the centuries.

FRENCH/ENGLISH MAPPING AND SURVEYING OF THE SOUTH CHINA SEA

Setting aside the quest for an Austral continent, the Drake and Le Maire-Shouten voyages revealed that the Pacific was no Spanish "lake." This was dramatically demonstrated with Dutch attacks on Spanish ships and outposts, as with Manila (and Macau), part of the Eighty Years War (1568–1648) against Hapsburg rule. This began with Olivier van Noort who, on December 14, 1600, engaged with the Spanish fleet under Antonio de Morga in Manila Bay leading to the loss of the richly laden and manned galleon *San Diego.* Van Noort, who returned to Holland, became the first Dutch to circumnavigate the world. Having expelled the Spanish stationed in northern Formosa in 1642, the Dutch became increasingly eager to attack Manila and the two adversaries engaged in with five naval battles in the nearby seas in the year 1646.

In one way or another, these European explorations, trading missions, and naval conflicts were reflected in the maps of the century. European rivalry over ownership of the South China Sea exemplifies this development. In this sense, we may envisage the cartography of the oceans as a form of "capture" of a bounded maritime space that would, in the future, be paraded, placed on display, and even parlayed into ownership. This was more difficult to achieve with the vast Pacific Ocean but a similar logic was also at work when it came to certain island groups.

By the time King Louis XIV's geographer Nicolas Sanson d'Abbeville produced his map of the East Indies in 1683, Borneo had been regularly visited by Dutch and English ships. D'Abbeville's map shows Pulo Condore off the southern tip of the Indochinese Peninsula along with islands to the north of Borneo, conceivably including the Spratlys. Known to Marco Polo in an earlier age, Pulo Condore was mapped, inter alia by Dampier in 1687–88. Maps of north-west Borneo and Sulu include that of official British hydrographer, Alexander Dalrymple, dating from ca. 1775 as a result of that nation's possession of Balambangan Island in 1763. Maps by Singapore-based James Horsburg also facilitated British control over pirate-infested north Borneo. A Spanish-era map of the Philippines including Palawan and northeast Borneo, namely the production of the Jesuit priest Pedro Murillo Verlarde who had the map published in Manila in 1734, is taken as the first scientific map of the Spanish colony bearing longitude and latitude along with rhumb lines (Nicholl 1986: 72–90).

Taking their name from *parcel,* the Portuguese term for reef, shoal or shallow place, the Paracel Islands (also known as Xisha in Chinese and Hoàng Sa in Vietnamese) sprawl south of Hainan, from latitude 17° N to 11° N. Because of the navigational dangers they presented, the Paracels were better known to early cartographers, as with Hondius's *InsulaeIndiae Orientalis* of ca. 1635. D'Anville's *Further India* map ca. 1755 also shows the Paracels in their approximate location as well as other islets and shoals in the approximate vicinity of the Spratlys. In 1735 coastal Cochinchina and Tonkin appeared in maps by Jacques Nicolas Bellin and, later, in those of the French Admiralty based upon observation by Vice-admiral de Buil and Rosili. Dalrymple, who made the journey from Balambangan to Cochinchina in 1760, also produced detailed charts of the coastal regions from 1780 to 1803 and charts that would be used by future navigators, including the La Pérouse expedition (see Gunn 1991: 266).

Detailed maps prepared by the French naval officer Jean Baptiste Marie Dayot of the Vietnamese coast in 1798 were published some years later by the French Government and by the British Admiralty (Fell 1988: 49–56; 86–93). As Pham (2016: 60) explains, having spent several years at the Nguyen court, Dayot also gleaned local knowledge from fishermen, sailors, and other informants. Despite his efforts to chart the Paracels and Spratlys, it would not be until Horsburgh succeeded Dalrymple as hydrographer of the East India Company that this project was completed. In particular, the passage between the Paracels and the Macclesfield Bank became safer with the survey conducted in 1807 by Lieutenant Daniel Ross of the Bombay Marine. In due time, Horsburgh's *India Directory* became the new model from which all subsequent pilots and sailing directions would descend (Pham 2016: 51; 52; 60). The point is that even if the Spratlys were not named in European maps at that point in time and while their survey awaited future navigators, their approximate location and the dangers they represented to navigation were well understood by Western seafarers. From the perspective of the long-distance traders of the sixteenth and seventeenth centuries, these shoals and islets were places to be avoided rather than invitations to get more closely involved (see Gunn 1991: 267).

Colonial Survey of the Spratlys or Nansha Islands

Setting aside modern-day diplomatic and international relations issues surrounding the disputed islands and maritime features, this section brings into context the diplomatic contest over islands in the South China Sea during the 1930s by the then two dominant imperial powers in the region, namely Great Britain and France, later to be joined by Japan. Aside from the questions of the right by discovery, the right of first occupant, colonial claim, the test of effective and continuous occupation, the discussion draws clear implications for the colonial successor state[s]. As Eric Tagliacozzo (2003: 107) has observed, the historiography of imperial mapping has traditionally focused on the land and not on the sea. Hydrographic mapping allowed Western empires to refashion the world into a template more suitable to colonial control. To be sure, as previously examined, much turned upon the Western legal concepts of *mare clausum* (closed sea) and *mare liberum* (open sea).

Fast forward to January 1988, following earlier clashes on land and sea during the Sino-Vietnamese war of 1979, Vietnam and China entered into bloody naval confrontation when Vietnamese soldiers died in an attempt to prevent Chinese forces occupying reefs in obscure islets in what Vietnam describes as its East Sea, and China, the Nan Hai or South Sea. This is a reference to what is known in English as the Spratly Islands dispute. A group of more than 759 reefs, islets, cays in the South China Sea, the archipelago is also known in Chinese as *Nánshā Qúndăo* and, in Vietnamese, *Quầnđảo Trường Sa*.

From a colonial-era perspective, the tussle over the so-named Spratly Islands conforms to the general thrust of world historical expansion by the colonial core countries into the periphery. With its origins back to the era of classic European colonialism, what is singular about the Spratlys dispute, however, has been its longevity especially in the way of reawakening ancient Asian notions of territoriality and even meta-geographical imagination. In line with the logic of colonialism the islands were *terra nullius*, unclaimed because no other colonial power had registered a claim. Their belated incorporation into a Wallersteinean "world-economy," however, was entirely in line with the making of the "world economy." Stated another way, the Spratly conflict between rival colonial claimants, as much the concerned local tributaries, respectively, China and Vietnam, reveals the clash of hegemonies through long historical time (*see* Gunn 1991: 263).

It is noteworthy that the official French chart of the period, *La Neptune Orientale*, published in 1745 with the support of the French Academy and the French East India Company, and credited to mapmaker Jean Baptiste Nicolas Denis d'Après de Mannevillette, came to British attention in an era of colonial competition over the South China Sea. Writing in February 1933 the British official J. S. Barnes found no islands or banks in this area whatsoever, with the exception of the Macclesfield Bank further north labeled "anglais." As noted, notwithstanding the "considerable interchange of ideas between the French and English hydrographers of the period," on the Horsburgh chart of the China Sea dated 1821 for use by the Bombay Marine, or East India Company Navy, the concerned islands were plainly given English names. As Barnes triumphantly recorded, there was "no indication" of French influence (Barnes cited in Gunn 1991: 268).

By way of summary, as R. T. Fell remarked, cartographers of many European nations were involved in the production of the first Western maps and charts of the South China Sea region, not to mention the local assistance of especially Filipinos and Malays. Moreover, "There were also good political reasons at that time for the British and the French to have more accurate information about a region in which both nations were taking an increasing interest." Sovereignty over the islands, however, was not yet an issue. (Fell 1998: 9 cited in Gunn 1991: 268).

THE "CLASH OF CONCEPTS" OVER THE SOUTH CHINA SEA

Writing in 1985, as the political geographer Victor Prescott advised, the warning "Dangerous Shoals" written across the charts of the Spratly Islands should equally be heeded by navigators and scholars. As he writes, "no other part of the world possesses to the same extent the twin difficulties of a plethora of claims and a lack of precise, basic geographical information." Geographically, the scattered archipelago of cays and low-lying reefs lying 400 kilometers east of the Vietnamese coast and 965 kilometers south of China's Hainan Island also presented very specific dangers to pioneer navigators. Lest we forget, this was still the age of sail, and alongside the best navigation, knowledge of winds, tides, seasons, and, in the South China and East China seas, typhoons was also crucial (Prescott 1985: 219 cited in Gunn 1991: 262).

The argument raised by former University of Hong Kong geographer R. D. Hill (1991: 1–3) as to giving weight to "ownership" in Asian historical terms bears scrutiny. In this discussion, he highlights the fact that power, whether stemming from king, emperor, or *devaraja,* concentrated the essence of state. In this formulation, power derived from control of people as opposed to terrain, whether sea or land. Whereas Romans, other Mediterranean peoples, and Arabs could "locate their territories in terms of the Ptolemaic grid, Southeast Asians could do little more than identify the center of power of their local rulers." With the Ptolemaic revival, he continues, "major corrections to the global graticule by Mercator (whose projection we still use for navigation purposes) in the sixteenth century and by Kepler, the great astronomer, in 1630, a thoroughly modern spatial framework was erected into which the lands and seas of many nations could be fitted."

Adding a note of caution, as Hill (1991: 1–3) writes, such a rendition of space did not hold for Asia. With few exceptions, as with Chinese maps indicating measurement in *li*, no traditional Asian state "seems to have had either defined boundaries or any indigenous spatial framework" into which measured territory could be slotted. While Chinese maps illustrate provincial boundaries it is unclear as to how "new discoveries" were fitted into the Chinese graticule. "For broadly-Confucian states such as China and Vietnam the oceans were *res nullius*—'nothings'—places beyond civilization and hence of little concern." Citing Dian Murray (1988: 4) he observes, "Confucians were more intent upon governing men than space." Nevertheless, it is precisely these two nations who *do* debate ownership of the sea with reference to traditional codes, maps, and imperial decrees. There is now a voluminous literature on ownership disputes in the South China (and East China Seas), certain scholarly and certain gushing from officialized publications and with each one trumping the other.[10] To me this sounds like a "clash of concepts" to parody Samuel Huntington's better known paradigm which also fits the bill.

Cartographic Evidences and Arguments

As Prescott notes, of all the claimants to the Spratly Islands only the Philippines could enclose them by archipelagic baselines. Otherwise the islands do not constitute part of the continental shelf of any country. More specifically, the group lies south of parallel 12° north and east of meridian 112° east. They exclude all islands within the archipelagic baselines of the Philippines and those which lie within 40 nautical miles of the coast of Borneo. Prescott identified the existence of twenty-six islands or cays in the Spratly Group plus seven sets of rocks standing above high water. Only fourteen of these "insular features" were occupied at the time he wrote (Prescott 1985: 219; 230 cited in Gunn 1991: 262).[11]

In the wake of armed clashes in the Spratly Islands occurring in the early 1990s, the Chinese media reiterated earlier claims that Chinese people actually began to settle in the Spratlys (Nansha Islands) since 200 BCE, in the time of Emperor Wu's reign during the Han dynasty. By Tang and Song times (618–1279), Chinese people were already living on the islands and engaged in fishing. By Ming and Qing times (1368–1911), the archipelago was placed under the jurisdiction of Qiongzhou Prefecture, Hainan Island, Guangdong province. To back these claims the Chinese purport to have discovered pottery shards and other archaeological evidence linking China with the early discovery and settlement of the islands (Yin Zhiping, 1988 cited in Gunn 1991: 263–64).

To be sure, we cannot ignore the role of Chinese trading junks in regional and long-distance trade which would have virtually made these waters a *Mar Sinensus*. Chinese navigators sailing to the south seas, however, avoided the center and either ran along the western and southwest margins of the South China Sea via the Hainan Paracel gateway and along the coast to Pulo Condore (Vietnam's Côn Đảo)—the favored course—or through the Palawan passage connecting with the Philippines and more southerly destinations. In any case, according to Murray, Qing maps of the Nanyang or China's southern ocean realm barely included such islands as the Paracels and Spratlys (Murray 1988: 4 cited in Gunn 1991: 264) (see Figure 8.3).

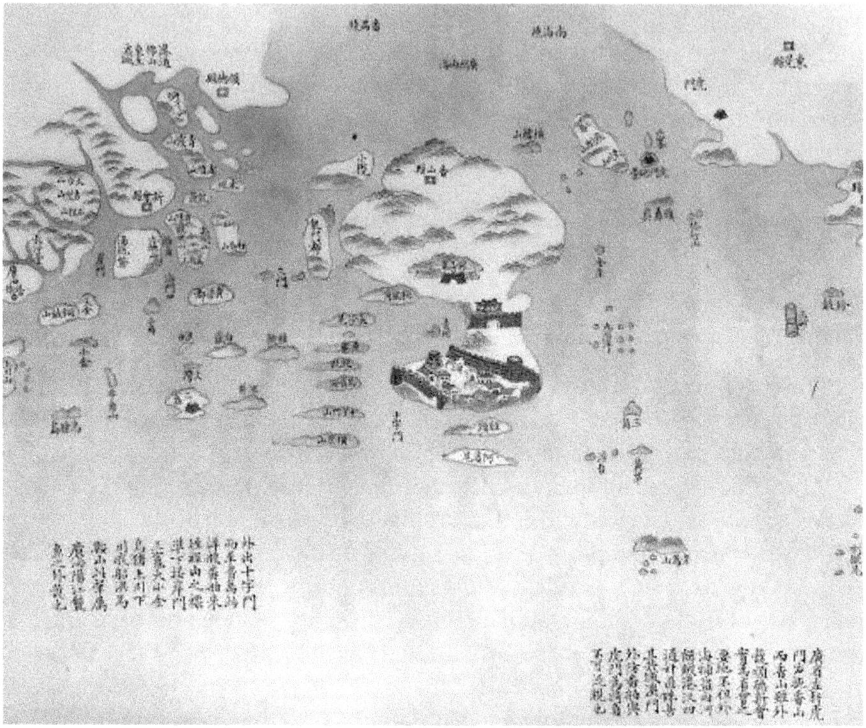

Figure 8.3 Coastal Map of China (Macau). Qi sheng yanhaitu. 七省沿海图 [Qing Guangxu 7 nian, 1881] Scale ca. 1:760,000. Late Qing coastal defense map. Area surrounding Macau including Pear River delta only. To note, Macau peninsula is falsely depicted as a detached, fortified island. *Source*: Retrieved from Library of Congress. Digital Id http://hdl.loc.gov/loc.gmd/g7822c.ct000747.

Separate chapters of this book have discussed Arab and subsequently European navigational technologies. We have also examined Chinese mapping, albeit setting aside the question of Chinese navigational skills. Allowing that this is not a well-researched subject, it is salutary to learn from the Ming-era *Shun Feng Xiang Song* (順風相送, *Fair Winds for Escort*), a major source on Chinese navigational routes during the fifteenth century, and certain navigational lore of the time. As J. V. Mills (1979: 72) summarizes, in specifying the stages of a sailing run, the usual formula in Chinese texts is "From place A steer x°: after y watches you mark place B." Time was measured in *kêng* (watch) of 2.4 hours; and when places could be identified with reasonable certainty, the distance was measured on a chart, allowing the speed to be calculated. On a long-distance voyage in the open seas the speed of a Chinese ship at this time averaged about four knots or 6.4 kilometers in 24 hours. As Mills adds, the book contains certain stellar altitudes but none further east than Pulo

We (Pulau Wey), 95° 15' E, the small volcanic island standing northwest of Sumatra. Nevertheless, we are surprised at both the spread of Chinese junk trade network from Japan in the northeast to Hormuz across the Indian Ocean, to Timor in the southeast, as well as to the scientific approach to navigation which, allowing for civilizational differences, was not altogether different from the European approach. In other words, as the *Shun Feng Hsiang Sung* suggests, China, or at least groups of navigators, did have a comprehensive understanding of the seas, their connectivity, their currents and winds, their navigational hazards, and their coastlines, and major features, many of them named. This is affirmed in the Selden map which also indicates major Chinese junk trade routes.

Vietnam's Historical Role in the Maritime World

There is no doubting Vietnam's historical role in the maritime world. Writing of the importance of the Dai Viet state in entering international trade between the tenth and fourteenth centuries, Japanese scholar Momoki Shiro (1998: 15) contends that Vietnam's coastal ports played a vital role in servicing commercial links between Southeast Asia and China, not excluding trade with the Ryukyu kingdom. Dai Viet was not just a trade broker, but its domestic production also supported an export industry. More modestly, the Vietnamese claim effective and continuous occupation of both the Paracels (Hoang Sa) and Spratlys (Truong Sa) since the seventeenth century. To this end, the Vietnamese back their claim with reference to historical texts. Among them is *Toàn tập Thiên Nam tứ chí lộ đồ thư* (Route Maps from the Capital to the Four Directions) a seventeenth-century map of Quang Nam province. Notes attached to the map refer to an elongated sandbank offshore known as the "golden sandbank" (referring to the Paracels and Spratlys) (Nguyen 2013: 42–51; 58). Additionally, other texts such as the *Phủ biên tạp lục* (1776) (Miscellaneous Records on the Pacification of the Frontiers), the *Đại Nam thực lục* (1844–1848) (Truthful Accounts about Dai Nam Former and Present Dynasties), and the *Đại Nam nhất thống chi* (1882) (Geography of Unified Dai Nam) all refer to the role of the state in exploiting the resources of the archipelago continuously under the reigns of the Nguyễn (1558–1786), the Tay Son (1786–1802), and Nguyễn (1802–1945). Specifically, in 1835–36 under Emperor Minh Mang (r. 1820–41), cartographic surveys of the surrounding seas were made and "sovereignty markers" were placed in the archipelago. In this account, evidence of the Vietnamese imperial connection with the islands can be seen on the islands in the form of vestiges of military camps, shrines and stellas dating from the time of Minh Mang or earlier. To this end, the Vietnamese approvingly cite French European sources, both textual and cartographic. For example, in 1634, Dutch merchants shipwrecked in the Hoang Sa (Paracels) dealt with Vietnamese not Chinese officials. Portuguese and, in turn, Dutch mapmakers duly recorded the navigational hazards (Nguyen 2013: 128–29).

In his study of China's 1979 cross-border war with Vietnam (sometimes glossed as the Third Indochina War), King C. Chen comments that historical records before the mid-nineteenth century prove that while both China and Vietnam had claimed ownership of the archipelagos neither was aware of the other's claim. In terms of evidence, he claims, China has more solid documentation than Vietnam to prove its first discovery and claims to ownership of the islands. But, even if first discovery of the Paracels can be substantiated, the evidence of prior discovery much less claim over the Spratlys is lacking (Chen 1987: 44 cited in Gunn 1991: 265). As the Korean specialist Park Choon-ho noted, while China is bound to win the "chronicle war" with Vietnam owing to its long history and its encyclopedic records, discovery is not a sufficient but only a necessary condition required by international law as one of the requirements of acquisition of territory. Additionally, "continuous and uninterrupted exercise of sovereignty" must be upheld (Park 1983: 216 cited in Gunn 1991: 265).

In any case, dynastic claims amounted to nothing in the era of European world historical expansion especially on the periphery of empire. For the European center, conflicting notions of power and sovereignty upheld by Asian tributary states were by-passed and have been by-passed in international law. For China, this meant the indignity of encroachment upon sovereignty and a system of unequal treaties. While China entered the world economy as semi-colony, Vietnam suffered a loss of all sovereignty to French imperial goals in the East Asia region. In any case, the Sino-French accords of 1887 and 1895 that established the land and offshore coastal boundary between France in Vietnam and the Qing dynasty was silent over the demarcation of the Paracels and Spratlys (Duiker 1986: 36–37 cited in Gunn 1991: 265).

In my article examining Franco-English colonial rivalry over the disputed islands (Gunn 1991: 279–280), I pointed out the seemingly unusual circumstance whereby post-independent South Vietnam looked to French colonial documentation to demonstrate prior claim or ownership of certain islands. More recently, as Alex Calvo (2015) has summarized, Vietnam relies on two kinds of Western sources; namely records of missionaries and traders, as well as official documents from the French colonial administration and other powers. Both the former Republic of Vietnam and its successor, the Socialist Republic of Vietnam, have followed the same line. Furthermore, Hanoi has been keen to emphasize that French domination over Indochina did not interrupt the exercise of sovereign powers but actually reinforced it, since the new authorities kept exercising sovereign rights. Vietnam's Eastern (China) Sea narrative also relies on international treaties signed by Paris, such as the 1887 Convention on the land border, signed by France and the Qing dynasty.

Citing an array of French authors and witnesses supporting Vietnamese claims, the Vietnamese author, Duy Chien (2014), references "Notes on the geography of Southern Vietnam" by Jean-Louis Taberd, the former apostolic vicar of Cochinchina, from 1824 to his death in 1840, and interpreter of Nguyễn Emperor Gia Long. Published in 1837–39 in the *Journal of the Royal Asiatic Society of Bengal*, Taberd describes the 1816 flag-raising ceremony on the Paracels by Gia Long thus: "Paracel or Hoang Sa islands is an area crisscrossed by small islands, reefs and sand, seems to be extended to 11° 0' North latitude and around 107° 0' longitude. . . . Although this is an archipelago covered by nothing else but islands and reefs, and the depth of the sea promises [more] inconveniences than advantages, Gia Long still thought that he had the right to expand his territory by that pathetic merger." The flag-raising ceremony was also confirmed by French missionary, Jean Baptiste Chaigneau writing in 1820. Taberd also published a map *An Nam Đại Quốc Họa Đồ* bearing the words "*Paracel seu Cát Vàng*" (1838), suggesting a Vietnam connection through propinquity.

As Calvo (2015) notes, among the strongest pieces of historical evidence in support of the Vietnamese advance is the Nguyễn dynasty's exercise of tax powers over foreign ships passing through the Paracels, as recorded in non-Vietnamese sources. Notably, this is recorded by the missionary Karl Gützlaff writing in the *Journal of the Royal Geographic Society* of London in 1849. Separately, both Vietnam and the Philippines ritually present Song, Ming, Qing, and even Republican-era maps (and there are many) revealing Hainan Island, as opposed to the marine features of the South China Sea, demarcates the southernmost part of China. Meanwhile, China maintains its long historical claim *tout court*.

CONCLUSION

As this chapter has revealed, a new conception of the *Sinus Magnus*, the sea gateway to China known to the Arabs since antiquity, came into being in the wake of the Magellan voyage especially around ownership. Treating the sea and maritime features within it as private domain, the Europeans pirated and plundered, just as they came into conflict with the Sinicized world of China-Vietnam-Korea-Japan along with the Ryukyu kingdom, the sultanates of Malacca, Brunei, Sulu, and those of the Moluccas as with Tidor and Ternate, and Java as with Banten, not to mention such Buddhist kingdoms as Siam and Cambodia, and the Hinduized Cham, all of them maritime in orientation, and each of them territorial within their own particular conceptions of space.[12]

Just as cadastral surveys on land served land alienation, exploitation of resources and the Western colonization mission at large, hydrographic surveys of the oceans were part and parcel of the imperial quest for naval and, hence, trade and political supremacy over local and European competitors alike. The hydrographic mapping of the coastal waters of China (such as around Macau and Hong Kong along with the Paracels and Spratlys and other strategic straits and waterways) had the effect of superseding local maps by incorporating even the most remote locations into the graticule of world mapping. The colonial survey of the southern islands of the South China Sea following on in quick succession to their "discovery" were no exception.

More than any other international dispute involving sea boundaries and claims, the South China Sea case illustrates the vastly different conceptions of space between those inherited and developed by the West versus concepts invoked by China and other Asian kingdoms and polities. This we called a "clash of concepts." As international relations specialist Katherine Morton (2016) expresses the matter, "On the side of the West stands the Grotian concept of free and open seas and, via its reversal under international consent, the modern-day Law of the Sea." On the side of China (and the "little dragon" of Vietnam) stands the dynastic historical record of ownership and concept of tributary. But for seaborne Portuguese, Spanish, Dutch, and English companies and seafarers arriving on the shores of the western Pacific Ocean in the early sixteenth and seventeenth centuries, whether sailing east or sailing west around the globe, the ocean seas were literally the sites of sea-battles and places to deny the objectives of their adversaries, mostly Muslim when they were not fighting or competing with each other.

But in entering the China seas, as we have argued, Europeans also had to radically reconfigure the old Ptolemaic template that somehow conflated the Gulf of Siam into the South China Sea, and otherwise reduced China to some peripheral appendage on the Eurasian landmass, coordinates unknown. Whereas the Ptolemaic template as passed down through *Geographia* represents a correlated assembly of toponyms and maps relating to a three-continental landmass, in the quest for appropriation—and possession—as this chapter has highlighted, the Spanish turned to the *mar Occéano*. Traveling westward willy-nilly—and with not a little design and planning—they also gained the measure of China and the China seas in a way unimaginable to the ancients. As also examined, later-arriving Europeans would discover the reality of a far smaller ice-dome covering the southern polar zone outside of the Ptolemaic imaginary of a vast encircling southern landmass.

NOTES

1. To this end, as Mundy (1996: 12) explains, from 1579 to 1585, cosmographer Juan Lopez de Velasco drew up the *Relacion geográficas,* a questionnaire posed to American-born Spanish, *mestizos,* and others in New Spain, Peru, and possibly even the Philippines.

2. It is important to note, as Gužauskyte (2014: 4) explains, that Columbus (and those in his wake) imposed an assembly of toponyms in his *Diarios*or voyage logs. In so doing, he thus followed the precedent set by Ptolemy's *Geographia.* But whereas Ptolemy dealt with landmasses, in an unprecedented departure, Columbus paired names with maps in the *mar Occéano.*

3. The historiography of Pacific Ocean exploration is obviously large, practically reaching back to Pigafetta, through Tasman, Cook, La Pérouse, and so on. Modern scholarship has tended to apply a Braudelian Mediterranean framework connecting up peoples and cultures (see Matsuda 2012).

4. Of great importance was the pioneering journey from Manila by Spanish Basque navigator, Andrés de Urdaneta, taking a northward journey across the Pacific to catch the prevailing winds and currents, thus establishing the wherewithal for seasonal return journeys. Known as the *tornaviaje* or return trip as Gordon and Morales (2017: 4; 14–19) explain, this could take up to six months relative to the outward journey from Mexico or Peru which seldom took longer than two months.

5. This important letter is lodged in the Bibliothèque National, Paris, and is reproduced in full by Randles (1984: 17–21).

6. Hernando de los Ríos Coronel, Memorial quedirigió al Rey desde Manila con fecha de 27 de junio de 1597 . . . Archivodel Museo Naval. Colección Fernández de Navarretehttp://www.upf.edu/asia/projectes/che/s16/rios.htm China en España Elaboración de un corpus digitalizado de documentos españolessobre China de 1555 a 1900.

7. Crossley (2011) has translated from Spanish to English de los Ríos'rutters of 1605, 1606, a Philippines rutter, and those of 1610 and 1611. These are extremely rich in revealing navigational lore, process, and techniques standard for the seventeenth century. http://www.csse.monash.edu.au/~jnc/Rios/

8. However, we do know that on December 11, 1685, an eclipse of the Moon was registered by German Jesuit Paul Clayn (Klein) at Manila (*Mémoires de l'Académie des Sciences*, vol. 7, 1725).

9. The Queiros historiography is long, in part because he produced a significant number of petitions or memorials directed at the Spanish king, one entering publication in English by Samuel Purchas. A fuller set of his voyages was published by Hakluyt. See also, Markham (1904, Vol. 1). Portuguese interest is implied in Lapa (1951). Australian interest in Queiros's colonization plans has also led to such works as Celsus (1965). In a more popular vein, see Estensen (2006).

10. An even newer literature has emerged around dispute resolution, as with Wu (2013), highlighting China's long historical claims and, inter alia, exposing weakness in Vietnam's claims.

11. The seascape continues to morph especially with the militarization of a number of the islands and the transformation through reclamation of features into islands. At least five landlords can be nominated. From 1968 the Philippines occupied ten features—seven islands and three reefs—including Loaita, Thitu, and Northwest Cay occupied and later West York Island, Flat Island, Nansha Island, and Lankiam Cay. From 1956, Taiwan occupied Itu Aba Island. Since 1978, Malaysia, militarily occupied Swallow Reef and erected obelisks on Louisa Reef and Commodore Reef (although the latter was destroyed by the Philippines authorities). Vietnam (until

1988) had forces on Spratly, Southwest Cay, Sin Cay, Namyit, and Amboyna Cay (Prescott 1985: 219). The fifth landlord, China, erupted on the scene in early 1988. On naming issues see Hayton (2014).

12. Understandings of Cham civilization, their Austronesian origins, and their seafaring traditions, have expanded greatly in recent times. For example, see Hayton (2014) who, inter alia, provides a historical overview of the South China Sea from prehistoric times to the early 2010s.

Conclusion

Post-Ptolemaic Constructions of Empire and Nation-State

Observing the glaring disconnect between the Ptolemy-revival atlases emerging in late Renaissance Europe and the reality announced by the "discoveries," this work set out to test the longevity of the Ptolemaic world image. This we did with specific reference to the macro-region termed *India Extra Gangem*, literally the space between India and China including Southeast Asia. The study also drew attention to the trap set by Ptolemy and Ptolemy-revival atlases for future navigators and cartographers seeking to better know the real world. In doing so, it looked back to classical points of reference, essentially to highlight the weight of Ptolemaic representations of Asia such as received in late medieval Europe from Arabic and other sources. As a corollary, the study also tracked the gradual erosion of Ptolemaic conceptions especially in the light of new empirical knowledge supplied by the Columbus, da Gama, and Magellan discoveries. One overarching conclusion emerged, namely that revisions to the Ptolemaic-revival template was a non-linear evolution suggesting advances and retreats. While Ptolemy's geocentric system succumbed to general irrelevance in the face of the revelations of Galileo, Copernicus, and others, we were nevertheless surprised that the Alexandrian inheritance doggedly survived into the early modern age. Not only did Ptolemy stand as a reference point or as a test of classical erudition as with the example of French mapmaker Jean Baptiste Bourguignon d'Anville, but his geographical conceptions became the yardstick against which French "scientific cartography" measured its success in global mapping even 200 years after Columbus.

To illustrate the contested constructions of place by European mapmakers and establishments, the study matched textual and cartographic adjustments to images of *India Extra Gangem,* especially during this period of European power projection. Where possible, as in the case of China, we gave full weight to local traditions of astronomy, renditions of space, and mapping, allowing as well complex East-West borrowings such as occasioned by the *pax Mongolica.*

SPATIALIZATION REVISITED

To heed the sage conclusions of the editors of *The History of Cartography, Cartography in the Traditional East and Southeast Asian Societies*, non-Western cartographies rival those of the West in length and variety. While we have not labored the subject, even before Ptolemy, Asian cartographers had produced maps "impressive for their mathematical accuracy." In this argument, there was no "normal" as to cartographic practice across Eurasia given civilizational and cultural difference. Neither does the displacement of traditional mapping by European mathematical cartography necessarily imply "progress," especially considering the loss incurred in aesthetic for example (Woodward et al. 1994: 844–845). In this vein, European "rationalism" was also crosscut with religious dogma, a predilection for astrology, superstition, and other medieval practices, some of it washing into cartographic representations and accompanying texts, a mind-set frequently serving as a decoy for discoverers as with the fetish for the biblical Ophir and lure of gold.

In the Introduction we have drawn attention to the now current acceptance on the part of geographers that maps stand as metaphors and not as immutable statements of truth. As Mitchell (2012: 16) summarizes, "In postmodern writing, therefore, emphasis is placed on the spatial, experimental, subjective, and cognitive nature of mapping and reading." Cartography, as explained, was also a harbinger of early arriving European colonialism. Especially we confirmed the Deleuze and Guattari (1972) thesis on territorialization sequential with the historical processes of European expansion and the colonizer's need to impose fixity, control, and order over conquered space and cultures, also implying disempowerment or loss. Although we have not hitherto explored the matter, not even the age of decolonization and the shaping of new nation-states could entirely escape the "classical" constructs.

More recently the question of scale has entered this discourse. Traditionally, as Robert B. McMaster and Eric Sheppard (2004: 2) write, "geographers thought about scale as predominately a cartographic concept, where scale associates a map distance of a feature to that feature's distance on the surface of the earth." As such, the definition of scale is mathematical and remains the focus of the cartographer. To make this point they offer the example of eighteenth-century France where the "science" of cartography emerged with the development of modern geodesy and the creation of the first state-sponsored national map—the Carte de Cassini (a reference to the four generations of the Cassini family and with their map finished in 1789). As they sum up, "the problem of measurement consumed the cartographic community," and would do so until relatively recent times.

However, in human geography scale has come to be understood not just as a matter of spatial resolution, but as a strategy and an outcome of political and social processes. Beginning in the 1970s such social theorists as Lefebvre (1974), Smith (1992), and Harley (2001) began to argue that space and, by implication, scale is socially constructed. In Neil Smith's (1992: 64) formulation, "Long conceived as fixed, dead and immobile, space is both available as foundational metaphor and at the same time being rediscovered as produced, mutable, an intrinsically complex expression of social relations." Because space is constructed and produced through political-economic processes it cannot be considered as purely exogenous or fixed. This understanding has led to further studies on politically constructed space. In the following section we seek to test this proposition with respect to various regions, new and old nations, and other constructed spaces.

VALORIZING "ASIA"

As Ptolemaic-revival atlases vividly portray, the eastern edge of the Eurasian landmass folds into a *Sinus Magnus* (or Great Gulf). Yet, this was not an open sea and the American continent was disallowed, especially as the Ptolemaic template also exaggerated the eastern extent of the Eurasian landmass. As our discussion revealed, Ptolemy's underestimation of the Earth's true circumference led to the misplacement of, especially, the Golden Chersonese configured significantly further east than the true position of the actual Malay Peninsula, even though distance from the Equator was better represented (Durand and Curtis 2013: 25).

The Ptolemaic heritage (at least as revised by medieval European visitors to the Mongol court) also carried over into late Renaissance map renditions of China, as with confounding Cathay and Mangi (or southern China), or China with Tartary, and even the name of the capital, Khanbalik. China, too vast to conquer or convert, was at least identified (and named). Reaching back to Ptolemy, China attracted a confusing ensemble of names, variously, Scythia, Seres (*Serica*) and Sina (*Sinae*). Not even the evidence of Muslim travelers arriving in Song China ports via the maritime silk roads fed back into European images of an open sea. Polo added information but also confounded. Knowledge derived from later-arriving Italian travelers was slow in being absorbed. Columbus's own denial of having discovered a new continent in deference to a belief in having reconnoitered the coast of

China (Mangi) or Vietnam (Champa) offers convincing evidence of the sway of Ptolemaic thinking at the outset of the age of discovery. As discussed, it would not be until the arrival of the first Jesuits in the Ming capital that a bounded sense of China and region would eventually appear in Western imagination. A Chinese capital—Beijing (alongside Nanjing)—rightly took its place on the map.

In the event, the Portuguese (and Spanish) voyages brought new empirical data on China and with portolan map renditions of coastal China first entering the *Livro* or Book of Francisco Rodrigues (ca. 1512). As we acknowledged, Jesuit knowledge of China and Japan also fed back to Europe. Reciprocally, Europe began to impose its own spatiality, nomenclature, and even values upon Asia, absolutely as with the Philippines (and Marianas), named after Spanish crowns, alongside other enclaves and, differentially, in India, Indochina, and the East Indies.

Not for want of trying, on January 23, 1576, Pope Gregory XII issued a bull setting up the diocese of Macau, whose jurisdiction grandiloquently included China, Japan, and adjacent islands, and with the formal spelling of Macau first making an appearance in Latin as "Machao." Although Tartary or "*Tartarie Chinoise*," as employed by d'Anville, lingered on, Russian expansion across the Eurasian landmass forced China under the Qing to think the unthinkable, namely to accept both embassies and fixed and measured boundaries.

The world historical import of this development cannot be underestimated, just as the measurements of longitude came to be scientifically conducted in India and China, respectively, through the seventeenth and eighteenth centuries. As the errors of the Ptolemaic template came to be exposed, so the eastern extent of the Eurasian landmass came to be measured, territorialized, and situated (paradoxically entering the world map graticule just as Ptolemy counseled). Almost everywhere, however, as with the Jesuits, as with the Malacca-based "cosmographer" Manuel Godinho de Erédia, and as with Crown projects such as that mounted by Louis XIV of France, Europeans drew upon indigenous Asian knowledge. But as Europe imposed its conceptual renderings of space upon "Asia," we also wonder as to local Asian cognizance of space and region or how "Asia" was valorized locally in China, Russia, Japan, India, and Southeast Asia?

The Sinic World View

What then distinguished the Sinic world view of "Asia/Eurasia" and distant Europe? According to Richard Smith (1996: 23), traditional Chinese maps reflected Sino-centric principles with China occupying the center of most renderings of cartographic space and with tributary relations never far from mind. From antiquity, a parallel secular Chinese mapmaking tradition developed adapted over time to encompass a wider world. Such, not only included Song-era maritime voyages to a southern sea (as with voyages to Timor) but, with the advent of the Mongol-Yuan dynastic power, new astronomical and geographical information derived from Muslim and other sources came to be incorporated. Lacking the mathematical precision or presumption of Ptolemy's *Geographia*, it has to be admitted that, as with the early Ming-era Zheng He voyages or even predecessors, China made up with its own chorography of description, or culture of data collection that went in hand with an ancient centralized literate record-keeping bureaucracy.

Newly arriving Jesuits, as with Matteo Ricci, sought to accommodate Chinese sensibilities as to their Middle Kingdom status by constructing world maps based upon the projection developed by Abraham Ortelius, but with a prime meridian running through China. This was the case of his *Kunyu Wanguo Quantu* (坤輿萬國全圖) printed at the request of the Wanli Emperor in 1602 in turn based upon a 1584 version. According to Smith (1996: 45), it was Ricci who (re)introduced the notion of the sphericity of the earth to Chinese literati (even if Persian scholars had attempted this much earlier). Importantly, he also introduced the concept of map coordinates. As Waley-Cohen (1999: 101–109) confirms, European astronomy was the "mainstay" of the Jesuit exchange with China, though they also stuck with geocentric theories against the evidence. As this author summarizes of the Jesuit mission to China, they undoubtedly made their impact in the field of mathematics, Euclidean geometry, and plain and spherical trigonometry, alongside astronomy, calendrical reform, and general science.

However, starting from 1700, it was the Jesuit mapping of China conducted under Emperor Kangxi which offered convincing empirical coordinate references, challenging but not actually overthrowing neo-Confucian and traditional cosmological views. Another example of the Chinese tradition carried through to Korea can be instanced with the "wheel maps" of the 1760s known for registering a large number of place names for "countries" (Smith 1996: 30). On the other hand, the mapping of the Qing empire, achieved on a province-level basis by Jesuit surveyors, fed directly back into the knowledge pool in Europe via d'Anville's map reproductions. By this age, the Chinese space vastly exceeded the scope of the Han dynasty confined to the Yellow River and with outliers in northern Vietnam. It also vastly exceeded the dimensions of late Song China such as attested by Marco Polo and others. While the Jesuit survey undoubtedly helped the Qing to validate its boundaries, the valorization of "Asia" in Chinese mapping would practically await the end of the dynasty.

The Russian Far East

But where did the expanded Qing empire stop in the vast central Asian steppes shading into Mongolia, not to mention the ocean? To be sure, as Nianshen Song (2005) has argued, the so-named Russian "Far East" betrays a deep-seated Eurocentrism (practically back to Ptolemy we may add). In Anglophone Asian studies scholarship, he points out, "northeast Eurasia" partly overlaps with "inner Asian frontiers," which includes (greater) Manchuria as well as (greater) Mongolia, Chinese Turkestan (Xinjiang), and Tibet. Until recently, national narratives from Russia, China, Korea, and Japan all describe indigenous peoples as "barbarian." By dividing the region into multiple subregions, as with the Russian Far East and Siberia, eastern Inner Mongolia, northeast Korea, eastern Mongolia, and northern Japan, such a view ignores or even denies autonomous history. "It also turns a blind eye to the *longue durée* development of this land by inhabitants of multiple cultures for thousands of years before the coming of modern imperial and national states."

Japan

But how was "Asia" perceived from the East? Writing on Meiji-era literati and reformer Fukuzawa Yukichi's late nineteenth-century program of "leaving Asia" and embracing Europe, Japan specialist Pekka Korhonen (2014) explains that political vocabulary and the meaning of words (including geographic names) change over time. As he writes, "Sometimes conceptual history even teaches us that specific concepts did not exist at a certain past moment, even though we believe so now." In Fukuzawa's time, he writes, the scope of Asia was much narrower than nowadays. (Even today, some of my Japanese students insist upon defining Asia in a restricted Buddhistic-Hindu sense, recalling ancient cosmographies.)

As Korhonen (2014) explains, perhaps the first instance when the name Asia appeared in a Japanese text is in a 1695 version of the Nishikawa Joken map. Similarly, his contemporary, the Confucian scholar and Tokugawa adviser, Arai Hakuseki, abstractly refers to *Ajia* (and the Italian pronunciation of the word in Japanese comes from him). However, the important geographic concepts that Arai used in his 1725 work *Sairan Igen* were taken from classical Chinese geography, namely *Toyo* (東洋) *or* Eastern Ocean, and *Seiyo* (西洋) or Western Ocean. Although borrowing from Chinese cosmology, he nevertheless based his maps upon European geographical concepts as with Ricci's world maps. Rather than referring to oceans, he denoted the countries bordering these oceans; and with *Seiyo* referring to the place where Europeans sailed from, and *Toyo* to the areas they came to in the east. The substantive analytical content differed from the two European geographical terms, Europe and Asia. As Korhonen elaborates, the usages of

Seiyo and *Toyo* would became established concepts in Japan in later times in referencing world cultural, economic, and political phenomena.

But for Fukuzawa, Asia was the part of the Eurasian continent that did not fit within the concept of Europe (actually reversing a European perception of Asia), spanning everything from Siberia southwards, and the Urals and Anatolia eastwards. His was a modern view, stripped of Chinese cosmology. He could foresee that practically all Asia was either colonized or in danger of becoming subjugated. Strikingly, Fukuzawa's double-hemispheric world map placed Japan and the Pacific at the center, outside of European conceptualizations which centered upon the Atlantic. To be sure, as Korhonen (2014) advises, he imported Western information, but reinterpreted it from an Eastern perspective.

The Indian Subcontinent

By the late eighteenth century, the Great Indian Trigonometric survey was already underway. In fact, as the official "Survey of India" (dating its origins to 1787) boasts on its website, "We know every inch of the nation, because we map every inch of it" (India Survey 2014). This is undoubtedly true in the positivistic tradition of d'Anville, but maps are not neutral either. While our analysis has resisted constructivist discourse, postmodern renditions of space and territorialization also see maps as totalizing and appropriating. In any case, British India soon fell in place and, as heirs to the Mughal empire and with the last remaining centers of Hindu power as with the southern Deccan-based Vijayanagara in decline, the raj was created. Ptolemy may not have been consciously consulted by British empire builders—and the matter does bear consideration—but Alexander the Great would have been proud with the British template spanning a swathe of territory from the Indus to the Ganges (*India Intra Gangem*) including Taprobane (Ceylon/Sri Lanka), and soon extended to embrace the Bay of Bengal, and Burma, literally establishing a frontier with China. Within decades of the launching of the Great Trigonometric Survey by the East India Company in 1802, the British had established outliers in the Malacca Straits, as with Penang and Singapore, and were pushing into northern Borneo and the China Sea or *India Extra Gangem* as well. To achieve this goal, not only did the British extinguish royal traditions in Burma, but they placed the ancient kingdom under rule from New Delhi (down until 1930). Only intra-colonial rivalry with the Dutch and French checkmated further British aggrandizement.

Mainland Southeast Asia

Reaching back to Ptolemy, a sense of mainland Southeast Asia or the space between India and China came to be consistently rendered as *India Extra*

Gangem. A region came to be credibly created, just as details of polities and peoples began to fall into place, especially entering the navigational lore of Arabs on the maritime route to China. In fact, the region was even better known to China as part of its tributary trade network and, as alluded, with the northern part of Vietnam occupied by China from Han times. Apart from introducing new material changes to trade protocols, Chinese knowledge of the Indian Ocean littoral also expanded dramatically with the seven voyages of the Zheng He fleet between 1403 and 1433. Such is attested though archaeological research at a range of sites, from Kota Cina in northeast Sumatra, to Banten on Java, to Brunei on Borneo not excluding marine archaeology (Miksic 2010).

Europe from the time of Alexander the Great better apprehended polities around the Bay of Bengal, morphing into the known world template of Ptolemaic representations. Arab knowledge of the maritime trade routes to China was quite detailed, such as reflected in the account of Ibn Battuta who traveled to Southeast Asia ca. 1345, as well as in Arab itineraries such as those translated by Renaudot. European traveler reports, some by the land route as with Niccolò de' Conti's account of 1444, began to reveal details on the charter kingdoms of Bengal, Aracan, Pegu, and Ava. As we observed, de' Conti's knowledge fed directly into the Frau Mauro world map of 1459, actually naming Melacha/Malacca at the base of a small peninsula, a first in Europe (Durand and Curtis 2013: 27).

As a geographical region, *India Extra Gangem* was relatively constant over the ages, although the progressive siltation and change of courses of river deltas as with the Brahmaputra River, the Mekong, the Red River did create topographic changes (actually measurable at the ancient Hoi An site, once connected with the ocean, or at Oc Eo now far inland, and at Palembang the presumed loci of Srivijaya astride the Musi River in southeastern Sumatra, now far inland from the open sea). *India Extra Gangem* was also coherent as a region, ambiguously embracing Tibet and the highlands but separated by mountain passes with neighboring Yunnan and Guangxi provinces of China but also neatly bisected by the elongated Golden Chersonese, with its many port cities open to oceanic trade. Politically, however, mainland Southeast Asia experienced great internal flux.

Modern historians of Southeast Asia divide the mainland into, respectively, western, central, and eastern (Lieberman 2003; 2009). All witnessed the rise and fall of civilizations, the most dramatic being the florescence and eclipse of Angkor in the central mainland, more or less contemporaneous with the European Middle Ages. But no less pivotal was the rise of Ayutthaya (Siam), communicable with the ocean at the expense of Angkor and its successor kingdoms in the lower Mekong. Near simultaneously, the Vietnamese Nguyen dynasty dominating the eastern mainland effectively eliminated the maritime trading kingdom of Champa (the Ciamba that Columbus looked to as a point of reference derived from Ptolemy), and encroached upon Cambodia when it pushed into the lower Mekong. Alongside Ayutthaya, another port-capital, namely Pegu in the western mainland also competed with the former. In the Burmese-Siamese War (1765–67), the Konbaung dynasty successfully sacked the Thai capital forcing a relocation and rebirth of dynasty downriver to the present-day Bangkok. A revived Chakkri dynasty centered on Bangkok, along with an invigorated Nguyen dynasty, both vied at Cambodia's expense, and with the Nguyen appropriating the lower Mekong delta.

French Indochina

Although French Jesuit missionaries arriving from Macau had made an early entry into Vietnam as with Alexandre de Rhodes, pushing as well into Cambodia and Laos, France itself was a late-coming colonizer in East Asia literally wresting a colonial empire by gunboat, as opposed to diplomacy and entreaty as demanded by Asian courts with China the obvious example. As dominant colonial powers on mainland Southeast Asia, England and France virtually determined the fate of the concerned nation-states, Siam included, which escaped direct colonization. Having annexed the empire of Annam with its seat of power in Hue, France then set about creating its own administrative region utterly contradictory to local history or culture, and entirely outside of local geomantic traditions or what Momiki (2010: 128) has described as "cultural territory." This was Indo-China, originally hyphenated to reveal its hybrid character, with Indo referring to the Indianized kingdoms of Laos and Cambodia, and with China referring to Sinicized Vietnam (in turn run as three administrative zones in a divide and rule tactic to rile would-be nationalists). As a construct, Indochina with its administrative capital in Hanoi never worked well and, by the mid-twentieth century, literally tore itself apart in a paroxysm of nationalist rebellion.

Commencing with Rhodes' *Regnu Annam* published in Rome in 1651, French mapping of Indochina commenced with the first missionary enterprise. As revealed, Vietnam was then divided between two rival dynasties, the Trinh in the north and the Nguyen in the center-south. The Hinduized Cham kingdom, prior to its destruction by the Vietnamese, is indicated along with Cambodia. Then unknown Laos is marked "*solitodo*." *Mar Sinicum* separates coastal Vietnam from China. Map coordinates are added. French mapping of Indochina progressed with conquest and "pacification." A two-century process of mapping and surveying, it was not until 1920 that the Service Géographique de l'Indochine produced its landmark *Atlas de l'Indochine* (Saigon) complete with forty-seven maps, variously, ethnographic, historic, administrative, communications, with city plans, and with coordinates added.

To this day, as mentioned in chapter 8, European maps alongside indigenous maps in the Sinic tradition are wielded by Vietnam to assert claims of ownership of islands in the Eastern/South China Sea (see Nguyen 2013).

Maritime Southeast Asia

The sense of a maritime Southeast Asia world has a long pedigree in European perception. An island realm off the eastern edge of the Eurasian continent is implicit in Ptolemy and Pliny and known to Arab navigators even prior to the advent of Islam. As we noted, Arab and/or Muslim traveler accounts eventually entering European language translations are replete with convincing empirical information on navigation route, dangers, ports, local customs, albeit highly generalized and sometimes fantasized. Marco Polo confirmed an island Southeast Asia feeding into Renaissance revival evocations of this region.

Indonesia

With independence declared on August 17, 1945, the Republic of Indonesia also claimed the boundaries of the former Dutch (sometimes the Netherlands) East Indies under the principle of state succession. Nevertheless, the matter was contested militarily as well as on the diplomatic front and with the Dutch retaining control over the stone-age culture on the eastern half of the island of New Guinea ostensibly pending separate preparation for independence. A Latinized term first announced by a German geographer-philosopher in the late nineteenth century, Indonesia meaning the "islands of India"—Ptolemaic in itself—was also a term adopted by early opponents to Dutch rule in the sprawling archipelagic colony. According to historian Bernard Vlecke (1965: 366), in colonial times "The use of the word 'Indonesia' was forbidden. Officially, there was no such territory on the map." To strike a neutral position Vlecke himself refers to the archipelago as Nusa Antara (or islands in-between), recalling a term used in fifteenth-century Javanese texts and, for that matter, by Erédia. Thus while the Latinized construct "Indonesia"—as with a number of other "-nesias" in the Western Pacific (Melanesia; Micronesia, Polynesia)—has no direct classical antecedents, the "Islands of India," actually conjure up the dimly comprehended mess of islands entering Pliny and other Roman histories and in a language they would have understood.

While the ancient Java-Sumatra-based Hindu-Buddhistic kingdoms of Srivijaya and Majapahit would provide valuable sustaining myths of ancient archipelagic unity for the Indonesian nationalist elite as with litterateur Muhammad Yamin (1903–62), the fact of the matter is that with over 2,000 islands, and hundreds of languages and ethnic groups, there was no natural or even historical unity. Neither did the kingdoms map their empires or, not at least outside of Hindu-Buddhistic cosmological concerns (Suárez 1999: 24–25). Paradoxically, it was Dutch social engineering and an opportunistic Japanese military endorsement of Indonesian independence that created a nation on paper. Moreover, the pro-independence elites in the Dutch colony tended to be Javanese or Sumatran, and with a wide gap between Javanese values and the peoples of the so-called outer islands. To this day, with their Melanesian traditions, Papuans resent incorporation into the Republic of Indonesia, such as facilitated by a flawed United Nations "plebiscite" of selected chiefs conducted in 1969 (Saltford 2002). In a United Nations-conducted poll of August 1999, the people of East Timor, reinventing their new nation as Timor-Leste, rejected incorporation into the Republic of Indonesia. In 2003, after a long and protracted war of secession punctuated by a massive and destructive tsunami of "biblical" proportions, the people of Aceh, the ancient port attested by Marco Polo, also entering into direct trade relations with the Ottoman empire, demanded and obtained a substantial degree of autonomy within the Republic, also introducing Shariah law.

Needless to say, from Ptolemy, Pliny, along with Arab and Chinese chronicle sources, arriving Iberians and northern Europeans found no political, social or religious unity in this watery region of "islands in-between." Rather, they found many polities, Hindu-Buddhistic (as on Bali or on the Cham coast of Vietnam), and Islamic, as with Pasai, Aceh, Brunei, Ternate, and Mataram on Java. They also encountered deeply ingrained local animist traditions, particularly in all the islands east of Bali, on Borneo, and in most of the hinterlands outside of major civilizational influences.

Far from perceiving unity, as with India or China, European mapmakers long struggled with, first; resolution of the Taprobane/Sumatra dislocation, and second; the denomination and even location of Java la Grande and/or Java Majore and Java Minore, not only confusing Sumatra and Java but, in some instances, even Borneo. Adding to the confusion, Ptolemy's Golden Peninsula plunging south of the Equator in many cartographic depictions, long impeded accurate conceptions of the straits and surrounding seas as with the shape, size, and location of the *Sinus Magnus*. Even as empirical information on the archipelago drawn from the Portuguese, Dutch, French, and English voyages entered European knowledge, the translation onto European maps was tardy. As we examined in some detail, Iberian competition over the spice islands led to a great deal of cartographic dissimulation as to the precise location of the Moluccas.

The Philippines

In the event, the Magellan circumnavigation brought back to Europe convincing empirical data of the island realm off the southeastern coast of China. Only vaguely apprehended in the classical template and reproduced in Ptolemy-revival atlases as a string of misshapen islands otherwise fitting Pliny's accounting of strange peoples on the edges of the world, even Columbus was deceived when first encountering the Arawak "Indians" of the Bahama Islands. Even if the demarcation question was not settled, Spanish priority in the so-named *Las Philippinas* was established. More so than other "regions" of Asia with established central kingdoms, the disparate islands were noted for their localization of power at the level of *barangay* or community, although Hinduized Cham may indeed have touched certain southern island groups and with other sites (as with Mindoro Island) connected up in active trade with China and external regions. In the event, however, the Spanish conquest of the Philippine islands, the implantation of a system of *fortaleza* from Manila to Zamboanga in the south, the privileging of Tagalog in print, alongside Spanish, and the mapping of the islands literally created a unity befitting a future nation where there had been none. Of course, it may also be argued that if it were not for the timing of the Spanish intervention and their naval supremacy, an alternative nation may indeed have gelled around the Sultanate of Sulu-Sultanate of Brunei with its pre-contact outliers in the Manila region. More bluntly, the irruption of Spanish navigators and Catholic missionaries into the Western Pacific, its colonization and evangelical project, matched by its maritime surveys and cartographic representations, would have 500 years ramifications.

Malaya and Borneo/Malaysia

As introduced, the map history of Malaya/Borneo by Durand and Curtis (2013) reveals not only the evolution of cartographic images of this region over time, but also exposes the enduring influence of the Ptolemaic template. In line with the master narrative of this book, representations of the Golden Chersonese, along with early modern depictions of the Malacca or "Malay" Peninsula, was literally a two millennium transition. Once again it would be Francisco Rodrigues's portolan-style mapping that would literally place the Malay Peninsula on the world map. He likewise produced convincing coastal renditions of parts of Borneo Island, including the Sultanate of Brunei. Within years of the Portuguese conquest of Malacca in 1511, the Malay Peninsula and Malacca town took graphic form—even illustrated—in such maps as those of Pedro Reinel dating from ca. 1517, and as in the Miller Atlas of 1519 with its Baroque embellishments. Portolan style and with practical navigation in mind, these maps lacked coordinates, although as described in chapter 8, later-arriving Dutch, British, and French would supply this missing element.

By this age, as Durand and Curtis (2013: 33) point out, the Portuguese began to improve their knowledge not only through firsthand surveys and observation, as with Rodrigues, but also through the incorporation of local knowledge, as with Reinel. The creation of "Malaysia" was as problematical as it was contested. As it played out, the "merger" of Malaya with Singapore and the British-controlled north Borneo states (minus Brunei) by the British Foreign Office was an end-of-empire gesture designed to neutralize "anti-merger" pro-communist threats and to appease Malay nationalists, a scheme also opposed by the radical anti-Western Republic of Indonesia under founder President Sukarno.

The South China Sea at The Hague

In July 2015, the Permanent Court of Arbitration at The Hague held its first hearing on the 2013 arbitration case filed by the Philippines against the People's Republic of China (PRC) questioning the legal validity of Beijing's "9-dash line" claim over the South China Sea.[1] In the run-up to the Award issued by the Permanent Court of Arbitration (The Hague) in July 2016 on the South China Sea Arbitration (the Republic of the Philippines vs. the People's Republic of China), the Philippines were not backward in touting map evidence to deny Chinese "ownership" of the Spratlys and Scarborough Shoal. For example, in November 2014, the University of the Philippines, Asian Center, Diliman, Quezon City mounted a map exhibition titled "Historical Truths and Lies: Scarborough Shoal in Ancient Maps."[2] The display featured reproductions of eighteen ancient Chinese maps, fifteen by Chinese officials and citizens, and three by non-Chinese. Especially, the exhibition made the point that, according to Chinese maps, Hainan Island-province always remained China's southern boundary and that the dynasties never included the Spratlys or Scarborough Shoal. In fact, Vietnam makes the same argument as for example with map displays strategically mounted in airports or museums.

On its part, China asserted that it would not accept or participate in arbitration initiated by the Philippines, facts reiterated in a position paper submitted to the Court in December 2014. Meantime, the Court agreed to take up seven of the fifteen submissions made by Manila, in particular the issue of whether or not Scarborough Shoal and low-tide areas like Mischief Reef can be considered islands.[3] In the same month, Vietnam endorsed the Court's jurisdiction in the proceedings. Finally, on July 12, 2016, the Court ruled in favor of the Philippines concluding that China has no legal basis to claim historic rights to the bulk of the South China hence there was "no legal basis for China to claim historic rights" over the nine-dash line. The Court also criticized

China's land reclamation projects and its construction of artificial islands in the Spratly Islands. It also characterized Ita Aba (Taiping Island), currently administered by Taiwan, and other features of the Spratlys as "rocks" under UNCLOS, and therefore not entitled to a 200 nautical mile exclusive economic zone. In the event, and in an obvious affront to proponents of "freedom of passage," the PRC rejected the ruling with President Xi Jinping declaring that "China will never accept any claim or action based on those awards." The Taipei government also rejected the ruling.[4]

ENVOI

In the introduction, I announced my ambition to problematize the creation of geographical knowledge especially in the wake of the discoveries. As this book has traced across many chapters, and tested against many regions of Asia, Ptolemy's ancient distinction between the science of representing space on a three-dimensional plane was one thing, versus the chorography of description, often second or third hand, lost in translation, subject to false accounting, and fantasized, on the other. I also advertised my intention to raise awareness as to the historical construction of region in Asia. As understood, this is no less problematical in view of the fact that political geography often conflates with the physical reality. Certainly, proximity leads to familiarity and obviously Venice knew the Mediterranean and "Near East" better than anyone. But with distance, as with India, Southeast Asia, and China, eventually only a massification of firsthand reports overcame the classic—actually medieval European—images of anthropophagi, humanoids, and faux islands and landmasses. As this work has highlighted, the framing of regions and subregions was the subject of many imaginings, contestations, and filters. To borrow a modern Japanese-English term often applied to rectifying historical truths in school textbooks, mapmaking schools, editors, and their sometimes powerful sponsors subjected empirical data to constant spin and "screening." But eventually outsiders could not neglect local images and perceptions as to who was Persian, Indian, Bengali, Burmese, Siamese (Thai), Malay, Formosan (Taiwanese), Japanese (easier in the case of islanders), Tartar, or Chinese (not unequivocal, as attested). But peoples (much less identities) and nations and their regions are not coterminous, and not even constant, with China a clear case in point. Hence mapmakers were both conservative in hewing to the Ptolemaic template and inventive when that image began to disaggregate in the light of superior knowledge.

Finally, I adverted to a more latent agenda, namely to sound an alert as to the perils of overdetermination in the construction of boundaries such as carried in nationalistic and neo-nationalist discourses of the present age.

With the East and South China Seas in mind, not to mention Antarctica and the Arctic, even the "ocean sea" is contested space today. All the world is "known" now, as the Amazon, the Congo, and Borneo, and other peripheral regions, yield to new conquistadors and profit-takers, and with competition for scarce resources no less keen, indeed massively more so than even the appetites of Constantinople and Venice combined. Today, geopolitics enters a new phase. "Clash of civilization" takes on a new dimension if we read back on the depredations wrought by the "new Mongols," read Washington's "war against terror" (latterly joined by Russia) as it played out in Afghanistan and even in the ancient old world heartland of Iraq, Syria, Egypt, Libya, and Tunisia. Across Eurasia today, and especially in the wake of the collapse of the Soviet Union, new nationalisms, new ethnic subnationalisms, alongside ethnic and religious identities tug across countries and regions up and down the ancient terrestrial silk roads. Even as I write, a new dynamic appears, namely the invocation of a revived Middle Eastern caliphate (with Southeast Asian offshoots) out of the chaos sown by Western assault upon Arab dictators in the ancient lands turned modern nation-states of Iraq, Syria, and Libya, respectively, and which, through the application of terror on an industrial scale, appears to have launched a frightening specter of medieval jihad as much as seeking to redraw the political map.

NOTES

1. Among the European maps assembled to prove Spanish-Philippine ownership of Scarborough include the Murillo map *Carta Hydrographica y Chorographica de lasYslas Filipinas* by the Jesuit Pedro Murillo, published in 1734 in Manila. This is the oldest map to give a name to "Panacot" (Scarborough) shoal. "Panacot" is Tagalog for "threat, danger" (see Tordesillas 2015).
2. Also hosted by the "Institute for Maritime and Ocean Affairs" the maps and their portrayals can be found athttp://www.slideshare.net/caraganews/catalogue-historical-truthlieslow
3. Permanent Court of Arbitration (PSA) Press Release, October 29, 2015.
4. As Hayton (2018) contends from an analysis of episodic Chinese interest in the South China Sea down to Nationalist China's fanciful postwar production of a nine-dash line chart, China has come to construct a veritable national "maritime geobody" or collective psychological attachment to the offshore islands.

Bibliography

Abu-Lughod, Janet L. *Before European Hegemony: The World System A.D. 1250–1350*. New York and Oxford: Oxford University Press, 1989.

Adler, Marcus Nathan. *The Itinerary of Benjamin of Tudela: Travels in the Middle Ages*. London: Oxford University Press, 1907.

Agência Nacional para a Cultura Científica, Ciência Viva, Descobertas e Invenções Portuguesas (in Portuguese) http://www.cienciaviva.pt/projectos/concluidos/inventions/instrumentos.asp

Albuquerque, Luis de, intro. Abraão Zacuto. *Almanach perpetuum de Abraão Zacuto*. Lisboa: Imprensa Nacional–Casa da Moeda, 1986.

Almagià, Roberto. *Monumenta Cartographica*. Vaticana: Biblioteca Apostolicae Vaticanae, 1944–1945.

Anthiaume, Albert-Marie-Ferdinand. *Cartes marines, constructions navale,voyages de découverte chez les normands, 1500–1650* (2 vols). Paris: Ernest Dumont, 1916.

Ariel, Avraham and Nora Ariel Berger. *Plotting the Globe: Stories of Meridians, Parallels, and the International Date Line*. Westport, CT: Praeger, 2005.

Armitage. David. Introduction to Grotius's The Free Sea, trans. Richard Hakluyt, with William Welwod'd Critique and Grotius's Reply, ed. David Armitage (Indianapolis: Liberty Fund), 2004. http://oll.libertyfund.org/pages/grotius-the-freedom-of-the-seas#

Ashmand, J. M., ed. *Ptolemy's Tetrabiblos*. London: Davis and Dickson, 1822. Reprinted, Bel Air, MD: Astrology Classics, 2002. http://www.sacred-texts.com/astro/ptb/

Augé, Marc. *Non-Places: Introduction to an Anthropology of Supermodernity*. London: Verso, 1995.

Auzout, Adrien. *Lettres de m. Auzout sur les grandes lunettes: Voyages de mm. Cassini en France, en Italie, en Flandre, en Hollande et en Angleterre. Observations envoyées des Indes et de la Chine à l'Academie Royale des Sciences*. Amsterdam: chez Pierre Mortier, 1736.

Bagrow, Leo. 1945. "The Origin of Ptolemy's Geographia." *Geografiska Annaler*, vol. 27 (1945): 318–387.

———. 1948. "The Maps from the Home Archives of the Descendants of a Friend of Marco Polo." *Imago Mundi*, vol. 5 (1948): 3–13.

———. *History of Cartography* (2nd ed.). (revised and enlarged by R.A. Skelton), New Brunswick, NJ: Transactions, 1951; 2010.

———. "The First Russian Maps of Siberia and Their Influence on the West-European Cartography of N. E. Asia." *Imago Mundi,* vol. 9 (1952): 83–93.

Barnes, Jerome R. "Giovanni Battista Ramusio and the History of Discoveries: An Analysis of Ramusio's Commentary, Cartography." Ph.D. thesis submitted to University of Texas at Arlington, 2007.

Barreto, Luis Filipe. "O problema do conhecimento na 'Sphaera' de D. João de Castro." *Revista da Universidade de Coimbra,* vol. 32 (1985): 387–397.

———. *Cartography of the East–West Encounter.* Macau: Territorial Commission for the Commemoration of the Portuguese Discoveries, n.d.

——— and Maria Manuela d'Oliveira Martins; trans. Clive Thoms. *Museum guide: Macao Scientific and Cultural Centre.* Lisbon: Macao Scientific and Cultural Centre, Ministry of Science and Technology, 1999.

Barros, João de. *Primeira Década da Ásia, dos feitos que os portugueses fizeram no descobrimento dos mares e terras do Oriente.* Lisbon: Germão Galherde,1552; 1553; 1563; 1615.

Batchelor, Robert K. 2013. "The Selden Map Rediscovered: A Chinese Map of East Asian Shipping Routes, c. 1619." *Imago Mundi,* vol. 65, no. 1 (2013): 37–63.

Berggren, J. Lennart and Alexander Jones. *Ptolemy's Geography: An Annotated Translation of the Theoretical Chapters.* Princeton, NJ: Princeton University Press, 2002.

Bibliothèque nationale de France (BnF). http://www.bnf.fr/fr/acc/x.accueil.html

Bigalow, John. "The so-called Bartolomew Columbus' Map of 1506." *Geographical Review*, vol. 25, no. 4 (1935): 643–656.

Blair, Emma Helen and James Alexander Robertson, Edited and annotated, with historical introduction and additional notes by Edward Gaylord Bourne. *The Philippine Islands, 1493–1803; explorations by early navigators, descriptions of the islands and their peoples, their history and records of the Catholic missions, as related in contemporaneous books and manuscripts, showing the political, economic, commercial and religious conditions of those islands from their earliest relations with European nations to the beginning of the nineteenth century.* Cleveland, Ohio: The A. H. Clark company, Volume I: 1493–1529, 1903–1907.

Bolghakov, J and A. Ahmad. In *Majallat ma'had al-makhtûtât al-'arabiya.* (Kuwait), vol. 8, (1962).

Bonnet-Bidaud, J. M., F. Praderie, and S. Whitfield. "The Dunhuang Chinese Sky: A Comprehensive Study of the Oldest Known Star Atlas." *Journal of Astronomical History and Heritage,* vol. 12, no. 1 (2009): 39–59.

Borschberg, Peter. "The Seizure of the *Sta. Catarina* Revisited: The Portuguese Empire in Asia, VOC Politics and the Origins of the Dutch-Johor Alliance (1602–ca. 1616)." *Journal of Southeast Asian Studies,* vol. 33 (2002): 31–62.

———. *Hugo Grotius, the Portuguese and Free Trade in the East Indies.* Leiden and Singapore: KITLV Press and NUS Press, 2011.

Boxer, C. R. *South China in the Sixteenth Century: Being the Narratives of Galeote Pereira, Fr. Gaspar Da Cruz, O.P. [and] Fr. Martín de Rada, O.E.S.A. (1550–1575).* London: Hakluyt Society, 1953.
Braddell, T. "The Ancient Trade of the Indian Archipelago." In J. R. Logan, ed., *The Journal of the Indian Archipelago and Eastern Asia.* New Series, vol. 2, no. 3 (1887): 238–277.
Braudel, Fernand. *The Mediterranean and the Mediterranean World in the Age of Philip II.* London: Fontana/Collins, 1975.
———. *The Perspective of the World*, vol. III of *Civilization and Capitalism.* New York: Harper & Row, 1984.
———. *The Wheels of Commerce (Civilization and Capitalism): 15th–18th Century (Vol. 2).* Berkeley, CA: University of California Press, 1992.
British Museum, 2014. http://www.britishmuseum.org/whats_on/past_exhibitions/2012/hajj/the_exhibition.aspx
Brown, Lloyd A. "The Longitude." In James Roy Newman, ed., *The World of Mathematics*, Volume 2. Parts V–VI, 780–819. New York: Simon & Schuster, 1956.
Buchmann, Hipolito. *The Esoteric Codex: Medieval Astrologers.* Lulu.com, 2015.
Bureau des longitudes. 2014. http://www.bureau-des-longitudes.fr/histoire.htm
Burkert, Walter. *The Orientalizing Revolution; Near Eastern Influence on Greek Culture in Archaic Age.* Cambridge, MA: Harvard University Press, 1992.
Buschman, Rainer F., Edward R. Slack Jr., and James B. Tueller. *Navigating the Spanish Lake: The Pacific in the Iberian World, 1521–1898.* Honolulu, HI: University of Hawai'i Press, 2014.
Calvo, Alex. "China, the Philippines, Vietnam, and International Arbitration in the South China Sea." *Japan Focus*, vol. 13, issue 43, no. 2 (2015). https://apjjf.org/-Alex-Calvo/4391
Camões, Luís Vaz de. *Os Lusíadas.* Lisbon: em casa de Antonio Gõçaluez, 1572.
Camões (Cameons), Luis de. *The Lusiad, or Portugals Historicall Poem; Written in the Portingall Language by Luis de Camoens, and now newly put into English by Richard Fanshaw, Esq.* London: printed for Humphrey Moseley, 1665.
Cams, Mario. "'Companions in Geography': The Sino-European Effort to Measure China, c.1685–1735." *Imago Mundi*, vol. 66, no. 1 (2013): 143–144.
———. "The China Maps of Jean-Baptiste Bourguignon d'Anville: Origins and Supporting Networks." *Imago Mundi,* vol. 66, no. 1 (2013): 51–69.
Cassini, Giovanni. "Reflections de M. Cassini sur la Longitude de la Côte Orientale de la Chine," 176–79. In *Mémoires de l'Académie royale des sciences depuis 1666 jusqu'en 1699 - 1729–1734* (1725).
Casson, L. *The Periplus Maris Erythraei: Text with Introduction, Translation, and Commentary.* Princeton, NJ: Princeton University Press, 1989.
Catz, Rebecca D., ed. trans. *The Travels of Mendes Pinto.* Chicago, IL: The University of Chicago Press, 1989.
Celsus, Kelly. *Calendar of documents: Spanish voyages in the South Pacific from Alvaro de Mendana to Alejandro Malaspina, 1567–1794, and the Franciscan missionary plans for the peoples of the Austral lands, 1617–1634: compiled from manuscripts and other documents in the archives and libraries of Spain, America,*

Rome, Paris, London, Sydney, etc. Madrid: Franciscan Historical Studies (Australia) in association with Archivo Ibero-Americano, 1965.

Chaigneau, Jean-Baptiste. "Mémoire sur la Cochinchine." *Bulletin des Amis du Vieux Hué*. Hanoi, 1820 [1923].

Chapman, Allan. "Uraniborg in Beijing: The Reconstruction of Tycho Brahe's Instruments in the Seventeenth-Century Jesuit Observatory," 270–77. In *Treasures from Imperial China: The Imperial City and the Royal Danish Court.* Christianborg Palace/Christianborg Stot: The Royal Silver Vault/ Det Kongelige Solukammer, 2006.

Chen, King C. *China's War with Vietnam.* Stanford, CA: Hoover Institution Press, 1987.

Christ, Georg. *Trading Conflicts: Venetian Merchants and Mamluk Officials in Late Medieval Alexandria.* Leiden: Brill, 2012.

Clancy, Robert Clancy, John Manning, and Henk Brolsma. *Mapping Antarctica: A Five Hundred Year Record of Discovery.* Dordrecht: Springer, 2014.

Claudii Ptolemaei Geographia. edidit Carolus Fridericus Augustus Nobbe Ptolemy, 2nd cent. Nobbe, Karl Friedrich August, Editio stereotypa 3 v.; 14 cm. Sumptibus et typis Caroli Tauchnitii Lipsiae, 1843–1845.

Connett, Christina Findlay. *Cartography as a Strategy of Empire in Historia general de los hechos de los castellanos en las Islas i Tierra Firme del Mar Oceano by Antonio de Herrera y Tordesillas.* Doctoral dissertation, Universitat de València, València, Spain, 2014.

Correira, Carlos Alberto Calinas. *A arte de navegar de Manoel Pimentel: as edições de 1699 e 1712.* Tese de mestrado, História dos Descobrimentos e da Expansão, Universidade de Lisboa, Faculdade de Letras, 2011.

Cortesão, Armando. *Cartografia e cartografos portugueses dos seculos XV e XVI: contribuição para um estudo completo.* Lisbon: Seara Nova, 1935.

Corvo, João de Andrade. João de Castro; *Roteiro de Lisboa a Goa.* Lisbon: Academia Real das Sciencias, 1882.

Crato, Nuno (Paulo de Sousa Arrobas). 2014. http://cvc.instituto-camoes.pt/ciencia_eng/e54.html

———. 2014. http://cvc.instituto-camoes.pt/ciencia_eng/e6.html

———. 2014. http://cvc.instituto-camoes.pt/ciencia_eng/e7.html

Crossley, John Newsome. *Hernando de los Ríos Coronel and the Spanish Philippines in the Golden Age.* Burlington, VT: Ashgate, 2011.

d'Anville, Jean-Baptiste Bourguignon. *Nouvel atlas de la Chine, de la Tartarie chinoise, et du Thibet.* The Hague: H. Scheurleer, 1737.

———. *Éclaircissemens géographiques sur la carte de l'Inde.* Paris: Impr. Royale, 1753.

———. *États formés en Europe après la chute de l'empire romain en occident.* Paris: Impr. Royale, 1771.

———. *Antiquité géographique de l'Inde et de plusieurs autres contrées de la Haute-Asie.* Paris: Impr. Royale, 1775.

———. *Mémoire et abrégé de géographie ancienne et générale.* Paris: Impr. Royale, 1834.

Dalché, Patrick Gautier. "The Reception of Ptolemy's Geography (End of the Fourteenth to Beginning of the Sixteenth Century)." 285–364. In David Woodward, ed. *The History of Cartography, Volume 3 Cartography in the European Renaissance.* Chicago, Il: University of Chicago Press, 2007.

Darwin, John. *After Tamerlane: The Rise and Fall of Global Empires, 1400–2000.* New York: Penguin, 2007.

Dawson, Christopher, ed. *The Mongol Mission: Narratives and Letters of the Franciscan Missionaries in Mongolia and China in the Thirteenth and Fourteenth Centuries.* London and New York: Sheed and Ward, 1955.

de Certeau, Michel; trans. Steven Rendall. *The Practice of Everyday Life.* Berkeley, CA: University of California Press, 1984.

De Romanis F. and A. Tchernia, eds., *Crossings. Early Mediterranean Contacts with India.* New Delhi: Manohar, 2005.

Deleuze, Giles and Felix Guattari. *Anti-Oedipus: Capitalism and Schizophrenia (L'anti-Oedipe).* London: Continuum, 1972; 2004.

Destombes, Marcel. "Les originaux chinois des plans de ville publiés par J. B. du Halde," 85–97. In Joseph Dehergne, ed., *Archives des jésuites, Actes du colloque international de sinologie*, Paris: Belles Lettres, 1976.

Diderot, Denis. *Encyclopédie, ou dictionnaire raisonné des sciences, des arts et des métiers.* Denis Diderot, 1713–84.

Dilke, O. A. W. "The Culmination of Greek Cartography in Ptolemy (with additional material supplied by the editors)," 177–200. In J. B. Harley and D. Woodward, eds., *The History of Cartography,* vol. 1, *Cartography in Prehistoric, Ancient and Medieval Europe and the Mediterranean.* Chicago, IL: University of Chicago Press, 1987.

Dozy, R. and M.J. de Goeje, notes and French translation. *Edrisi, Description de l'Afrique et de l'Espagne.* Leiden: E. J. Brill, 1866.

Ducène, Jean-Charles. 2017. "Formes de l'océan Indien dans la cartographie arabe." 57–72. In Emmanuelle Vagnon and Éric Vallet, eds., *La fabrique de l'océan Indien. Cartes d'Orient et d'Occident (Antiquité-XVIe siècle).* Paris: Publications de la Sorbonne.

Du Halde, Jean-Baptiste. *Description Geographique, Historique, Chronologique, Politique, et Physique de l'Empire de la Chine et de la Tartarie Chinoise.* Paris: P.-G. le Mercier, Four vols, 1735.

———. *The General History of China Containing a Geographical, Historical, Chronological, Political and Physical Description of the Empire of China, Chinese-Tartary, Corea, and Thibet; Including an Exact and Particular Account of Their Customs, Manners, Ceremonies, Religion, Arts, and Sciences (3rd ed.).* London: J. Watts, 1741.

Duiker, William J. *China and Vietnam: The Roots of Conflict.* Berkeley, CA: University of California Press, 1986.

Durand, Frederic and Richard Curtis. *Maps of Malaya and Borneo: Discovery, Statehood and Progress.* Singapore: Editions Didier Miller, 2013.

Duteil, Jean-Pierre. *Le mandat du ciel: le rôle des Jésuites en Chine, de la mort de François- Xavier à la Dissolution de la Compagnie de Jesus (1552–1774).* Paris: AP Editions Arguments, 1994.

Duy Chien. "Western Witnesses of Vietnam's Sovereignty over Hoang Sa." *VietNamNet Bridge,* June 26, 2014. http://english.vietnamnet.vn/fms/special-reports/105809/western-witnesses-of-vietnam-s-sovereignty-over-hoang-sa.html
Edgerton, Samuel Y. *The Renaissance Discovery of Linear Perspective.* New York: Basic, 1975.
Edson, Evelyn. *The World Map, 1300–1492: The Persistence of Tradition and Transformation.* Baltimore, MA: The John Hopkins University Press, 2007.
Encyclopedia Britannica. "Marinus of Tyre," Vol. 17, 1919.
Encyclopedia Britannica. 2014a. http://global.britannica.com/EBchecked/topic/471291/portolan-chart
Encyclopedia Britannica. 2014b. http://global.britannica.com/EBchecked/topic/14638/Alfonsine-Tables
Encyclopedia Iranica. http://www.iranicaonline.org/articles/portugal-i.
Estensen, Miriam. *Terra Australia Incognita: The Spanish Quest for the Great South Land.* Sydney: Allen & Unwin, 2006.
Falchetta, Pierro. *Fra Mauro's World Map.* Turnhout, Belgium: Brepols, 2006.
Fausett, David. *Writing the New World: Imaginary Voyages and Utopias of the Great Southern Land.* New York: Syracuse University Press, 1993.
Fell, R. T. 1988. *Early Maps of South-East Asia.* Singapore: Oxford University Press, 1988.
Fernández-Armesto. Felipe. *Before Columbus: Exploration and Colonization from the Mediterranean to the Atlantic, 1229–1492.* Philadelphia, PA: University of Philadelphia Press, 1987.
———. *Amerigo: The Man Who Gave His Name to America.* New York: Random House, 2007.
Ferraro, Giuseppe. *Relazione Delle Scoperte Fatte Da C. Colombo, Da A. Vespucci E Da Altri Dal 1492 Al 1506.* Bologna: Romagnoli, 1875.
Finocchiaro, Maurice A. *Galileo on the World Systems: A New Abridged Translation and Guide.* Berkeley, CA: University of California Press, 1997.
Flinders, Mathew. *A Voyage to Terra Australis.* London: Nicol, 1814.
Florovsky, Anthony. "Maps of the Siberian Route of the Belgian Jesuit, A. Thomas (1690)." *Imago Mundi,* vol. 8 (1951): 103–108.
Foss, Theodore N. "Jesuit Cartography: A Western Interpretation of China." *Review of Culture,* 21 (2nd series) (1994): 133–156.
Francis, Louis. ed. *Ptolemy's Geographia* [Electronic resource]: Books I & II, University of Oxford Text Archive, 1994. http://ota.ahds.ac.uk/desc/2422
Freeman, Michael. "Luis (Vaz) de Camões (Or de Camoens) c. 1524–1580: Portuguese Poet and Dramatist." 222–223. In L. Classe, ed., *Encyclopedia of Literary Translation Into English: A-L.* London: Taylor & Francis, 2000.
Frolova, L., compiler. "Siberia Mapping," Exhibition, National Library of Russia, 1999. *http://nlr.ru/eng/exib/siberia/sib01.htm*
Gallica, a web portal of the BNF. via http://gallica.bnf.fr/?lang=EN
Gaspar, Joaquim Alves. "The Myth of the Square Chart." e-Perimetron, vol. 2, no. 2 (2007): 66–79. https://www.academia.edu/1560621/The_Myth_of_the_Square_Chart
———. "Blunders, Errors and Entanglements: Scrutinizing the Cantino Planisphere with a Cartometric Eye." *Imago Mundi*, vol. 64, part 2 (2012): 181–200.

———. "From the Portolan Chart to the Latitude Chart: The Silent Cartographic Revolution." *Cartes & Géomatique* (June 2013): 67–77. https://www.academia.edu/5146610/From_the_Portolan_Chart_to_the_Latitude_Chart_The_silent_cartographic_revolution

Gaubil, Antoine. *Traité de l'Astronomie Chinoise.* Paris: chez Rollin, 1739.

Giraldez, Arturo. *The Age of Trade: The Manila Galleons and the Dawn of the Global Economy.* Lanham, MD: Rowman & Littlefield Publishers, 2015.

Gislén, Lars. "Analysis of the Astronomical Information in Tachard: 'Voyage to Siam in the Year 1685.'" *Centaurus,* vol. 46, no. 2 (September 2004): 133–144.

Glasner, Ruth. *Averroes' Physics: A Turning Point in Medieval Natural Philosophy.* Oxford: Oxford University Press, 2009.

Godhino de Eredia, Manuale. *Malaca, l'Inde méridionale et le Cathay: Manuscrit original autographe de Godinho de Eredia,* reproduit en fac-simile et traduit par Léon Janssen. Bruxelles: C. Muquardt, 1882.

Goodchild, Richard. *Cyrene and Apollonia: An Historical Guide (4th ed).* Tripoli: Department of Antiquities, 1981.

Goody, Jack. *Renaissances: One or the Many?* Cambridge: Cambridge University Press, 2010.

Gordon, Peter and Juan José Morales. *The Silver Way: China, Spanish America and the Birth of Globalization, 1565–1815.* Sydney: Penguin Random House Australia, 2017.

Grafton, Anthony with April Shelford and Nancy Siraisi. *New Worlds, Ancient Texts: The Power of Tradition and the Shock of Discovery.* Cambridge, MA: Harvard University Press, 1995.

Grégoire, Henri. *Instruction publique. Rapport sur l'établissement du bureau des longitudes.* Paris: Imprimèrie Nationale: Messidor, Year III of the Republic, 1795.

Grotius, Hugo. *De iure prae dae.* Leiden, 1604.

Grotius, Hugo. *Mare Liberum.* Leiden, 1609.

Guerreiro, Inácio. "Reflexos Geo-económico na Catrografia do Oceano Indico do Extremo Oriente dos Séculos XIV a XVI." 453–468. In *As Relações entre a India Portuguesa, a Asie do Sueste e do Extremo Oriente,* Actas do VI Semanário Internacional a História Indo-Portuguesa, 1993.

Guillemard, Francis H. *The Life of Ferdinand Magellan.* New York: AMS Press, 1971.

Gunn, Geoffrey C. 1991. "Anglo-French Rivalry over the Spratlys (1930–1937)." 262–283. In R. D. Hill, Norman G. Owen and E. V. Roberts, eds., *Fishing in Troubled Waters: Proceedings of an Academic Conference on Territorial Claims in the South China Sea.* Hong Kong: Centre of Asian Studies, University of Hong Kong, 1991.

———. *First Globalization: The Eurasian Exchange 1500–1800.* Lanham, MD: Rowman & Littlefield, 2003.

———. *History without Borders: The Making of an Asian World Region, 1000–1800.* Hong Kong: Hong Kong University Press, 2011.

———. "East-Southeast Asia in World History." *TRaNS: Trans – Regional and – National Studies of Southeast Asia,* vol. 2, no. 1 (2014): 5–30.

———. "The Timor-Macao Sandalwood Trade and the Asian Discovery of the Great South Land?'" *Review of Culture* (Macau), vol. 53 (2016): 125–148.

Gužauskyte, Evilina. *Christopher Columbus's Naming in the 'Diarios' of the Four Voyages (1492–1504): A Discourse of Negotiation.* Toronto: University of Toronto Press, 2014.

Haguet, Lucille. "Les cartes de la Description de la Chine de Jean-Baptiste Du Halde: quelques remarques sur les sources et les originaux." December 28, 2011. http://danville.hypotheses.org/1266

Hall, Joseph. *Mundus alter et idem, siue Terra Australis ante hac semper incognita longis itineribus peregrini Academici nuperrime lustrata. Auth: Mercuric Britannico.* Francofurti [Frankfurt am Main], Apud haeredes Ascanij de Riniasme. London: H. Lownes, 1605.

Halma, Nicolas. *Composition mathématique de Claude Ptolémée, ou astronomie ancienne, traduite pour la première fois du gec en français sur les manuscrits de la bibliothèque du roi.* tome 2, Paris: H. Grand, 1813–16.

Halsberghe, Nicole. "Sources and Interpretation of Chapters One to Four in Ferdinand Verbeist's *Xin Zhi Lingtai Yixiang Zhi* (Discourse on the Newest Built Astronomical Instruments in the Observatory), Beijing, 1674." *Review of Culture* (Macau), no. 21 (2nd series) (1994): 213–234.

Hamilton, H. C. and W. Falconer. *The Geography of Strabo*, 3 vols. (Google eBook) ... H. G. Bohn, 1854–57.

Han Qi. 2001. "Sino-French Scientific Relations through the French Jesuits and the Académie Royale des Sciences in the Seventeenth and Eighteenth Centuries." 137–147. In Stephen Uhalley, Jr. and Xiaoxin Wu, eds., *China and Christianity: Burdened Past, Hopeful Future.* London: Armonk, 2001.

Harley, J. B. *The New Nature of Maps: Essays in the History of Cartography.* Baltimore, MA: Johns Hopkins University Press, 2001.

Harley, J. B. and David Woodward. *The History of Cartography, Volume 2, Book 2 Cartography in the Traditional East and Southeast Asian Societies.* Chicago, IL: Chicago University Press, 1994.

Harris, John. *Navigantium atque itinerantium bibliotheca, or, A Compleat Collection of Voyages and Travels (3rd ed).* London: Printed for T. Osborne, 1705; 1764.

Harvey, David. *The Condition of Postmodernity. An Enquiry into the Origins of Cultural Change.* Oxford: Blackwell, 1991.

Hayton, Bill. *The South China Sea: The Struggle for Power in Asia.* New Haven, CT: Yale University Press, 2014.

———. "The Modern Origins of China's South China Sea Claims: Maps, Misunderstandings, and the Maritime Geobody." *Modern China* (May 2018): 1–44.

Herrera y Tordesillas, Antonio de. *Historia General De Los Hechos De Los Castellanos En Las Islas I Tierra Firme Del Mar Oceano.* Madrid: la Emplenta Real, 1601–1615.

———. "*Descripción de las indias del Poniente," Descripción de las Indias.* Madrid: En la Emplenta Real, 1601.

Heylyn, Peter. *Cosmographie in Four Books, Containing the Chorographie and Historie of the Whole World.* London: Printed for Henry Seile, 1652.

———. *Cosmography in Four Books. Containing the Chorography and History of the Whole World, and all the Principal Kingdoms, Provinces, Seas, and Isles thereof.* London: Anne Seile and Philip Chetwind, 1657; 1673, 1674; 1677; 1682.

Hill, R. D. "Fishing in Troubled Waters: Some Thoughts on Territorial Claims in the South China Sea." 1–8. In R. D. Hill, Norman G. Owen and E. V. Roberts, eds., *Fishing in Troubled Waters: Proceedings of an Academic Conference on Territorial Claims in the South China Sea.* Hong Kong: Centre of Asian Studies, University of Hong Kong, 1991.

Hirth, Friedrich and W. W. Rockhill (trans; annot). *Chau Ju-kua: His Work on the Chinese and Arab Trade in the Twelfth and Thirteenth Centuries, entitled Chu-fan-chï.* St. Petersburg: Printing Office of the Imperial Academy of Sciences, 1911.

Hobson, John M. *The Eastern Origins of Western Civilization.* Cambridge: Cambridge University Press, 2004.

Holland, Philemon. *The Historie of the World, Commonly Called, The Naturall Historie of C. Plinius Secundus.* London, 1601.

Howard, Deborah. "Venice as an 'Eastern City'." 59–71. In Stefano Carboni, ed., *Venice and the Islamic World, 828–1797.* New York: Metropolitan Museum of Art/New Haven, NJ: Yale University Press, 2007a.

———. "Venice and the Mamlucks." 73–89. In Stefano Carboni, ed., *Venice and the Islamic World, 828–1797.* New York: Metropolitan Museum of Art/New Haven, NJ: Yale University Press, 2007b.

India Survey. 2014. http://www.surveyofindia.gov.in/

Infopédia. 2003–2014. Porto: Porto Editora http://www.infopedia.pt/$maquina-do-mundo [consult. 2014–12–25 03:26:24].

Instituto Camões. 2014. http//cvc.instituto-camoes.pt/cienncia_eng/e42.htmL

Institute for Maritime and Ocean Affairs. http://www.slideshare.net/caraganews/catalogue-historical-truthlieslow

Jack-Hinton, Colin. "The Political and Cosmographical Background to the Spanish Incursion into the Pacific in the Sixteenth Century." *Journal of the Malaysian Branch of the Royal Asiatic Society,* vol. 37, no. 2 (1964): 125–161.

Jamil, Ali. *The Determination of the Coordinates of Positions for the Correction of Distances between Cities.* Beirut: The American University of Beirut, 1967.

Jarric Pierre de. *Histoire des choses plus memorables advenues tant ez Indes orientales, que autres païs de la descouverte des Portugois*, 3 vols. Bordeaux: Millanges, 1610–14.

Jaubert, P. Amédée, trans. *Géographie d'Édrisi traduite de l'arabe en français d'après deux manuscrits de la Bibliothèque du roi et accompagnée de notes* (2 vols). Paris: L'imprimerie Royale, 1836–40.

Jin Guoping (金國平). 吳志良著,《過十字門》 澳門:澳門成人教育學會, 2004.

———. "Some Reflections on Mendes Pinto's Knowledge of the Chinese Language and the Conceivable Chinese Vocabulary of the Peregrinação." 319–330. In Jorge Santos Alves, ed., *Fernão Mendes Pinto and the Peregrinação: Studies, Restored Portuguese text, Notes and Indexes*, Lisbon: Fundação Oriente, 2010.

Jones, Alexander. "Marinus of Tyre." *Complete Dictionary of Scientific Biography,* 2008. http://www.encyclopedia.com/doc/1G2–2830905894.html

Jonkers, A. R. T. "Parallel Meridians: Diffusion and Change in Early Modern Oceanic Reckoning." 7–42. In J. Parmentier, ed., *Noord-Zuid in Oostindisch Perspectief*, The Hague: Walburg, 2005.

The Hague, and G. Forstner, *Längenfehler und Ausgangsmeridiane in alten Landkarten und Positionstabelle* (dissertation). Munich, 2004.

Kameswara-Rao, N., A. Vagiswari, and C. Louis. "Father J. Richaud and Early Telescope Observations in India." *Bulletin of the Astronomical Society of India,* vol. 12, no. 1 (1984): 81–85.

Karamustafa, Ahmet T. "Introduction to Ottoman Cartography." 206–227. In J. B. Harley and David Woodward, eds., *The History of Cartography, Volume 2, Book 1 Cartography in the Traditional Islamic and South Asian Societies.* Chicago, IL: Chicago University Press, 1992.

Keuning, Johannes. "Nicolaas Witsen as a Cartographer." *Imago Mundi,* vol. 11 (1954): 95–110.

King, David A. "Two Iranian World Maps for Finding the Direction and Distance to Mecca." *Imago Mundi,* vol. 49 (1997): 62–82.

———, and Richard P. Lorch. "Qibla Charts, Qibla Maps, and Related Instruments." 189–205. In J. B. Harley and David Woodward, eds., *The History of Cartography, Volume 2, Book 1 Cartography in the Traditional Islamic and South Asian Societies.* Chicago, IL: University of Chicago, 1992.

Kircher, Athanasius. *China monumentis.* Amsterdam: Jansson, 1667.

Kizilov, Mikhail. "Noord en Oost Tartarye by Nicolaes Witsen: The First Chrestomathy on the Crimean Khanateand its Sources." 169–187. In Denise Klein ed., *The Crimean Khanate between East and West (15th-18th Century)* (=Forschungen zur osteuropäischen Geschichte 78). Wiesbaden: Harrassowitz, 2012.

Korhonen, Pekka. "Leaving Asia? The Meaning of Datsu-A and Japan's Modern History." *The Asia-Pacific Journal,* vol. 12, issue 9, no. 3, March 3, 2014. http://www.japanfocus.org/-Pekka-Korhonen/4083#sthash.VcICWfvf.dpu

Lach, Donald F. *Asia in the Making of Europe, Volume I: the Century of Discovery.* Book 1–2. Chicago, IL: University of Chicago Press, 1965.

———, and Edwin J. Van Kley. *Asia in the Making of Europe* (vol. III, Book 1). Chicago, IL: University of Chicago Press, 1993.

Landry-Derron, Isabel. "Les mathémeticiens envoyés en Chine par Louis XV en 1685." *Arch. Hist Exact Sci.*, vol. 55 (2001): 423–463.

Lapa, Albino. *Pedro Fernandes de Queirós: o último navegador português.* Lisbon: Agência Geral do Ultramar, 1951.

Larner, John. *Marco Polo and the Discovery of the World.* New Haven, CT: Yale University Press, 1999.

Laurencich-Minelli, Laura. *Un "giornale" del Cinquecento Sulla Scoperta dell'America. Il Manoscritto Di Ferrara.* Cisalpino-Goliardica, 1985.

Le Gobien, Charles. *Histoire des Isles Marianes, Nouvellement converties à la Religion Chrestienne; & de la mort glorieuse des premiers Missionnaires qui y ont prêché la Foy.* Paris: Nicolas Pepie, 1700.

Lee Yu-Chung and José Luis Caño Ortigosa. "Los Inicios del Conocimento Mutuo entre China Y Espana." *Tinta China-e*, no. 9 (2015): 36–41. Accessed 19 March 2017 https://studylib.es/doc/8194671/tinta-china---ministerio-de-educaci%C3%B3n--cultura-y-deporte

Lefebvre, Henri. *La Production de l'espace.* Paris, Éditions Anthropos, 1974. (English edition, 1991).

Lester, Toby. *The Fourth Part of the World: An Astonishing Epic of Global Discovery, Imperial Ambition, and the Birth of America.* New York: Free Press, 2010.

Lewis, Martin and Kären E Wigen. *The Myth of Continents.* Berkeley, CA: University of California Press, 1997.

Li, Narangoa and Robert Cribb. *Historical Atlas of Northeast Asia, 1590–2010: Korea, Manchuria, Mongolia.* New York: Columbia University Press, 2014.

Lieberman, Victor. *Strange Parallels: Southeast Asia in Global Context, c. 800–1830. Volume 1: Integration on the Mainland.* Cambridge: Cambridge University Press, 2003.

———. *Strange Parallels: Southeast Asia in Global Context, c. 800–1830. Volume 2: Mainland Mirrors: Europe, Japan, China, South Asia, and the Islands.* Cambridge: Cambridge University Press, 2009.

Lieske, Jay, H. "A Collection of Gallilean Satellite Eclipses 1652–1982." 51–60. In Vassilis Markellos and Yoshihide Kozai, eds., *Dynamical Trapping and Evolution in the Solar System: Proceedings*, vol. 106. Dordrecht: D. Reiedel Publishing Co., 1983.

Lilley, Keith D. ed. *Mapping Medieval Geographies: Geographical Encounters in the Latin West and Beyond, 300–1600.* Cambridge: Cambridge University Press, 2013.

Lim Jongtae. "Western Astronomy vs. Korean Geography: Intellectual Exchanges Between a Korean and the Jesuits as Seen From Yi Kiji's 1720 Beijing Travelogue." 205–309. In Luís Saraiva, ed., *Europe and China: Science and Arts in the 17th and 18th Centuries.* Singapore: World Scientific, 2013.

Liu Jing Jing, "Michael Baum and Athanasius Kircher – Sinomanic promoter." 181–85. In Eugenio Lo Sardo, A. Parisi and R. Pittella, eds., *Journey to the Ends of the world: Michele Ruggieri and Jesuits in China.* Macau: Instituto Cultural do Governo da R.A.E. de Macau-Museu de Macau (vols. II), 2013.

Livieratos, Evangelos. "Ptolemy." 1996. *http://cartography.web.auth.gr/Livieratos/fil/Ptolemy.html*

Lockhart, Bruce and Tran Ky Phuong, eds. *The Cham of Vietnam: History, Society and Art.* Singapore: National University of Singapore Press, 2011.

Loubère, Simon de la. *Du Royaume de Siam.* Paris: Jean Baptiste Coignard, 2 vols, 1691.

Loureiro, Rui Manuel, ed. *O Manuscrito de Lisboa da "Suma Oriental" de Tomé Pires: Contribução por Uma Edição Crítica.* Macau: Instituto Português do Oriente, 1996.

———. "Mission Impossible: In Search of the Sources of Fernão Mendes Pinto's Peregrinação." 235–55. In Jorge Santos Alves, ed., *Fernão Mendes Pinto and the Peregrinação: Studies, Restored Portuguese Text, Notes and Indexes.* Lisbon: Fundação Oriente, 2010.

Lucas, Henry S. *The Renaissance and Reformation, (2nd ed).* New York: Harper & Row, 1960.

Macías, Luis A. Robles. "Alessandro Zorzi's sketch maps (text) Presentation at the Brussels Circle Map Afternoon (MAPAF), Royal Library, Brussels, 22 April 2017." https://hal.archives-ouvertes.fr/…/Presentation%20Zorzi%20maps%20for%20Mapaf%.

McIntosh, Gregory C. *The Piri Reis Map of 1513.* Athens, GA: University of Georgia Press, 2000.

McLaughlin, Raoul. *Rome and the Distant East: Trade Routes to the Ancient Lands of Arabia, India and China.* London: Continuum, 2010.

McMaster, Robert B. and Eric Sheppard. "Introduction: Scale and Geographic Inquiry." 1–22. In Eric Sheppard and Robert B. McMaster, eds., *Scale and Geographic Inquiry: Nature, Society, and Method.* Malden: Blackwell, 2004.

Mapp, Paul W. *The Elusive West and the Contest for Empire, 1713–1763.* Chapel Hill, NC: University of North Carolina Press, 2011.

MapForum.com. The Development of the Printed Atlas Part 2: Ptolemaic Atlases, 2014 http://www.mapforum.com/02/ptolemy1.htm

Markham, Clements R. *The Letters of Amerigo Vespucci: and Other Documents Illustrative of His Career.* London: Hakluyt Society, 1894.

———. ed. *The Voyages of Pedro Fernandez de Quiros, 1595, 1606* (Vol. 1). London: Hakluyt Society, 1904.

Marques, Alfredo Pinheiro. *Origem e desenvolvimento da cartografia portuguesa na epoca dos descobrimentos.* Lisbon: Imprensa Nacional, Casa da Moeda, 1987.

Martini, Martino. *Bellum Tartaricum: Or, The Conquest of the Great and most Renowned Empire of China, by the Invasion of the Tartars.* London: Printed for John Crook, 1654.

Matsuda, Mat K. *Pacific Worlds: A History of Seas, Peoples, and Cultures.* Cambridge: Cambridge University Press, 2012.

Meltzer, Milton. *Ferdinand Magellan: First to Sail Around the World.* New York: Marshall Cavendish/Benchmark Books, 2001.

Mémoires de l'Académie royale des sciences depuis 1666 jusqu'en 1699. Paris: La Compagnie des Libraries, 1729–34.

Mendes Pinto, Fernão. *The voyages and adventures of Ferdinand Mendez Pinto, a Portugal.* London: Printed for Richard Bently, Jacob Tonson, Francis Saunders, and Tho. Bennet, 1614; 1653.

Michaud, Louis-Gabrie. *Biographie universelle ancienne et moderne ou histoire par ordre alphabétique, de la vie privée et publique de tous les hommes qui se sont distingués par leurs écrits, leurs actions, leurs talents.* Chez Michaud frères, 1816.

Miksic, John N. *Archaeological research on the "Forbidden Hill" of Singapore: excavations at Fort Canning.* Singapore: National Museum, 1984.

———. "Before and after Zheng He: Comparing some Southeast Asia Archaeological Sites of the 14th and 15th Centuries." 384–408. In Geoff Wade and Sun Laichen, eds., *Southeast Asia in the Fifteenth Century: The China Factor.* Singapore: National University of Singapore Press, 2010.

Mills, J. V. trans.; annot. *Ying-yai Sheng-lan/The Overall Survey of the Ocean's Shores (1433).* Bangkok: White Lotus Press, 1979; 1997.

Mitchell, Peta. *Cartographic Strategies of Postmodernity: The Figure of the Map in Contemporary Theory and Fiction.* New York: Routledge Studies in Twentieth-Century Literature, 2012.

Mohammed Hamidullah. 1968. "Muslim Discovery of America before Columbus." *Journal of the Muslim Students' Association of the United States and Canada,* vol. 4, no. 2 (1968): 7–9.

Momoki Shiro. "Dai Viet and the South China Sea Trade: from the 10th to the 15th Century." *Crossroads: An Interdisciplinary Journal of Southeast Asian Studies,* vol. 12, no. 1 (1998): 1–34.

———. "Nation and Geobody in Early Modern Vietnam: A Preliminary Study through Sources of Geomancy." 126–53. In Geoff Wade and Sun Laichen, eds., *Southeast Asia in the Fifteenth Century: The China Factor.* Singapore: National University of Singapore Press, 2010.

Montalboddo, Francazano. *Paesi nouamente retrouati et Nouo Mondo da Alberico Vesputio Florentino intitulato.* Vicenza, 1508.

Morton, Katherine. "China's Ambition in the South China Sea: Is a Legitimate Maritime Order Possible?" *International Affairs,* vol. 92 (2016): 909–940. http://onlinelibrary.wiley.com/doi/10.1111/1468–2346.

Moseley, C. W. R. D. *The Travels of Sir John Mandeville.* London: Penguin, 1983.

Mosley, Adam. *Bearing the Heavens: Tycho Brahe and the Astronomical Community of the Late Sixteenth Century.* Cambridge: Cambridge University Press, 2007.

Moule, Arthur Christopher and Paul Pelliot. *Marco Polo: The Description of the World.* Vol. II. London: George Routledge and Sons, Ltd., 1938.

Mundy, Barbara E. *The Mapping of New Spain: Indigenous Cartography and the Maps of the Relaciones Geograficas.* Chicago, IL: University of Chicago Press, 1996.

Murray, Diane. *Conflict and Co-existence: The Sino-VietnameseMaritime Boundaries in Historical Perspective.* Wisconsin University Papers on Southeast Asia, Occasional Paper, no.13, 1988.

Muslim Heritage, 2014. http://muslimheritage.com/topics/default.cfm?ArticleID=690#ftnref14 Accessed March 18, 2017.

Muzaffar Alam and Sanjay Subrahmanyam. *Indo-Persian Travels in the Age of Discoveries 1400–1800.* Cambridge: Cambridge University Press, 2007.

Needham, Joseph. *Science and Civilisation in China,* Volume 3: *Mathematics and the Sciences of the Heavens and the Earth.* Cambridge: Cambridge University Press, 1959.

Newbold, Thomas John. *Political and Statistical Account of the British Settlements in the Straits of Malacca, viz. Pinang, Malacca, and Singapore: With a History of the Malayan States on the Peninsula of Malacca.* vol. 2. London: J. Murray, 1839.

Nicholl, Robert. "A Note on the Verlade Map." *Brunei Museum Journal,* vol. 6, no. 2 (1986): 72–76.

Nicolet, Claude. *Space, Geography, and Politics in the Early Roman Empire.* Anne Arbor, MI: University of Michigan Press, 1991.

Nguyen Q. Thang. *The Hoang Sa and Truong Sa Archipelagos: Part of Vietnam's Territory from the Standpoint of International Law.* Ho Chi Minh City: Ho Chi Minh City General Publishing House, 2013.

Nordenskiöld, Adolf Erik. *Facsimile-atlas to the Early History of Cartography with Reproductions of the Most Important Maps Printed in the XV and XVI centuries.* Stockholm: P. A. Norstedt & Söner, 1889.

Nowell, Charles E., ed. *Magellan's Voyage Around the World: Three Contemporary Accounts.* Evanston, IL: Northwestern University Press, 1962.

Nunn, George E. *The Columbus and Magellan Concepts of South American Geography.* Glenside, Pennsylvania: Privately printed, 1932.

———. "The Imago Mundi and Columbus." *The American Historical Review,* vol. 40, no. 4 (1935): 646–661.

———. "Marinus of Tyre's Place in the Columbus Concepts." *Imago Mundi,* vol. 2, no. 1 (1937): 27–35.

———. "The Three Maplets attributed to Bartholomew Columbus." *Imago Mundi,* vol. 9, no. 1 (1952): 12–22.

Odifreddi, Piergiorgio. "An Emblematic Case of the Heterogenesis of the Ends." 105–108. In *The Master from the West; Commemorating the 400th Anniversary of the Passing of Matteo Ricci.* Supplementary Catalogue, Macau: Museu de Arte de Macau, 2010.

Oliveira, Isócrates de. "Itinerário de Camões no Extremo Oriente." *Boletim da Sociedade de Geografia de Lisboa,* 1–12 (1975): 71–89.

Ota, Emma. "Localities of Mediation: Deterritorialization and Embeddedness in the Mediated experience of Place." www.eonsbetween.net/localities_of_mediation.pdf? Accessed March 23, 2018.

Park Choon-ho. *East Asia and the Law of the Sea.* Seoul: Seoul National University, 1983.

Park, Hyunhee. *Mapping the Chinese and Islamic Worlds: Cross-Cultural Exchange in Pre-Modern Asia.* Cambridge: Cambridge University Press, 2012.

Parks, George B., compiler. *The Contents and Sources of Ramusio's Navigationi.* New York: New York Public Library, 1955.

Permanent Court of Arbitration (PCA) Press Release, October 29, 2015. www.pcacases.com/web/sendAttach/1503. Accessed March 23, 2018.

Peto, Christine Marie. *When France Was King of Cartography: The Patronage and Production of Maps in Early Modern France (Toposophia: Sustainability, Dwelling, Design).* Lanham, MD, Lexington Books, 2007.

Pham, Minh Hà, Charlotte. "European Navigation, Nautical Instructions and Charts of the Cochinchinese Coast (16th-19th Centuries)." *Moussons,* vol. 27 (2016): 101–129. Accessed June 11, 2017 http://moussons.revues.org/3544

Pigafetta, Antonio. *The Voyage of Magellan, The Journal of Antonio Pigafetta,* Engelwood Cliffs, NJ: Prentice Hall (facsimile of *Le voyage et navigation fait par les espignolz es isles Mollucques.* Paris 1525), 1969.

Pieterse, Jan Nederveen. *Ethnicities and Global Multiculture: Pants for an Octopus.* Lanham, MD: Rowman & Littlefield, 2007.

Pimentel, Manuel. *Arte pratica de navegar e regimento de pilotos repartido em duas partes…* Lisbon: Luis Serrão Pimentel, 1681.

———. *Arte de navegar: emque se ensinam as regras praticas, e o modo de cartear pela carta plana, & reduzida, o modo de graduar a balestilha por via de … Orientaes, agora novamente emendado, & accrescentadas muitas derrotas novas.* Lisbon: Na Officina Real Deslandesiana, 1712.

Pires, Benjamin Vieira. *A Viagem de Comércio Macau-Manila nos séculos XVI a XIX.* Macau: Centro de Estudos Marítimos de Macau, 1994.

Pires, Tomé; trans. A. Cortesão. *The Suma Oriental of Tomé Pires, (1515).* London: Hakluyt Society, 1944.

Portuondo, Maria M. *Secret Science: Spanish Cosmography and the New World.* Chicago, IL: University of Chicago Press, 2009.

Prescott, J. V. R. *The Maritime Political Boundaries of the World.* London: Methuen, 1985.

Purchas, Samuel. *Hakluytus Pothumus or, Purchas his Pilgrimes.* London, 1625.

Rackham, Horace. *Pliny's Natural History.* Translated by H. Rackham (vols. 1–5, 9) and W.H.S. Jones (vols. 6–8) and D.E. Eichholz (vol. 10). Cambridge, MA: Harvard University Press, 1949–54.

Ragep, Jamil, F. "Copernicus and his Islamic Predecessors: Some Historical Remarks." *History of Science*, vol. xlv (2007), 65–81.

———. "Islamic Reactions to Ptolemy's Imprecisions." 121–34. In Alexander Jones, ed., *Ptolemy in Perspective: Use and Criticism of his Work from Antiquity to the Nineteenth Century.* New York: Springer, 2010.

Raman, Shankar. *Framing "India": The Colonial Imaginary in Early Modern Culture.* Stanford, CA: Stanford University Press, 2001.

Ramusio, Giovanni Battista. *Delle Navigationi Et Viaggi (Band 1): In Molti Lvoghi Corretta, Et Ampliata, Nella Qvale Si Contengono La Descrittione Dell'Africa, & del paese del Prete Ianni, con varij viaggi ... et la Nauigatione attorno il Mondo ...* Venice, 1554.

———. *Delle Navigationi Et Viaggi* (Band 3): *Nel Quale Si Contengono Le Nauigationi al Mondo Nuouo, alli Antichi incognito, fatte da Don Christoforo Colombo Genouese, che su il fu il Primo a scoprirlo a'i Re Catholici, detto hora le Indie occidentali.* Venice, 1556.

———. *Delle Navigationi Et Viaggi (Band 2): Nel Qvale Si Contengono L'Historia delle cose de Tartari, & diuersi fatti de loro Imperatori, descritta da M. Marco Polo Gentilhuomo Venetiano, & da Hayton Armeno; varie descrittioni* ..Venice, 1559.

———. *Navigationi et Viaggi.* Venice 1563–1606. First Volume: With an introduction by R.A. Skelton, and an analysis of the contents by Prof. George B. Parks First Series - Volume II-IV., Amsterdam: Theatrum Orbis Terrarum, 1970.

Randles, W. G. M. "Portuguese and Spanish Attempts to Measure Longitude in the 16th Century." *Bol. Bibl Univ. Coimbra*, vol. 39 (1984): 143–60.

———. *Pedro Nunes and the Discovery of the Loxodromic Curve: Or How, in the Sixteenth Century, Navigating with a Globe Had Failed to Solve the Difficulties Encountered with the Plane Chart.* Coimbra: Imprensa de Coimbra, 1990.

Raynal, Guillaume Thomas. *Histoire Philosophique et Politique des Etablissements et du Commerce de Européens dans les deux Indes.* Paris, np., 1782.

———. *A Philosophical and Political History of the Settlements and Trades of the Europeans in the East and West Indies. By the Abbe Raynal. Newly translated from the French, by J. O. Justamond,* Six vols. Dublin: John Exshaw and Luke White, 1784.

Reinaud, J. T. *Mémoire sur le Périple de la mer Érythrée et sur la navigation des mers orientales au milieu du troisième siècle de l'ère chrétienne, d'après les témoignages grecs, latins, arabes, persans, indiens et chinois.* Paris: Academie des inscriptions et belles-lettres, 1864.

———. trans. notes. *Géographie d'Aboulféda.* Paris: Imprimèrie nationale, 1848–1883.

Reiske, Johann Jakob. *Abulfedae annales Moslemici. Latinos ex arabicis fecit Io. Iacobus Reiske.* Leipzig, 1754.

Renaudot, Eusèbe. *Anciennes Relations des Indes et de la Chine de deux voyageurs mahométans qui y allèrent dans le neuvième siècle; traduites d'arabe: avec des Remarques sur les principaux endroits de ces relations.* Paris: Jean-Baptiste Coignard, 1718.

Renshaw, Steve and Saori Ihara. "Kitora Kofun: A Detailed Astronomical Star Chart in an Ancient Japanese Tomb." 2002. http://www2.gol.com/users/stever/kitora.htm. Accessed March 23, 2018.

Rochberg-Halton, Francesca. "Nabu-Rimani." 749–752. In Frank Northen Magill, ed., *Dictionary of World Biography, Volume 1: The Ancient World.* Pasadena, CA: Salem Press, 1998.

Romer, F. E. *Pomponius Mela's Description of the World.* Ann Arbor, MI: University of Michigan Press, 1998.

Roque de Oliveira, Francisco. 2012. "Seventeenth Century Surveys for a Secure Overland Route from Europe to China." 468–503. In Antonio Vasconcelos de Saldanha and Artur K. Wardega, eds., *In the Light and Shadow of an Emperor: Tomás Pereira, SJ (1645–1708), the Kangxi Emperor and the Jesuit Mission in China.* Newcastle upon Tyne: Cambridge Scholars Publishing, 2012.

Rule, Paul. "François Noël, SJ and the Chinese Rites Controversy." 137–165. In W. F. Vande Walle and Noël Golvers, eds., *The History of the Relations between the Low Countries and China in the Qing Era (1644–1911).* Leuven, 2003.

Ruscelli, Girolamo. *La Geographi di Claudio Tolomeo.* Venice, 1561.

Sachau, Edward, ed. *Albêrûnî's India, an Account of the Religion, Philosophy, Literature, Chronology, Astronomy, Customs, Laws and Astrology of India about A. D. 1030.* Edited in the Arabic original by Dr Edward Sachau,... London: Trübner, 1887.

Saliba, George. "Al-Biruni," 248–52. In Joseph Strayer, ed., *Dictionary of Middle Ages.* Vol 2. New York: Charles Scribner's Sons, 1982.

———. "Revisiting the Astronomical Contacts between the World of Islam and Renaissance Europe: the Byzantine Connection." 361–373. In Paul Magdalino and Maria Mavroudi, eds., *The Occult Sciences in Byzantium.* Sofia: Editions La Pomme d'or S.A, 2006.

———. *Islamic Science and the Making of the European Renaissance.* Cambridge, MA: MIT Press, 2007.

Saltford, John. *The United Nations and the Indonesian Takeover of West Papua, 1962–1969: Anatomy of a Betrayal.* London: Routledge, 2002.

San Antonio G.Q. de. *Brève et véridique relation des événements du Cambodge. Nouvelle édition du texte espagnol, avec une traduction et des notes par Antoine Cabaton.* Paris, Ernest Leroux, 1914.

Santarém, Manuel Francisco de Barros and Sousa de Mesquita de Macedo Leitão e Carvalhosa. *Essai sur l'histoire de la cosmographie et de la cartographie pendant le moyen-âge: et sur les progrès de la géographie après les grandes découvertes du XVe siècle....* Paris: Impr. de Maulde et Renou, 1848–1852.

Santarém, Visconde de. *Atlas composé de cartes des XIVe, XV, XVI et XVII siécles [Material cartográfico]: pour la plupart inédites, et devant seuvir de preuves a l'ouvrage sur la priorité de la découverte de la Côte Occidentale d'Afrique au dela du Capo Bojador par les portugais / recueillies et gravées sous la direction du Vicomte de Santarém.* Paris, 1841.

Santos Lopes, Marília dos. "Peregrinação by Fernão Mendes Pinto: A Source of Knowledge," 257–70. In Jorge Santos Alves, ed., *Fernão Mendes Pinto and the Peregrinação: Studies, Restored Portuguese text, Notes and Indexes.* Lisbon: Fundação Oriente, 2010.

Sardo, Eugenio Lo. "Michelle Ruggieri's Atlas: Re-Edition," 59–72. Ruggieri, Michele. 大明國圖志：羅明堅中國地圖集 = Atlas da China de Michele Ruggieri = Atlas of China by Michele Ruggieri, Macau: Cultural Affairs Bureau of the Macao SAR Government and the Archivio di Stato di Roma, 2013.

Sasaki, Chikara. *Descartes's Mathematical Thought.* Boston, MA: Springer, 2003.

Sauer, Carl Ortwin. *The Early Spanish Main.* Cambridge: Cambridge University Press, 1966.

Savage, Victor R. *Western Impressions of Nature and Landscape in Southeast Asia.* Singapore: Singapore University Press, 1984.

Savage-Smith, Emilie. "Celestial Mapping." 256–62. In J. B. Harley and David Woodward, eds., *The History of Cartography in Volume 2, Book 1 Cartography in the Traditional Islamic and South Asian Societies.* Chicago, IL: University of Chicago, 1992.

Schaeffer, Bradley E. "The Epoch of the Constellations of the Farnese Atlas and their Origin in Hipparchus's Lost Catalogue." 2005. http://www.phys.lsu.edu/farnese/JHAFarneseProofs.htm. Accessed March 23, 2018.

Schoff, William H., trans. and ed. *The Periplus of the Erythraean Sea: Travel and Trade in the Indian Ocean by a Merchant of the First Century.* London, Bombay & Calcutta, 1912.

Seland, Eivind Heldaas. *Ports and Political Power in the Periplus: Complex Societies and Maritime Trade on the Indian Ocean in the First Century AD.* British Archaeological Reports, Oxford: Archaeopress, 2010.

Selden, John. *Mare clausum sev de dominio maris Libri Duo.* London: William Stansbury for Richard Meighen, 1635.

Sepinwall, Alyssa Goldstein. *The Abbé Grégoire and the French Revolution: The Making of Modern Universalism.* Berkeley, CA: University of California Press, 2005.

Sezgin, Fuat. "World: Historian Reveals Incredible Contributions Of Muslim Cartographers." 2004. https://www.rferl.org/a/1055354.html. Accessed February 10, 2018.

Service Geographique de l'Indochine. *Atlas de l'Indochine.* Saigon, 1920.

Sharratt, Michael. *Galileo: Decisive Innovator.* Cambridge: Cambridge University Press, 1996.

Sheehan, Kevin Joseph. *Iberian Asia: The Strategies of Spanish and Portuguese Empire Building, 1540–1700.* Berkeley, CA: University of California, 2008.

Short, John Rennie. *Alternative Geographies.* London: Routledge, 2000.

Silveira, Luis, ed. *Livro das plantas das fortalezas, cidades e povoaçoes do Estado da India Oriental com as descriçoes do maritimo dos reinos e provincias onde estao situadas e outros portos principais daquelas partes: contribuição para a historia das fortalezas dos portugueses no ultramar.* Lisbon: Ministerio do Planeamento e da Administração do Territorio, Secretaria de Estado da Ciencia e Tecnologia, Instituto de Investigação Cientifica Tropical, 1991.

Skelton, R. A. trans., ed. *Antonio Pigafetta, Magellan's Voyage: A Narrative Account of the First Navigation.* New York: Dover, 1969.

Small, Margaret. "Displacing Ptolemy: The Textual Geographies of Ramusio's Navigazioni e viaggi." 152–172. In Keith D. Lilley, ed. *Mapping Medieval Geographies: Geographical Encounters in the Latin West and Beyond, 300–1600.* Cambridge: Cambridge University Press, 2013.

Smith, Adam. *The Wealth of Nations, Books IV-V.* London: Penguin, 1776 (1999).

Smith, Neil. "Geography, Difference and the Production of Scale." 57–78. In Joe Donerty et al., eds., *Postmodernism and the Social Sciences.* London: Macmillan, 1992.

Smith, Richard J. *Chinese Maps: Images of "All Under Heaven."* Hong Kong: Oxford University Press, 1996.

Smith, Peter J. and Joseph Needham. "Magnetic declination in Medieval China." *Nature,* vol. 214 (17 June 1967): 1213–1214.

Smithies, Michael. "Introduction." In Marcel Le Blanc, S. J. ed., *History of Siam in 1688.* Chiang Mai: Silkworm Books, 2003.

Sobel, Dava. *Longitude: The True Story of a Lone Genius Who Solved the Greatest Scientific Problem of His Time.* London: Fourth Estate, 1996.

Soll, Jacob. *The Information Master: Jean-Baptiste Colbert's Secret State Intelligence System.* Ann Arbor, IL: University of Michigan, 2009.

Song Liming. "The Maps of China by Michelle Ruggieri and Matteo Ricci." 148–57. In Eugenio Lo Sardo, A. Parisi and R. Pittella eds., *Journey to the Ends of the world. Michele Ruggieri and Jesuits in China.* Macau: Instituto Cultural do Governo da R.A.E. de Macau-Museu de Macau. (vol. II), 2013.

Song Nianshen. "Northeast Eurasia as Historical Center: Exploration of a Joint Frontier." The *Asia-Pacific Journal,* Vol. 13, issue 43, no. 1 (2015). -Nianshen-Song/4392/article.html

Spence, Jonathan. *The Memory Palace of Matteo Ricci.* New York: Penguin, 1985.

———. *The Chan's Great Continent: China in Western Minds.* New York: Penguin, 1998.

Spooner, Paul B. "Geopolitical Lens: The Turkish Invasions of Europe and the Portuguese Expansion to Asia." *Revue of Culture,* vol. 42 (2013): 7–19.

Stallard, Avan Judd. "Antipodes to Terra Australis." doctoral dissertation, School of History, Philosophy, Religion and Classics, University of Queensland, 2010. https://espace.library.uq.edu.au/view/UQ:243057/s4092025_phd_finalthesis.pdf

———. *Antipodes: In Search of the Southern Continent.* Monash, Clayton: Monash University Press, 2016.

Stanley, Henry Edward John, ed., *Pigafetta's Treatise on Navigation.* Cambridge University Press, 1874 (2010).

Steinhardt, Nancy Shatzman. "China's Earliest Mosques." *Journal of the Society of Architectural Historians,* vol. 67, no. 3 (2008): 339–341.

Stanley, Henry Edward John, ed. *First Voyage Round the World by Magellan: Translated from the Accounts of Pigafetta and Other Contemporary Writers.* Cambridge: Cambridge University Press, 2010.

Stefoff, Rebecca. *Ferdinand Magellan and the Discovery of the World Ocean.* New York: Chelsea House, 1990.

Stephenson, F. Richard. "Chinese and Korean Star Maps and Catalogs," 511–78. In J. B. Harley and David Woodward, eds., *The History of Cartography,* Volume 2, Book 2, Cartography in the Traditional East and Southeast Asian Societies. Chicago, IL: University of Chicago Press, 1994.

Suárez, Thomas. *Early Mapping of Southeast Asia.* Singapore: Periplus, 1999.

———. *Early Mapping of the Pacific: The Epic Story of Seafarers, Adventurers, and Cartographers Who Mapped the Earth's Greatest Ocean.* Singapore: Periplus, 2004.

Szykuła, Krystyna. "Unexpected 16th Century Finding to Have Disappeared Just After Its Printing – Anthony Jenkinson's Map of Russia, 1562." In Carlos Bateira, ed., *Cartography: A Tool for Spatial Analysis.* InTech: Rijeka, Croatia, 2012. http://dx.doi.org/10.5772/50224

Szcześniak, Boleslaw. "The Antoine Gaubil Maps of the Ryukyu Islands and Southern Japan." *Imago Mundi,* vol. 12 (1955): 141–149.

Taberd, Jean-Louis. *Dictionarium Anamitico-Latinum et Latino-Anamiticum.* 1838.

———. "Notes on the Geography of Cochin China." *Journal of the Asiatic Society of Bengal,* vol. 6–7, 1837–1839.

Tachard, Guy. *Voyage de Siam des peres Jesuit Envoyez par le ROY aux Indes & à la Chine. Avec leurs Observations Astronomiques, et leurs Remarques de physique, de Géographie, d'Hydrographie, & d'Histoire.* Paris: Chez Arnould Seneuze et Daniel Horthemels, 1686.

———. *A Relation of the Voyage to Siam Performed by Six Jesuits.* Bangkok: White Lotus, 1999 (1688).

Tagliacozzo, Eric. "Hydrography, Technology, Coercion: Mapping the Sea in Southeast Asian Imperialism, 1850–1900." *Archipel,* vol. 65 (2003): 89–107.

Tang, Kaijian. *Setting Off from Macau: Essays on Jesuit History during the Ming and Qing.* Leiden: Brill, 2015.

Thatcher, John Boyd. *Christopher Columbus: His Life, His Work, His Remains.* New York: G.P. Putnam's Sons, Vol. II, 1903.

Thongchai Winichakul. *Siam Mapped: A History of the Geo-Body of a Nation.* Honolulu, HI: University of Hawaii Press, 1994.

Thrower, Norman J. W. *Maps & Civilization: Cartography in Culture and Society, (3rd ed).* Chicago, IL: The University of Chicago Press, 2008.

Tibbetts, Gerald R. "The Role of Charts in Islamic Navigation in the Indian Ocean." 256–62. In J. B. Harley and David Woodward, eds., *The History of Cartography, Volume 2, Book 1 Cartography in the Traditional Islamic and South Asian Societies.* Chicago, IL: University of Chicago, 1992.

Tolmacheva, Marina. "The Indian Ocean in Arab Geography." *Terra Brasilis (Nova Série)* [Online], 6, 2015. http://journals.openedition.org/terrabrasilis/1585; DOI: 10.4000/terrabrasilis.1585

Tooley, Ronald Vere. *Tooley's Dictionary of Mapmakers.* New York: A. R. Liss, 1979.

Tordesillas, Ellen T. "Philippines to Submit 300–Year-Old Map to UN in Case vs China." *The Philippine Star,* June 8, 2015. *http://www.philstar.com/headlines/*

2015/06/08/1463593/philippines-submit-300-year-old-map-un-case-vs-china. Accessed August 3, 2015.

Tsukamoto, Kazuto. "Star chart in Kitora Tomb likely depicts the heavens observed in ancient China." *Asahi Shimbun*, July 16, 2015. *http://ajw.asahi.com/article/behind_news/social_affairs/AJ201507160056*

Vagnon, Emmanuelle and Éric Vallet, eds. *La fabrique de l'océan Indien. Cartes d'Orient et d'Occident (Antiquité-XVIe siècle).* Paris: Publications de la Sorbonne, 2017.

Van Gent, R. H. and A. Van Helden. "Lunar, Solar, and Planetary Representations to 1650." 123–34. In David Woodward, ed., *The History of Cartography, Volume 3 Cartography in the European Renaissance.* Chicago, IL: University of Chicago Press, 2007.

van den Broecke, Marcel. *Ortelius Atlas Maps. An Illustrated Guide.* Second revised edition. Houten, 2011.

Vlekke, Bernard H. *Nusantara: A History of Indonesia.* The Hague: van Hoewe, 1965.

Vogel, Hans Ulrich. *Marco Polo Was in China: New Evidence from Currencies, Salts and Revenues.* Leiden: Brill, 2013.

Waley-Cohen, Joanna. *The Sextants of Beijing: Global Currents in Chinese History.* New York: Norton & Co., 1999.

Wallis, Helen. 1992. "What Columbus Knew." *History Today*, vol. 42 (May 1992): 17–23.

———. and E. D. Grinstead. "A Chinese Terrestrial Globe, A.D. 1623." *The British Museum Quarterly,* vol. 25, no. 3/4 (June 1962): 83–91.

Washburn, Wilcomb E. "A Proposed Explanation of the Closed Indian Ocean on Some Ptolemaic Maps of the Twelfth-Fifteenth centuries." *Revista da Universidade de Coimbra*, vol. 33 (1985): 431–441.

Wey Gómez, Nicolás. *The Tropics of Empire: Why Columbus Sailed South to the Indies.* Cambridge, MA: MIT Press, 2008.

White, Landeg, trans., intro. *The Lusiads.* New York: Oxford University Press, 1997; 2001.

Whitney, Elspeth. *Medieval Science and Technology.* Westport, CA: Greenwood Press, 2004.

Wieser, Franz R. von, 1893. *Die Karte des Bartolomeo Colombo über die vierte Reise des Admirals.* Innsbruck: Wagner, 1893.

Wilson, Eric. "Making the World Safe for Holland: 'De Indis' of Hugo Grotius and International Law as Geoculture." *Review (Fernand Braudel Center),* vol. 32, no. 3 (2009): 239–287.

Winter, Heinrich. "Francisco Rodrigues' Atlas of ca. 1513." *Imago Mundi,* vol. 6 (1949): 20–26.

Witsen, Nicolaes. *Noord en Oost Tartarye* [North and East Tartary], 2 vols. Amsterdam: François Halma, 1705; repr.

Wong, Katrine K. 2014. "Introduction." xxiii–xxxiv. In Katrine K. Wong and C. X. George Wei, eds., *Macao-Cultural Interaction and Literary Representations.* London: Routledge, 2014.

Wood, Frances. *Did Marco Polo Go to China?* London: Secker & Warburg, 1995.

———. "Did Marco Polo go to China?" *Asian Affairs*, vol. 27, no. 3 (1996): 296–304.

Woodward, David. "Medieval Mappaemundi." 285–370. In J. B. Harley and David Woodward, eds., *The History of Cartography, Volume 1, Cartography in Prehistoric, Ancient, and Medieval Europe and the Mediterranean.* Chicago, IL: University of Chicago Press, 1987.

Woodward, David, Cordell D. K. Yee and Joseph E. Schwartzberg. "Concluding Remarks." 843–849. In J. B. Harley and David Woodward, eds., *The History of Cartography,* Volume 2, Book 2, *Cartography in the Traditional East and Southeast Asian Societies.* Chicago, IL: University of Chicago Press, 1994.

Wu, Shicun. *Solving Disputes for Regional Cooperation and Development in the South China Sea: A Chinese Perspective.* Oxford: Chandos Publishing, 2013.

Yin Zhiping, "China's Sovereignty Over the Nansha Islands Indisputable," *Beijing Review*, 23–29 May, 1988.

Yule, Henry, ed. *Cathay and the way thither: being a collection of medieval notices of China* (2 vols). London: Hakluyt Society, 1866.

Yule, Henry and A.C. Burnell. *Hobson-Jobson: A Glossary of Colloquial Anglo-Indian Words and Phrases.* London: J. Murray, 1886; 1903. http://dsal.uchicago.edu/dictionaries/hobsonjobson

Yunli, Shi. "Eclipse observations made by Jesuit astronomers in China: A reconsideration." *Journal for the History of Astronomy*, vol. 31, part 2 (2000): 135–147.

Index

About the Author

Geoffrey C. Gunn is a graduate of Melbourne, Queensland, and Monash universities. He has taught in universities in North Africa, Australia, Southeast Asia (Singapore; Brunei), and China (Macau). He has published a number of dedicated country studies on East-Southeast Asia (Brunei, Laos, Vietnam, Cambodia, East Timor, and Macau), which have been translated into variously Portuguese, Indonesian, and Chinese. He is also the author of several world regional history texts, namely, *First Globalization: The Eurasian Exchange, 1500-1800* (Rowman & Littlefield); *History without Borders: The Making of an Asian World Region, 1000-1800* (University of Hong Kong Press); and *World Trade Systems of the East and West: Nagasaki and the Asian Bullion Trade Networks* (Brill). He is Emeritus Professor of Nagasaki University and an adjunct professor of the Center of Macau Studies, University of Macau.

www.ingramcontent.com/pod-product-compliance
Ingram Content Group UK Ltd.
Pitfield, Milton Keynes, MK11 3LW, UK
UKHW021906150926
475560UK00003B/202

9 781498 590136